AN INTRODUCTION TO ORDINARY DIFFERENTIAL EQUATIONS

This refreshing, introductory textbook covers standard techniques for solving ordinary differential equations, as well as introducing students to qualitative methods such as phase-plane analysis. The presentation is concise, informal yet rigorous; it can be used for either one-term or one-semester courses.

Topics such as Euler's method, difference equations, the dynamics of the logistic map and the Lorenz equations, demonstrate the vitality of the subject, and provide pointers to further study. The author also encourages a graphical approach to the equations and their solutions, and to that end the book is profusely illustrated. The MATLAB files used to produce many of the figures are provided in an accompanying website.

Numerous worked examples provide motivation for, and illustration of, key ideas and show how to make the transition from theory to practice. Exercises are also provided to test and extend understanding; full solutions for these are available for teachers.

AN INTRODUCTION TO ORDINARY DIFFERENTIAL EQUATIONS

JAMES C. ROBINSON

CAMBRIDGE
UNIVERSITY PRESS

CAMBRIDGE
UNIVERSITY PRESS

University Printing House, Cambridge CB2 8BS, United Kingdom

Cambridge University Press is part of the University of Cambridge.

It furthers the University's mission by disseminating knowledge in the pursuit of
education, learning and research at the highest international levels of excellence.

www.cambridge.org
Information on this title: www.cambridge.org/9780521533911

© Cambridge University Press 2004

First published 2004
8th printing 2013

A catalogue record for this publication is available from the British Library

Library of Congress Cataloguing in Publication data

Robinson, James C. (James Cooper), 1969–
An introdusction to ordinary differential equations / James C. Robinson.
p.   cm.
Includes bibliographical references and index.
ISBN 0 521 82650 1 (hardback) – ISBN 0 521 53391 1 (paperback)
1. Differential equations.   1. Title

QA372.R77 2003
512´.352–dc21   2003055186

ISBN 978-0-521-82650-1 Hardback
ISBN 978-0-521-53391-1 Paperback

To
Mum and Dad,
for all their love, help and support.

Contents

Some of the chapters, and some sections within other chapters, are marked with an asterisk (*). These parts of the book contain material that either is more advanced, or expands on points raised elsewhere in the text.

Preface

The aim of this book is to deal with all of the elementary methods for obtaining explicit solutions of ordinary differential equations, and then to introduce the ideas of qualitative analysis using phase plane techniques. Simple difference equations are also included, since their methods of solution are similar to those for linear differential equations. As well as being, I hope, an internally consistent choice of material, this selection of topics also has the advantage of preparing a student for a basic course on dynamical systems.

The book arose from my unsuccessful efforts to find a suitable text to recommend when I taught the first year Warwick differential equations course. Although there are a number of well-established and successful textbooks that treat this subject (these are discussed, along with other possibilities for further reading, in the final chapter), they seem either to include a large amount of additional material, or to concentrate only on the more advanced topics. I therefore produced a detailed set of lecture notes, which, with the encouragement of Alan Harvey and David Tranah, and most significantly Kenneth Blake at Cambridge University Press, eventually became this book. My thanks here to all those students who made useful suggestions while this book was still at the lecture note stage.

Part I contains an informal discussion of the issues of existence and uniqueness of solutions, and treats the standard classes of first order differential equations that can be solved explicitly, as well as covering exact equations and substitution methods.

The first chapter of Part II shows that two linearly independent solutions are needed in order to solve the general homogeneous problem, and also contains a brief treatment of the Wronskian. The remainder of this section treats equations with constant coefficients, concentrating for the most part on the second order case, with higher order equations discussed briefly at the end.

Second order equations with non-constant coefficients are treated in Part III, which covers reduction of order, the method of variation of constants, and series solutions.

Part IV turns aside from differential equations, motivating the study of difference equations by discussing Euler's method of numerical solution. Constant coefficient linear difference equations are covered, and then there are two chapters devoted to nonlinear difference equations. One of these goes beyond the confines of an introductory course and discusses the dynamics of the logistic map in some detail.

Part V treats coupled systems of two linear differential equations, starting with the substitution method that reduces the problem to a second order differential equation in one variable, the most reliable way to find explicit solutions. The remainder of this portion of the book deals with the matrix approach, showing how a calculation of the eigenvalues and eigenvectors of an appropriate matrix is enough to draw the phase portrait. This is done by changing to a coordinate system in which the equation is put into a standard form, providing an illustration of the Jordan canonical form of a matrix.

Part VI uses the methods from Part V in order to draw the phase plane diagrams for a variety of nonlinear systems, with examples taken from mathematical ecology and simple one-dimensional particle systems, including the pendulum. The book ends with a brief discussion of Dulac's criterion and the Poincaré–Bendixson Theorem, a chapter that investigates the complicated dynamics of the Lorenz Equations, and suggestions for further reading.

In addition to those already mentioned above I would like to thank various people who have contributed to this book. I first learned much of the material here from Tristram Jones-Parry at Westminster School, to whom much belated thanks for all his fine teaching many years ago. I also owe a debt of gratitude to all those who taught the course at Warwick before me, shaping its contents and therefore those of this book; in particular, I had useful guidance from the course notes of Alan Newell and Claude Baesens. I am most grateful to Andrew Stuart, who, in encouraging me to emphasise the links with linear algebra, made me fond of a subject that I still remembered with a shudder from my own undergraduate days. Thanks too to James Macdonald, whose 'Swarm of flies' program for his MMath project on the Lorenz equations was the inspiration behind Figure 37.8.

Over the past two months I have been able to think of little except phase planes and drawing figures in MATLAB: my wife, Tania Styles, has managed to endure my many variations on 'come and see this picture of a washing machine' with a smile. Heartfelt thanks to her for this, and, of course, for everything.

Finally, I would particularly like to thank my Ph.D. student, Oliver Tearne, and my father, John, both of whom read this book extremely carefully and made a number of very helpful comments. For whatever imperfections remain, my apologies to them and to my readers.

Introduction

Differential equations date back to the mid-seventeenth century, when calculus was discovered independently by Newton (*c.* 1665) and Leibniz (*c.* 1684). Modern mathematical physics essentially started with Newton's *Principia* (published in 1687) in which he not only developed the calculus but also presented his three fundamental laws of motion that have made the mathematical modelling of physical phenomena possible.[1]

Historically, advances in the theory of differential equations have come from the insights gained when trying to treat specific physical models. Despite this somewhat piecemeal development, the subject has become a well-defined and coherent area of mathematics. This book adopts a theoretical point of view, developing the theory to the point at which it can no longer be described as 'basic differential equations' and is about to become entangled with more advanced topics from the theory of dynamical systems. Of course, applications are used throughout to serve as motivation and illustration, but the emphasis is on a clean presentation of the mathematics.

You may find that some of the problems covered in the first few chapters are already familiar. The methods of solving these problems are well established, and you may be well practised at applying them. However, we will take care here to show why these methods work; giving proper justification of the methods can take some time, but as mathematicians we should not be satisfied merely with a set of 'recipes'. Nevertheless, knowing something about the details should not stop you from applying the methods you know already; rather you should be able to use them with more confidence.

Some of the chapters, and some sections within other chapters, are marked with an asterisk (*). These parts of the book contain either material that is more advanced, or material that expands on points raised elsewhere; while they could be omitted in the interests of brevity, they are intended to give some indication of the richness of the subject beyond the confines of an introductory course.

---

[1] Various modern editions of this work are available, translated from its original Latin.

There are three appendices, covering background material that is necessary at various points in the book. While some of this is elementary and may already be familiar (Appendix A recalls some notation and various facts about real and complex numbers that will be used throughout the book) some is a little more advanced. Problems with timetabling often mean that certain undergraduate courses have to rely on material that is yet to be taught in others, hence there are appendices on matrices, eigenvalues and eigenvectors (Appendix B) and on derivatives, partial derivatives and Taylor series (Appendix C). The calculation of eigenvalues and eigenvectors is treated in detail in the main part of the book.

The use of mathematical computer packages is now a standard part of the undergraduate curriculum, and an important tool in the armoury of practising mathematicians, scientists and engineers. Although the emphasis in the text is on pencil and paper analysis, and the book in no way relies on the availability of such software, some topics, particularly the treatment of coupled nonlinear equations using phase plane ideas in Chapters 28–37, can benefit greatly from the graphical possibilities modern computers provide. Almost all of the figures in this book have been generated using MATLAB, and very occasionally particular MATLAB commands are mentioned in the text. Nevertheless, it should be possible to carry out the numerical exercises suggested here using any of the major commercially available mathematical packages; and with a little more ingenuity using any programming language with graphical capabilities. The MATLAB files used to produce some of the figures, and mentioned in certain of the exercises, are available for download from the web at `www.cambridge.org/0521533910`.

There is no better way to learn this material than by working through a selection of examples. One set of examples is included in what is, I hope, a natural way in the text, with the end of each worked solution marked with a box ($\square$). Another set of examples is given in the exercises that end each chapter, and these should be considered an integral part of the book. The majority consist of sample problems that can be treated with the methods of the chapter – in order to give teachers a reasonable choice of problems, there are intentionally more of these than you could reasonably be expected to do. Others, labelled with a 'T', are more theoretical and designed to give an indication of some of the mathematical issues raised, but not treated in detail, in the text. Finally, those exercises labelled with a 'C' are intended to encourage the use of the computer to perform routine calculations and investigate equations and their solutions graphically. Those involved in teaching courses based on this book may obtain copies of solutions to these exercises by applying to the publisher by email (`solutions@cambridge.org`).

I would welcome any comments or suggestions, either by post to the Mathematics Institute, University of Warwick, Coventry, CV4 7AL, U.K. or by email to `jcr@maths.warwick.ac.uk`; any errata that arise will be posted on my own website `www.maths.warwick.ac.uk/~jcr/IntroODEs.html`.

Part I

First order differential equations

1

Radioactive decay and carbon dating

Before we start our formal treatment of the subject we will look at a very simple example that nonetheless exhibits the power of differential equations as models of reality. One point to bear in mind in this chapter is the distinction to be made between finding the solution of a differential equation, and interpreting this solution.

1.1 Radioactive decay

Let $N(t)$ denote the number of radioactive atoms in some sample of material at time t. Then with $k > 0$ the equation

$$\frac{dN}{dt} = -kN \tag{1.1}$$

is a very good model for the way that the number of radioactive atoms decays (see Exercise 1.1).

Although we will see later how to solve this equation, for now we will assume that when there are N_s atoms at time s, the solution is

$$N(t) = N_s e^{-k(t-s)}. \tag{1.2}$$

You can check that we really do have the solution: when $t = s$ the formula in (1.2) gives $N(s) = N_s$, while we have

$$\frac{d}{dt} N(t) = -kN_s e^{-k(t-s)} = -kN(t),$$

and so the differential equation (1.1) is satisfied.

It follows from (1.2) that the number of radioactive atoms decays exponentially to zero. Graphs of the solution for various values of $N(0)$, showing this decay, are plotted in Figure 1.1.

The *half-life* of a particular radioactive isotope is the time it takes for half of the radioactive atoms to decay, and this is related to the constant k that appears in

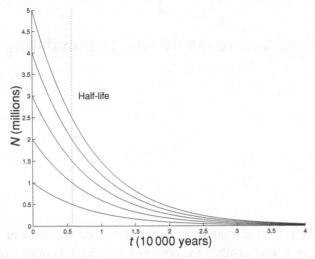

Fig. 1.1. Graph showing the number $N(t)$ of radioactive atoms falling off as a function of time, for a number of different values of N_0; the constant k is that for radioactive carbon 14. The half-life, approximately 5700 years, is marked by a dashed vertical line.

the equation. To find this relationship, suppose that there are N_0 radioactive atoms at time $t = 0$. Then the solution of (1.1) is

$$N(t) = N_0 e^{-kt}.$$

Half of the atoms will have decayed by time t_{half} when $N(t_{half}) = \frac{1}{2}N_0$, i.e.

$$N_0 e^{-k t_{half}} = \frac{1}{2}N_0 \qquad \Rightarrow \qquad e^{-k t_{half}} = \frac{1}{2}.$$

Taking the (natural) logarithm of both sides gives

$$-k t_{half} = -\ln 2,$$

and so the half-life is given by $t_{half} = (\ln 2)/k$. Note that this time does not depend on the initial number of radioactive atoms.

1.2 Radiocarbon dating

The solution (1.2) forms the basis of the technology of radiocarbon dating. The essence of the method is as follows. Living matter is constantly taking up carbon from the air. The result is that within such material the ratio of the number of isotopes of radioactive carbon 14 (^{14}C) to the number of isotopes of stable carbon 12 (^{12}C) is essentially constant. Once the specimen is dead (for example, a tree is cut down for its wood, or cotton is harvested for weaving), the radioactive ^{14}C atoms begin to decay according to the model (1.1). Since the half-life of carbon 14 is approximately 5700 years, we need to take the constant k in (1.1) to be

$$k = \frac{\ln 2}{5700} \approx 1.216 \times 10^{-4}.$$

By examining the ratio of the number of isotopes of carbon 12 to carbon 14 in a sample of the material that we want to date, it is possible to work out the proportion remaining of the ^{14}C atoms that were initially present. Suppose that the sample stopped taking up carbon from the air when $t = s$, and that the number of ^{14}C atoms present then was N_s. If we know that the sample now (at time t_0) contains only a fraction p of the initial level of ^{14}C, then $N(t_0) = pN_s$.

Using our explicit solution $N(t) = N_s e^{-k(t-s)}$, we should have

$$pN_s = N(t_0) = N_s e^{-k(t_0 - s)}.$$

Cancelling the factor of N_s in the two outside terms yields the equation

$$p = e^{-k(t_0 - s)}.$$

Taking logarithms of both sides we have

$$\ln p = -k(t_0 - s),$$

and so the year s from which the sample dates is given by

$$s = t_0 + \frac{\ln p}{k}. \tag{1.3}$$

In 1988, the Shroud of Turin (see Figure 1.2) was dated by three independent groups of scientists from Arizona, Oxford and Zurich. Fibres from the shroud were

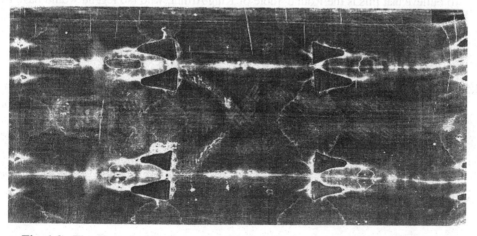

Fig. 1.2. The Shroud of Turin: carbon dated to the fourteenth century. Photograph © 1978 Barrie M. Schwortz (his website at www.shroud.com is well worth a visit).

found to contain about 92% of the level in living matter.[1] Using the expression in (1.3) shows that the Shroud therefore dates from

$$s = 1988 + \frac{\ln 0.92}{0.000\ 121\ 6} \approx 1302,$$

putting its origin squarely in the Middle Ages.

Exercises

1.1 Radioactive isotopes decay at random, with a fixed probability of decay per unit time. Over a time interval Δt, suppose that the probability of any one isotope decaying is $k\Delta t$. If there are N isotopes, how many will decay on average over a time interval Δt? Deduce that

$$N(t + \Delta t) - N(t) \approx -Nk\Delta t,$$

and hence that $dN/dt = -kN$ is an appropriate model for radioactive decay.

1.2 Plutonium 239, virtually non-existent in nature, is one of the radioactive materials used in the production of nuclear weapons, and is a by-product of the generation of power in a nuclear reactor. Its half-life is approximately 24 000 years. What is the value of k that should be used in (1.1) for this isotope?

1.3 In 1947 a large collection of papyrus scrolls, including the oldest known manuscript version of portions of the Old Testament, was found in a cave near the Dead Sea; they have come to be known as the 'Dead Sea Scrolls'. The scroll containing the book of Isaiah was dated in 1994 using the radiocarbon technique;[2] it was found to contain between 75% and 77% of the initial level of carbon 14. Between which dates was the scroll written?

1.4 A large round table hangs on the wall of the castle in Winchester. Many would like to believe that this is the Round Table of King Arthur, who (so legend would have it) was at the height of his powers in about AD 500. If the table dates from this time, what proportion of the original carbon 14 would remain? In 1976 the table was dated using the radiocarbon technique, and 91.6% of the original quantity of carbon 14 was found.[3] From when does the table date?

1.5 Radiocarbon dating is an extremely delicate process. Suppose that the percentage of carbon 14 remaining is known to lie in the range $0.99p$ to $1.01p$. What is the range of possible dates for the sample?

[1] P. E. Damon et al., 'Radiocarbon dating of the Shroud of Turin', Nature **337** (1989), 611–615.
[2] A. J. Jull et al., 'Radiocarbon dating of the scrolls and linen fragments from the Judean Desert', Radiocarbon **37** (1995), 11–19.
[3] M. Biddle, King Arthur's Round Table (Boydell Press, 2001).

2

Integration variables

Because of the intimate relationship between differentiation and integration (discussed in more detail in the next chapter) there will be many integrals in this book, and it is worth pausing now in order to make sure that we have an appropriately unambiguous notation.

Although in theory mathematicians make careful distinctions between 'the function f' and '$f(x)$', the value that f takes at a particular point x, this distinction is rarely maintained in day-to-day informal discussions.

Usually this does not cause any trouble. However, consider the following problem, posed in 'everyday' language:

Find the area under the graph of $f(x)$ between a and x.

Although the meaning of this is clear, 'find the shaded area in Figure 2.1', there is some potential for confusion when we try to write this down mathematically, since there are too many xs around. Converting the English into symbols gives

$$\int_a^x f(x) \, dx, \tag{2.1}$$

and it should be clear that this is not satisfactory, since the symbol x is used in two different ways: once as the upper limit of the range of integration ($\int_a^x$), and once as the variable that is being integrated over (dx).

When we integrate a function between two limits, for example[1]

$$\int_a^b f(x) \, dx,$$

the variable that we are integrating over is a 'dummy' variable. It is just there to tell us how to do the integration, and plays no rôle in the final answer, which will

---

[1] Observe that there is no need to change our notation for this particular definite integral, since no confusion can arise as to the rôle of x.

9

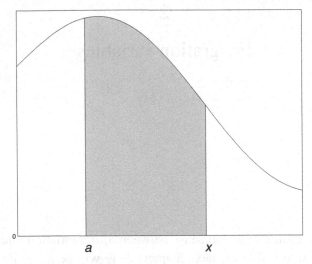

Fig. 2.1. 'Find the shaded area'.

only depend on a and b. So

$$\int_a^b f(x)\,\mathrm{d}x = \int_a^b f(\theta)\,\mathrm{d}\theta = \int_a^b f(\aleph)\,\mathrm{d}\aleph.$$

(We can change the name of the dummy variable with no effect on the integral.)

The obvious solution, then, is to change the integration variable in (2.1) to something other than x. However, changing the variable to something completely different from x is likely to be confusing. The approach we will adopt will be to add a tilde $\sim$ to the integration variable, so that instead of (2.1) we will write

$$\int_a^x f(\tilde{x})\,\mathrm{d}\tilde{x}. \tag{2.2}$$

All being well this should keep things 'clean' but should not be too jarring.

We will also do something similar when evaluating integrals where x is an upper limit, i.e.

$$\int_a^x f(\tilde{x})\,\mathrm{d}\tilde{x} = \left[\, F(\tilde{x}) \,\right]_{\tilde{x}=a}^x,$$

when $F' = f$.

Of course, very few people are this careful when they are doing calculations and the backs of mathematicians' envelopes are full of things like (2.1) rather than the pedantic (2.2).

3

Classification of differential equations

Before we begin we need to introduce a simple classification of differential equations which will let us increase the complexity of the problems we consider in a systematic way.

3.1 Ordinary and partial differential equations

The most significant distinction is between ordinary and partial differential equations, and this depends on whether ordinary or partial derivatives occur.

Partial derivatives cannot occur when there is only one independent variable. The independent variables are usually the arguments of the function that we are trying to find, e.g. x in $f(x)$, t in $x(t)$, both x and y in $G(x, y)$. The most common independent variables we will use are x and t, and we will adopt a special shorthand for derivatives with respect to these variables: we will use a dot for $\mathrm{d}/\mathrm{d}t$, so that

$$\dot{z} = \frac{\mathrm{d}z}{\mathrm{d}t} \quad \text{and} \quad \ddot{z} = \frac{\mathrm{d}^2 z}{\mathrm{d}t^2};$$

and a prime symbol for $\mathrm{d}/\mathrm{d}x$, so that

$$y' = \frac{\mathrm{d}y}{\mathrm{d}x} \quad \text{and} \quad y'' = \frac{\mathrm{d}^2 y}{\mathrm{d}x^2}.$$

Usually we will prefer to use time as the independent variable.

In an ordinary differential equation (ODE) there is only one independent variable, for example the variable x in the equation

$$\frac{\mathrm{d}y}{\mathrm{d}x} = f(x),$$

11

specifying the slope of the graph of the function y; the variable t in

$$m\ddot{\mathbf{x}} = \mathbf{f}(t)$$

which we could solve for the position $\mathbf{x}(t) = (x(t), y(t), z(t))$ of a particle at time t moving under the action of a force $\mathbf{f}(t)$ (the equation is Newton's second law of motion, $F = ma$); or x in

$$-\frac{\hbar^2}{2m}\frac{d^2\psi}{dx^2} + V(x)\psi = E\psi$$

where $\psi(x) = \alpha(x) + i\beta(x)$ is complex (this is the Schrödinger equation from quantum mechanics).

In a partial differential equation there is more than one independent variable and the derivatives are therefore *partial derivatives*, for example the heat in a rod at position x and time t, $h(x, t)$, obeys the heat equation

$$\frac{\partial h}{\partial t} = k\frac{\partial^2 h}{\partial x^2}.$$

A much more complicated example is given by the Navier–Stokes equations used to determine the velocity of a fluid

$$\mathbf{u}(x_1, x_2, x_3, t) = (u_1(x_1, x_2, x_3, t), u_2(x_1, x_2, x_3, t), u_3(x_1, x_2, x_3, t))$$

(think of $x_1 = x$, $x_2 = y$, and $x_3 = z$), which are:[1]

$$\rho\left[\frac{\partial u_j}{\partial t} + \left(\sum_{i=1}^{3} u_i \frac{\partial u_j}{\partial x_i}\right)\right] - \mu\left[\frac{\partial^2 u_j}{\partial x_1^2} + \frac{\partial^2 u_j}{\partial x_2^2} + \frac{\partial^2 u_j}{\partial x_3^2}\right] + \frac{\partial p}{\partial x_j} = f_j. \quad (3.1)$$

(one for each component, $j = 1, 2, 3$) and

$$\frac{\partial u_1}{\partial x_1} + \frac{\partial u_2}{\partial x_2} + \frac{\partial u_3}{\partial x_3} = 0. \quad (3.2)$$

In this book we will consider only ordinary differential equations.

---

[1] It is possible to write these two equations much more concisely using vector calculus notation. Imagine that ∇ represents a vector of partial derivatives, $\nabla = (\partial/\partial x_1, \partial/\partial x_2, \partial/\partial x_3)$, which can be manipulated like a normal vector. Then, for example, Equation (3.2) is just $\nabla \cdot \mathbf{u} = 0$. Defining also $\Delta = \nabla \cdot \nabla$ (the sum of all second derivatives) we can rewrite (3.1) as

$$\rho\left[\frac{\partial \mathbf{u}}{\partial t} + (\mathbf{u} \cdot \nabla)\mathbf{u}\right] - \mu\,\Delta\mathbf{u} + \nabla p = \mathbf{f}. \quad (3.3)$$

3.2 The order of a differential equation

The *order* of a differential equation is the highest order derivative that occurs: the equation

$$\frac{dy}{dx} = f(x)$$

specifying the slope of a graph is first order, as is the following equation expressing energy conservation,

$$\tfrac{1}{2}m\dot{x}^2 + V(x) = E$$

($\tfrac{1}{2}m\dot{x}^2$ is the kinetic energy while $V(x)$ is the potential energy at a point x); Newton's second law of motion

$$m\frac{d^2x}{dt^2} = F$$

is second order; the equation

$$\psi''' + \tfrac{1}{2}\psi\psi' = 0$$

(which occurs in the theory of fluid boundary layers) is third order (recall that ψ''' is shorthand for $d^3\psi/dx^3$).

To be more formal, an nth order ordinary differential equation for a function $y(t)$ is an equation of the form

$$\mathcal{F}\left(\frac{d^n y}{dt^n}, \frac{d^{n-1} y}{dt^{n-1}}, \ldots, \frac{dy}{dt}, y, t\right) = 0. \tag{3.4}$$

(Of course we want $d^n y/dt^n$ to occur in $\mathcal{F}$; if $\mathcal{F}(\ddot{y}, \dot{y}, y, t)$ is $y - t$ then the resulting equation ($y - t = 0$) is not a differential equation at all.) If t does not occur explicitly in the equation, as in

$$\frac{dy}{dt} = f(y),$$

then the equation is said to be *autonomous*.

3.3 Linear and nonlinear

Another important concept in the classification of differential equations is linearity. Generally, linear problems are relatively 'easy' (which means that we can find an explicit solution) and nonlinear problems are 'hard' (which means that we cannot solve them explicitly except in very particular cases).

An nth order ODE for $y(t)$ is said to be linear if it can be written in the form

$$a_n(t)\frac{d^n y}{dt^n} + a_{n-1}(t)\frac{d^{n-1} y}{dt^{n-1}} + \cdots + a_1(t)\frac{dy}{dt} + a_0(t)y = f(t), \qquad (3.5)$$

i.e. only multiples of y and its derivatives occur. Such a linear equation is called *homogeneous* if $f(t) = 0$, and *inhomogeneous* if $f(t) \neq 0$.

3.4 Different types of solution

When we try to solve a differential equation we may obtain various possible types of solution, depending on the equation. Ideally, perhaps, we would find a *fully explicit solution*, in which the dependent variable is given explicitly as a combination of elementary functions of the independent variable, as in

$$y(t) = 3\cos 5t + 8\sin t. \qquad (3.6)$$

We can expect to be able to find such a fully explicit solution only for a very limited set of examples.

A little more likely is a solution in which y is still given directly as a function of t, but as an expression involving an integral, for example

$$y(t) = 1 + \int_0^t e^{-s^2}\,ds. \qquad (3.7)$$

Here y is still an explicit function of t, but the integral cannot be evaluated in terms of elementary functions.

Sometimes, however, we will only be able to obtain an *implicit form* of the solution; this is when we obtain an equation that involves no derivatives and relates the dependent and independent variables.[2] For example, the equation

$$\ln y + 4\ln x - y - 2x + 4 = 0 \qquad (3.8)$$

relates x and y, but cannot be solved explicitly for y as a function of x.

All these types of solution will occur in what follows.

There are many situations, however, in which it is not possible to obtain any useful expression for the solution. For some equations it is still possible to understand

---

[2] We could also have an implicit solution containing integrals that cannot be evaluated in terms of elementary functions. For example, we will see that the equation $dx/dt = f(x)g(t)$ has solution

$$\int \frac{dx}{f(x)} = \int g(t)\,dt,$$

which in general gives such an implicit solution.

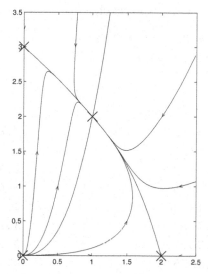

Fig. 3.1. A qualitative, graphical solution of the coupled system of equations (3.9). The axes are x (horizontally) and y (vertically), and it is safe to assume that this is the case for any unlabelled axes in the rest of the book.

the *qualitative* behaviour of the solutions, i.e. to describe how the solutions behave, even though we cannot specify them exactly. This is the approach we will take in Chapter 7, and throughout Chapters 32–37. Such a description is often best expressed graphically. For example, Figure 3.1 shows the *phase diagram* (or *phase portrait*) for the solutions of the equations

$$\dot{x} = x(4 - 2x - y)$$
$$\dot{y} = y(9 - 3x - 3y). \tag{3.9}$$

The diagram is a plot of sample curves traced out by solutions $(x(t), y(t))$ labelled with arrows indicating the direction in which t increases. The crosses show points at which the solutions of this equation are constant. We can tell from this diagram that every solution with $x(0)$, $y(0) > 0$ eventually approaches the point $(1, 2)$ [i.e. $x(t) \to 1$ and $y(t) \to 2$ as $t \to +\infty$], even though we do not have any form of explicit solution for (3.9).

For some equations all our analytical tools may fail, and in this case we can often use a computer to approximate the solution. A 'numerical solution' of a differential equation is usually only an approximation, and the initial result of such a calculation will not be an expression for x in terms of t, say, but a list of times, t, and corresponding approximate values for $x(t)$. Using MATLAB's ODE solving routine, ode45, to solve the equation

$$\frac{dx}{dt} = t - x^2 \qquad x(0) = 0$$

between times $t = 0$ and $t = 5$, yields such a list:

```
>> xdot=inline('t-x^2','t','x');
>> [t x]=ode45(xdot,[0 5],0);
>> [t x]
```

ans =	0	0		
	0.1250	0.0078	2.6250	1.4921
	0.2500	0.0312	2.7500	1.5407
	0.3750	0.0700	2.8750	1.5864
	0.5000	0.1235	3.0000	1.6299
	0.6250	0.1907	3.1250	1.6721
	0.7500	0.2700	3.2500	1.7127
	0.8750	0.3591	3.3750	1.7515
	1.0000	0.4555	3.5000	1.7891
	1.1250	0.5563	3.6250	1.8261
	1.2500	0.6585	3.7500	1.8621
	1.3750	0.7596	3.8750	1.8969
	1.5000	0.8574	4.0000	1.9310
	1.6250	0.9505	4.1250	1.9646
	1.7500	1.0377	4.2500	1.9976
	1.8750	1.1187	4.3750	2.0297
	2.0000	1.1935	4.5000	2.0612
	2.1250	1.2628	4.6250	2.0925
	2.2500	1.3268	4.7500	2.1231
	2.3750	1.3856	4.8750	2.1531
	2.5000	1.4403	5.0000	2.1826

We will discuss one simple method of numerical approximation in Chapter 21.

Exercises

3.1 Classify the following equations as ordinary or partial, give their order, and state whether they are linear or nonlinear. In each case identify the dependent and independent variables.

(i) Bessel's equation (v is a parameter)

$$x^2 y'' + xy' + (x^2 - v^2)y = 0,$$

(ii) Burger's equation (v is a parameter)

$$\frac{\partial u}{\partial t} - v\frac{\partial^2 u}{\partial x^2} + u\frac{\partial u}{\partial x} = 0,$$

(iii) van der Pol's equation (m, k, a and b are parameters)

$$m\ddot{x} + kx = a\dot{x} - b\dot{x}^3,$$

(iv) $dy/dt = t - y^2$,

(v) the wave equation (c is a parameter)

$$\frac{\partial^2 y}{\partial t^2} = c^2 \frac{\partial^2 y}{\partial x^2},$$

(vi) Newton's law of cooling (k is a parameter and $A(t)$ is a specified function)

$$\frac{dT}{dt} = -k(T - A(t)),$$

(vii) the logistic population model (k is a parameter)

$$\frac{dp}{dt} = kp(1 - p),$$

(viii) Newton's second law for a particle of mass m moving in a potential $V(x)$,

$$m\ddot{x} = -V'(x),$$

(ix) the coupled equations in (3.9)

$$\dot{x} = x(4 - 2x - y)$$
$$\dot{y} = y(9 - 3x - 3y),$$

and

(x)

$$\frac{dx}{dt} = \mathbb{A}x,$$

where x is an n-component vector and $\mathbb{A}$ is an $n \times n$ matrix.

4

*Graphical representation of solutions using MATLAB

The list of numbers that formed the example of a numerical solution at the end of the previous chapter indicates how useful a graphical representation of solutions can be. In fact MATLAB's default presentation of a numerical solution of a differential equation is as a graph: the commands

```
>> xdot=inline('t-x^2', 't', 'x');
>> ode45(xdot, [0 5], 0)
```

produce the graph shown in Figure 4.1 (only the axis labels have been added).

Whichever kind of solution we manage to obtain for our equation, the graphical capabilities provided by modern computer packages enable us to visualise these solutions and so obtain a much better understanding of their behaviour. All the solutions in Section 3.4 benefit from a graphical presentation. In this section we briefly discuss the main MATLAB commands that can be used to visualise and solve a variety of equations.

Almost all of the figures in Parts I, II, and III of this book are the graphs of explicit solutions; these are very easy to produce with MATLAB. For example, to plot $y(t) = 3\cos 5t + 8\sin t$ against t for $0 \le t \le 20$, the three lines

```
t=linspace(0,20);
y=3*cos(5*t)+8*sin(t);
plot(t,y)
```

produce Figure 4.2.

If the solution is given as an integral that cannot be evaluated explicitly, like (3.7),

$$y(t) = 1 + \int_0^t e^{-s^2}\, ds,$$

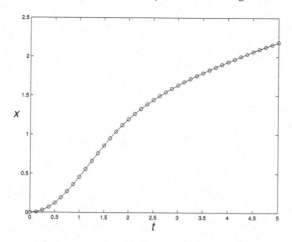

Fig. 4.1. The solution of $\dot{x} = t - x^2$ with $x(0) = 0$, as produced by the MATLAB ode45 command. The individual pairs (x, t) are represented by the circles, and are joined to produce an approximation to the solution $x(t)$ of the original equations.

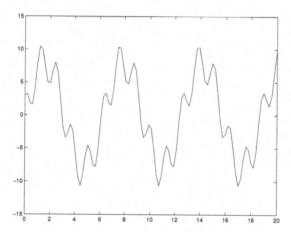

Fig. 4.2. The graph of $y(t) = 3\cos 5t + 8\sin t$ (y against t).

then we can find the value of y at any given value of t by approximating the integral; this is something that computers are very good at. The integral of e^{-t^2} between 0 and 2 (for example) can be evaluated by defining an 'inline function' $f(t) = \exp(-t^2)$ and then using the quad command:

```
>> f=inline('exp(-t.^2)','t')

f = Inline function:

  f(t) = exp(-t.^2)
```

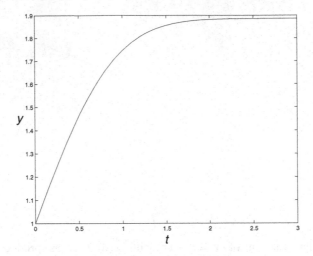

Fig. 4.3. The graph of $y(t) = 1 + \int_0^t e^{-s^2}\,ds$.

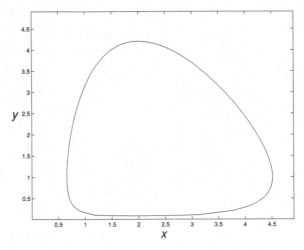

Fig. 4.4. The curve $\ln y + 4\ln x - y - 2x + 4 = 0$.

```
>> quad(f,0,2)
```

```
ans = 0.8821
```

The graph of y against t can be plotted using the short M-file

```
f=inline('exp(-t.^2)','t');
  for i=0:30
   t(i+1)=i/10;
   y(i+1)=1+quad(f,0,t(i+1));
end
plot(t,y)
```

Given an implicit formula like (3.8),

$$\ln y + 4 \ln x - y - 2x = -4,$$

we can notice that x and y lie on a curve that makes

$$F(x, y) = \ln y + 4 \ln x - y - 2x$$

constant. The 'contour plot' of the level set $F(x, y) = -4$,

```
>> [x, y]=meshgrid(.01:.1:5, .01:.1:5);
>> z=log(y)+4*log(x)-y-2*x;
>> contour(x,y,z,[-4 -4])
```

is shown in Figure 4.4.

Exercises

4.1 (C) Plot the graphs of the following functions:
    (i) $y(t) = \sin 5t \sin 50t$ for $0 \le t \le 3$,
    (ii) $x(t) = e^{-t}(\cos 2t + \sin 2t)$ for $0 \le t \le 5$,
    (iii)

$$T(t) = \int_0^t e^{-(t-s)} \sin s \, ds \qquad \text{for} \qquad 0 \le t \le 7,$$

    (iv) $x(t) = t \ln t$ for $0 \le t \le 5$,
    (v) plot y against x, where

$$x(t) = Be^{-t} + Ate^{-t} \qquad \text{and} \qquad y(t) = Ae^{-t},$$

    for A and B taking integer values between -3 and 3.

4.2 (C) Draw contour plots of the following functions:
    (i)

$$F(x, y) = x^2 + y^2 \qquad \text{for} \qquad -2 \le x, y \le 2;$$

    (ii)

$$F(x, y) = xy^2 \qquad \text{for} \qquad -1 \le x, y \le 1,$$

    with contour lines where $F = \pm 0.1, \pm 0.2, \pm 0.4$, and ± 0.8;

    (iii)

$$E(x, y) = y^2 - 2 \cos x \qquad \text{for} \qquad -4 \le x, y \le 4;$$

    (iv)

$$E(x, y) = x - \tfrac{1}{3}x^3 + \tfrac{1}{2}y^2(x^4 - 2x^2 + 2)$$

    for $-2 \le x \le 4$ and $-2 \le y \le 2$, showing contour lines where $E = 0, 0.5$, 0.8, 1, 2, 3 and 4;

    (v)

$$E(x, y) = y^2 + x^3 - x \qquad \text{for} \qquad -2 \le x, y \le 2.$$

5

'Trivial' differential equations

In this chapter we consider the simplest possible kind of differential equation, one that can be solved directly by integration. Although the problem is relatively straightforward, it will serve to introduce several important ideas.

You have probably already met and solved one simple kind of differential equation:

$$\frac{dy}{dx}(x) = f(x). \tag{5.1}$$

Viewed as an equation to solve for $y(x)$, this asks us to find the function whose graph has slope $f(x)$ at the point x. So in order to solve this equation we 'just' have to find a function whose derivative is $f(x)$.

5.1 The Fundamental Theorem of Calculus

Any function F that satisfies $F' = f$ is called an *anti-derivative*[1] of f. Clearly if F is an anti-derivative of f then so is $F(x) + c$ for any constant c.

This terminology allows us to distinguish between reversing the process of differentiation (finding an anti-derivative) and integration (finding the area under a curve). Put like this it becomes possible, perhaps, to appreciate how remarkable it is that these two concepts are so intimately related. This is formalised in the Fundamental Theorem of Calculus (FTC).

Essentially this theorem says that differentiation reverses the action of integration, and that if we know an anti-derivative of f we can calculate the area under the graph of f between any two points; it is easy to forget that the FTC is a major result because we use it so frequently in order to calculate integrals.

In the statement of the theorem we use $\mathbb{R}$ to denote the set of all real numbers, and $[a, b]$ denotes the closed interval $a \leq x \leq b$.

---

[1] The more puzzling word 'primitive' is sometimes used instead of 'anti-derivative'.

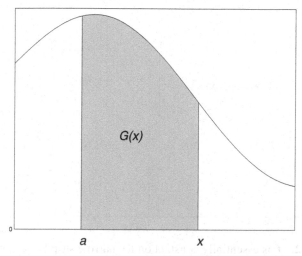

Fig. 5.1. $G(x)$ is the area under the graph of f between a and x.

Theorem 5.1 *Suppose that* $f : [a, b] \to \mathbb{R}$ *is continuous, and for* $a \le x \le b$ *define*

$$G(x) = \int_a^x f(\tilde{x}) \, d\tilde{x} \tag{5.2}$$

(the integral $G(x)$ *is the area under the graph of* f *between* a *and* x, *see Figure 5.1). Then*

$$\frac{dG}{dx}(x) = f(x),$$

and furthermore

$$\int_a^b f(x) \, dx = F(b) - F(a) \tag{5.3}$$

for any anti-derivative F *of* f *(i.e. for any* F *with* $F' = f$).

We often write (5.3) in the more convenient shorthand

$$\int_a^b f(x) \, dx = \left[F(x) \right]_{x=a}^b. \tag{5.4}$$

Proof (Sketch) If we calculate dG/dx using the formal definition of the derivative as a limit (see Appendix C),

$$\frac{dG}{dx}(x) = \lim_{\delta x \to 0} \frac{G(x + \delta x) - G(x)}{\delta x}, \tag{5.5}$$

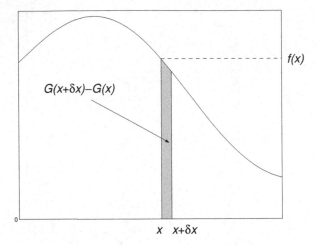

Fig. 5.2. f is essentially constant on the narrow strip $[x, x + \delta x]$.

then we get

$$\frac{\mathrm{d}G}{\mathrm{d}x}(x) = \lim_{\delta x \to 0} \frac{1}{\delta x} \left(\int_a^{x+\delta x} f(\tilde{x}) \, \mathrm{d}\tilde{x} - \int_a^x f(\tilde{x}) \, \mathrm{d}\tilde{x} \right)$$

$$= \lim_{\delta x \to 0} \frac{1}{\delta x} \int_x^{x+\delta x} f(\tilde{x}) \, \mathrm{d}\tilde{x}.$$

The expression

$$\int_x^{x+\delta x} f(\tilde{x}) \, \mathrm{d}\tilde{x} \tag{5.6}$$

represents the area in the little strip between x and $x + \delta x$ (see Figure 5.2). Since $f(\tilde{x}) \approx f(x)$ for this range of $\tilde{x}$ the value of (5.6) is roughly $\delta x \, f(x)$, and so we get

$$G'(x) \approx \lim_{\delta x \to 0} \frac{1}{\delta x} \delta x \, f(x) = f(x);$$

in other words $G(x)$ is an anti-derivative of $f(x)$. This argument can be made precise if f is continuous (see Exercise 5.9).

We now show how we can use an anti-derivative in order to calculate a definite integral (between two fixed limits) as in (5.3). If F is any anti-derivative of f then $(\mathrm{d}/\mathrm{d}x)(F - G) = F' - G' = 0$ and so F and G can only differ by a constant,

$$F(x) = G(x) + c.$$

Since $G(a) = 0$ (from its definition in (5.2)) we have $F(a) = c$, and so

$$\int_a^b f(x)\,dx = G(b)$$
$$= F(b) - c$$
$$= F(b) - F(a),$$

which is (5.3). $\qquad\qquad\qquad\qquad\qquad\qquad\qquad\qquad\qquad\qquad\square$

Because of the relationship between anti-derivatives and integrals, the notation

$$\int f(x)\,dx, \qquad\qquad (5.7)$$

(note the lack of limits on the integral) is often used as a shorthand to mean 'an anti-derivative of f'. We will use this notation at times, but when we need to be more careful we will explicitly use a particular choice of anti-derivative $F(x)$.

5.2 General solutions and initial conditions

Now let us return to our simple differential equation[2]

$$\frac{dy}{dx} = f(x). \qquad\qquad (5.8)$$

Any anti-derivative F of f is a solution of this equation ($y(x) = F(x)$); hence we could simply write

$$y(x) = \int f(x)\,dx.$$

If we choose one particular anti-derivative F, then we know that not only is $y(x) = F(x)$ a solution, but also $y(x) = F(x) + c$ for any c. So as it stands (5.8) has many solutions. We say that

$$y(x) = F(x) + c \qquad\qquad (5.9)$$

is the *general solution* of the equation (5.8), since any possible solution of (5.8) can be obtained by choosing c appropriately. It should not be a surprise that there are many possible solutions; we can move a graph 'up and down' and not change its slope – all the curves in Figure 5.3, which differ by only a constant, have the same slope at any given x value.

As illustrated in the figure, one way to pick out a particular solution is to give a point that must lie on the graph of y, in other words to specify the value $y(x_0)$ of

---

[2] Usually we will not make explicit the dependence of dy/dx on x (and similarly for other derivatives).

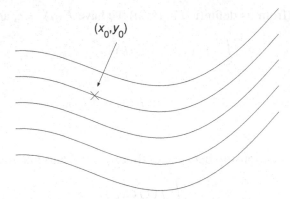

Fig. 5.3. Adjusting the constant c in (5.9) corresponds to moving the graph of F 'up and down' and does not affect the slope. An initial condition (x_0, y_0) will pick out one of the curves.

y at some particular x value, x_0. We refer to such a restriction

$$y(x_0) = y_0$$

as an *initial condition*, the idea being that we could construct the solution of (5.8) starting at (x_0, y_0) and then drawing the graph by using the information about the derivative of y contained in (5.8).

There are two ways to find the solution of (5.8) that satisfies $y(x_0) = y_0$. You should make sure that you understand what follows, since we will use similar reasoning very often throughout the rest of the book.

For the first method we do a little more than we have to: we find the general solution, and then solve a very simple algebraic equation to find the correct constant. We have seen that if we can find one anti-derivative F of f then the general solution of (5.8) is

$$y(x) = F(x) + c.$$

The particular solution that we want has $y(x_0) = y_0$, and so we need

$$y_0 = y(x_0) = F(x_0) + c \quad \Rightarrow \quad c = y_0 - F(x_0).$$

Thus the solution with $y(x_0) = y_0$ is

$$y(x) = y_0 + F(x) - F(x_0). \tag{5.10}$$

The alternative is to proceed more directly, and integrate both sides of

$$\frac{dy}{dx} = f(x) \tag{5.11}$$

between x_0 and x. Then we get

$$\int_{x_0}^x \frac{dy}{dx}(\tilde{x})\,d\tilde{x} = \int_{x_0}^x f(\tilde{x})\,d\tilde{x},$$

which gives

$$\left[\, y(\tilde{x})\,\right]_{\tilde{x}=x_0}^x = \int_{x_0}^x f(\tilde{x})\,d\tilde{x}.$$

Putting in the limits on the left-hand side this is

$$y(x) - y(x_0) = \int_{x_0}^x f(\tilde{x})\,d\tilde{x}. \tag{5.12}$$

You should make sure that you are happy going straight from (5.11) to (5.12); we will generally skip the two intermediate steps.

Since $y(x_0) = y_0$, (5.12) gives the solution in the form

$$y(x) = y_0 + \int_{x_0}^x f(\tilde{x})\,d\tilde{x}. \tag{5.13}$$

(The FTC shows that we do indeed have $y'(x) = f(x)$, and clearly $y(x_0) = y_0$ as required.) If we know that F is an anti-derivative of f then we can use the FTC,

$$\int_a^b f(x)\,dx = F(b) - F(a),$$

(this was (5.3), and is just the usual rule for evaluating integrals) to write the solution more explicitly as

$$y(x) = y_0 + F(x) - F(x_0).$$

Of course, this is the same expression that we obtained above in (5.10). Note, however, that in some cases the integral form in (5.13) may be the best that we can do, if it is not possible to find an explicit anti-derivative of f.

We now look at some simple examples.

Example 5.2 *Find the general solution of the equation*

$$\frac{dy}{dx} = x + 10\sin x. \tag{5.14}$$

What is the equation of the graph with slope $x + 10\sin x$ passing through the point $(\pi, 0)$?

In order to find the general solution we have to find an anti-derivative of $x + 10\sin x$. Using standard integrals this is $\frac{1}{2}x^2 - 10\cos x$, and so the general

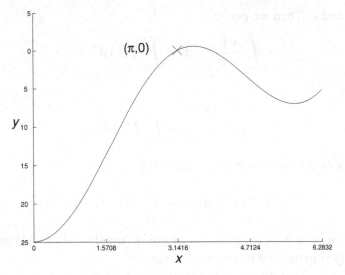

Fig. 5.4. The graph of $y(x) = \frac{1}{2}(x^2 - \pi^2) - 10(1 + \cos x)$. It passes through the initial condition $(\pi, 0)$, which is marked by a cross.

solution of (5.14) is

$$y(x) = \tfrac{1}{2}x^2 - 10\cos x + c$$

for any c. The one solution that passes through $(\pi, 0)$ must have

$$0 = y(\pi) = \tfrac{1}{2}\pi^2 - -10 + c \qquad \Rightarrow \qquad c = -\tfrac{1}{2}\pi^2 - 10,$$

and so

$$y(x) = \tfrac{1}{2}(x^2 - \pi^2) - 10(1 + \cos x).$$

The graph of y against x is shown in Figure 5.4, along with the initial condition.

$\square$

Example 5.3 *A curve passing through the point* $(1, 0)$ *has slope* $\ln x$. *What is the equation of the curve?*

We have to solve the equation

$$\frac{dy}{dx} = \ln x.$$

Since we have already used the 'long-winded' method in the first example, let us

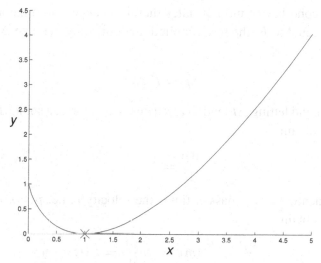

Fig. 5.5. The graph of $y(x) = 1 + x \ln x - x$, with the initial condition $(1, 0)$ marked by a cross.

do this one directly. We integrate both sides of the equation between 1 and x to give

$$y(x) - y(1) = \int_1^x \ln \tilde{x} \, d\tilde{x}$$

$$= \left[\tilde{x} \ln \tilde{x} - \tilde{x} \right]_{\tilde{x}=1}^x$$

$$= (x \ln x - x) - (0 - 1)$$

and so (since we want $y(1) = 0$)

$$y(x) = 1 + x \ln x - x.$$

This solution is shown in Figure 5.5. $\qquad\square$

In the next section we will give some more examples, this time more practically based, using Newton's second law of motion ($F = ma$).

5.3 Velocity, acceleration and Newton's second law of motion

Newton formulated the calculus, and his theory of differential equations, in order to be able to write down and solve the mathematical models that resulted from his laws of motion. Since derivatives are essentially the 'rate of change', questions concerning velocities (the rate of change of position) and acceleration (the rate of change of velocity) are most naturally framed as differential equations.

Newton's second law of motion states that the change Δp in the momentum p of an object is equal to F, the force applied, multiplied by the time Δt over which the force acts,

$$\Delta p = F \, \Delta t.$$

Dividing by Δt and letting Δt tend to zero (this is, of course, a somewhat imprecise derivation) we obtain

$$\frac{dp}{dt} = F(t).$$

Since the momentum is the mass m times the velocity v, i.e. $p = mv$, if the mass is constant we obtain

$$\frac{dp}{dt} = \frac{d}{dt}(mv) = m\frac{dv}{dt} = F(t).$$

The rate of change of v, dv/dt, is precisely what we mean by the acceleration, and so this equation is the familiar formula '$F = ma$' written another way.

Example 5.4 *A car of mass m is travelling at a speed v_0 when it suddenly has to brake; the brakes apply a constant force k until the car comes to rest. How long does it take the car to stop, and how far does it travel before it comes to rest?*

Using Newton's second law we have

$$m\frac{dv}{dt} = -k,$$

since the force acts to oppose the motion of the car. Rewriting this as $\dot{v} = -k/m$ and integrating both sides between times 0 and t we get

$$v(t) - v_0 = -\int_0^t \frac{k}{m} \, d\tilde{t},$$

or

$$v(t) - v_0 = -\frac{kt}{m},$$

and so

$$v(t) = v_0 - \frac{kt}{m},$$

see Figure 5.6. The car stops when $v(t_c) = 0$; this implies that $t_c = mv_0/k$.

Since the velocity v is the time derivative of the position x, $v = \dot{x}$ and we have

$$\frac{dx}{dt} = v(t) = v_0 - \frac{kt}{m}.$$

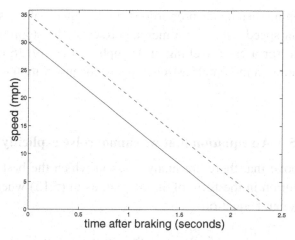

Fig. 5.6. The speed of a car that suddenly brakes: $v(t) = v_0 - (kt/m)$, with $m = 1000$ kg and $k = 6500$ N and for initial speeds of 30 mph (solid line) and 35 mph (dashed line).

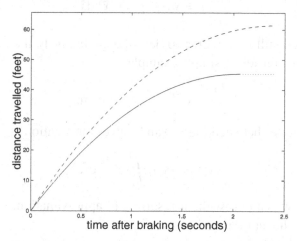

Fig. 5.7. The distanced moved by the car once the brakes have been applied. The choice of k and m is as for Figure 5.6, and again the solid line is for an initial speed of 30 mph and the dashed line is for an initial speed of 35 mph.

Integrating both sides between $t = 0$ and $t = t_c$ (when the car stops) we get

$$x(t_c) - x(0) = \int_0^{t_c} \left(v_0 - \frac{kt}{m} \right) \, \mathrm{d}t = \left[v_0 t - \frac{kt^2}{2m} \right]_{t=0}^{t_c} = v_0 t_c - \frac{kt_c^2}{2m}.$$

Substituting for t_c we have

$$x(t_c) - x(0) = \frac{m v_0^2}{2k},$$

as shown in Figure 5.7.

Since the stopping distance is proportional to the square of the speed, relatively small increases in speed will have a marked effect on the stopping distance. The stopping distance for a car travelling at 35 mph will be $49/36$ of that for a car travelling at 30 mph, almost half as much again for just 5 mph extra speed,[3] see Figure 5.11.   □

5.4 An equation that we cannot solve explicitly

We remarked above that there are many cases in which the best that we can do is to find the solution in the form of an integral, as in (5.13) where we wrote the solution of the general equation

$$\frac{dy}{dx} = f(x) \qquad \text{with} \qquad y(x_0) = y_0$$

as

$$y(x) = y_0 + \int_{x_0}^{x} f(\tilde{x}) \, d\tilde{x}. \qquad (5.15)$$

However, it can still be possible to describe qualitatively the behaviour of the solution. Here we consider a simple example,

$$\frac{dx}{dt} = e^{-t^2} \qquad x(0) = x_0.$$

Integrating both sides between times 0 and t gives the solution

$$x(t) = x_0 + \int_{0}^{t} e^{-\tilde{t}^2} \, d\tilde{t}. \qquad (5.16)$$

This is as far as we can go without resorting to approximation, since there is no explicit form for the anti-derivative of e^{-t^2}.

However, it is known[4] that

$$\int_{0}^{\infty} e^{-t^2} \, dt = \sqrt{\pi}/2.$$

---

[3] Realistic values of m and k are $m = 1000$ kg and $k = 6500$ N (one newton is one kg m/s^2) which means that the stopping distances at 30 mph (≈ 13.4 m/s) and 35 mph (≈ 15.6 m/s) are 13.8 m and 18.7 m respectively (to fall prey completely to the British imperial/metric confusion, that is roughly 40 feet and 60 feet respectively). An extra 10 mph on the motorway means that the stopping distance at 80 mph is $64/49$ of the stopping distance at 70 mph, over a quarter as much again: 225 feet (≈ 75 m) at 70 mph (≈ 31 m/s) and 300 feet (≈ 100 m) at 80 mph (≈ 36 m/s).

[4] This 'Gaussian integral' arises frequently, making it very frustrating that it cannot be evaluated explicitly. In particular the normal distribution, which is fundamental in the theory of statistics, is described by a bell-shaped curve whose equation is $e^{-x^2/2}/\sqrt{2\pi}$; statistical tables for the normal distribution are essentially based on evaluating the integral in (5.16) numerically.

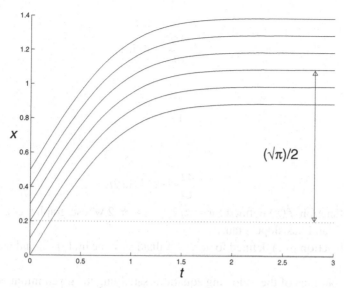

Fig. 5.8. We cannot solve the equation explicitly, but we know that the solution always increases and tends to a value $\sqrt{\pi}/2$ greater than its initial value.

So we can say, since $\dot{x} = e^{-t^2}$ is always strictly greater than zero, that $x(t)$ increases as t increases, and that

$$x(t) \rightarrow x_0 + \sqrt{\pi}/2$$

as $t \rightarrow \infty$, see Figure 5.8. Even though we cannot write down an explicit form for the solution, we can still say exactly what happens 'eventually'. In this way we can still understand something about the behaviour of the solution. This 'eventual' behaviour is often referred to as the long-time, or time asymptotic, behaviour.

Exercises

5.1 Find the general solution of the following differential equations, and in each case find the particular solution that passes through the origin.

(i)

$$\frac{d\theta}{dt} = \sin t + \cos t,$$

(ii)

$$\frac{dy}{dx} = \frac{1}{x^2 - 1}$$

(use partial fractions)

(iii)

$$\frac{dU}{dt} = 4t \ln t,$$

(iv)

$$\frac{dz}{dx} = xe^{-2x},$$

and

(v)

$$\frac{dT}{dt} = e^{-t} \sin 2t.$$

5.2 Find the function $f(x)$ defined for $-\pi/2 < x < \pi/2$ whose graph passes through the point $(0, 2)$ and has slope $-\tan x$.

5.3 Find the function $g(x)$ defined for $x > -1$ that has slope $\ln(1 + x)$ and passes through the origin.

5.4 Find the solutions of the following equations satisfying the given initial conditions:

(i)

$$\dot{x} = \sec^2 t \qquad \text{with} \qquad x(\pi/4) = 0,$$

(ii)

$$y' = x - \tfrac{1}{3}x^3 \qquad \text{with} \qquad y(-1) = 1,$$

(iii)

$$\frac{d\theta}{dt} = 2 \sin^2 t \qquad \text{with} \qquad \theta(\pi/4) = \pi/4,$$

(iv)

$$x\frac{dV}{dx} = 1 + x^2 \qquad \text{with} \qquad V(1) = 1,$$

and

(v)

$$\frac{d}{dt}\left[x(t)e^{3t}\right] = e^{-t} \qquad \text{with} \qquad x(0) = 3,$$

5.5 The Navier–Stokes equations that govern fluid flow were given as an example in Chapter 3 (see equations (3.1) and (3.2)). It is not possible to find explicit solutions of these equations in general. However, in certain cases the equations reduce to something much simpler.

Suppose that a fluid is flowing down a pipe that has a circular cross-section of radius a. Assuming that the velocity V of the fluid depends only on its distance from the centre of the pipe, the equation satisfied by V is

$$\frac{1}{r}\frac{d}{dr}\left(r\frac{dV}{dr}\right) = -P,$$

where P is a positive constant.

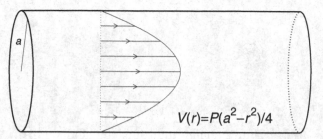

Fig. 5.9. The quadratic velocity profile in a circular pipe.

Multiply by r and integrate once to show that

$$\frac{dV}{dr} = -\frac{Pr}{2} + \frac{c}{r}$$

where c is an arbitrary constant. Integrate again to find an expression for the velocity, and then use the facts that (i) the velocity should be finite at all points in the pipe and (ii) that fluids 'stick' to boundaries (which means that $V(a) = 0$) to show that

$$V(r) = \frac{P}{4}(a^2 - r^2),$$

see Figure 5.9. (This is known as Poiseuille flow.)

5.6 An apple of mass m falls from a height h above the ground. Neglecting air resistance its velocity satisfies

$$m\frac{dv}{dt} = -mg \qquad v(0) = 0,$$

where $v = \dot{y}$ and y is the height above ground level. Show that the apple hits the ground when

$$t = \sqrt{\frac{2h}{g}}.$$

5.7 An artillery shell is fired from a gun, leaving the muzzle with velocity V. If the gun is at ground level and inclined at an angle θ to the horizontal then the initial horizontal velocity if $V \cos \theta$, and the initial vertical velocity is $V \sin \theta$ (see Figure 5.10). The horizontal velocity remains constant, but the vertical velocity is affected by gravity,

Fig. 5.10. Firing a shell at muzzle velocity V at an angle θ to the horizontal. The shell follows a parabolic path.

Fig. 5.11. A recent UK campaign to persuade drivers to cut their speed in town from 35 mph to 30 mph. The film at www.thinkroadsafety.gov.uk/ slowdown/download/slowdown.mpg makes the point more forcefully.

and obeys the equation $v = -g$. How far does the shell travel before it hits the ground? (Give your answer in terms of V and θ.)

5.8 In Dallas on 22 November 1963, President Kennedy was assassinated; by Lee Harvey Oswald if you do not believe any of the conspiracy theories. Oswald fired a Mannlicher–Carcano rifle from approximately 90 m away. The sight on Oswald's rifle was less than ideal; if the bullet travelled in a straight line after leaving the rifle (at a velocity of roughly 700 m/s) then the sight aimed about 10 cm too high at a target 90 m away. How much would the drop in the trajectory due to gravity compensate for this? (The initial vertical velocity v is zero, and satisfies the equation $\dot{v} = -g$, while the horizontal velocity is constant if we neglect air resistance.)

5.9 (T) This exercise fills in the gaps in the proof of the Fundamental Theorem of Calculus. Suppose that f is continuous at x, i.e. given any $\epsilon > 0$, there exists a $\delta = \delta(\epsilon)$ such that

$$|\tilde{x} - x| \leq \delta \quad \Rightarrow \quad |f(\tilde{x}) - f(x)| \leq \epsilon.$$

By writing

$$f(x) = \frac{1}{\delta x} \int_x^{x+\delta x} f(x) \, d\tilde{x}$$

show that for all δx with $|\delta x| \leq \delta(\epsilon)$

$$\left| f(x) - \frac{1}{\delta x} \int_x^{x+\delta x} f(\tilde{x}) \, d\tilde{x} \right| \leq \epsilon,$$

and hence that

$$\lim_{\delta x \to 0} \frac{1}{\delta x} \int_x^{x+\delta x} f(\tilde{x}) \, d\tilde{x} = f(x).$$

You will need to use the fact that

$$\left| \int_a^b g(x) \, dx \right| \leq \int_a^b |g(x)| \, dx \leq (b - a) \max_{x \in [a,b]} |g(x)|.$$

6

Existence and uniqueness of solutions

Because we are going to spend some time trying to solve equations like

$$\frac{dx}{dt} = f(x, t) \tag{6.1}$$

we need to be sure that such equations will actually have solutions. Clearly it is a hopeless task to search for a solution of (6.1) if the solution does not exist (hunting for a unicorn will take you a very long time).

6.1 The case for an abstract result

It is quite easy to write down equations that do not have any solutions, for example

$$x^2 + t^2\frac{dx}{dt} = 0 \qquad x(0) = c$$

does not have any solutions if $c \neq 0$: if $t = 0$ then the second term of the differential equation disappears and we must have $x(0) = 0$.

We have already seen that there are many possible solutions when we want to find a function whose graph has a particular slope; the question of the uniqueness of solutions of a differential equation is somewhat subtle. However, we saw that by specifying a particular initial condition we could tie down one particular solution. So the problem that we will consider for our general theory will be the *initial value problem* (IVP), consisting of the differential equation supplemented by an initial condition,

$$\frac{dx}{dt} = f(x, t) \qquad x(t_0) = x_0. \tag{6.2}$$

If we suppose for a moment that the independent variable is time, then the requirement of uniqueness has a physical interpretation. Suppose that we specify an initial condition 'now', i.e. at time $t = 0$; then the existence of a unique solution

38

means that we can use the equation to predict the future, since the solution is uniquely determined for $t > 0$. In this context, uniqueness of solutions is equivalent to the requirement that our model be deterministic.

Uniqueness is also useful since occasionally we may be able to guess what the solution of an equation is. If we substitute this guess in and it works, then it must in fact be *the* solution since we know that there is no other. We have already used this implicitly in Chapter 1 when we just checked that our solution $N(t) = N_s e^{k(t-s)}$ worked, and then assumed that it must be the only solution.

As with existence, uniqueness is not automatic. The innocuous looking IVP

$$dx/dt = \sqrt{x} \qquad x(0) = 0 \qquad (6.3)$$

has an infinite number of solutions. The 'obvious' solution is $x(t) = 0$ for all $t \geq 0$. But if you choose any value of $c > 0$, the function

$$x_c(t) = \begin{cases} 0 & t \leq c \\ (t-c)^2/4 & t > c \end{cases}$$

also satisfies the equation. Here the solution 'waits around' at $x = 0$, before eventually 'deciding' (at time $t = c$) to wander off to infinity. Some of the solutions of (6.3) are shown in Figure 6.1.

The issues of existence and uniqueness are real, and it is possible to come up with very simple equations in which they fail. The good news is that there is a very general theorem guaranteeing existence and uniqueness, with a hypothesis which is very simple to check.

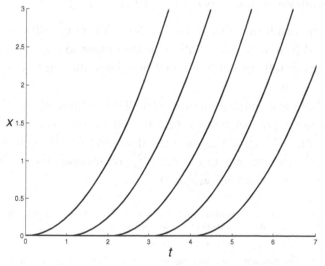

Fig. 6.1. A number of solutions of equation (6.3).

6.2 The existence and uniqueness theorem

The proof of the general existence and uniqueness theorem is beyond the scope of this book, and we will just state the result. However, if you are interested Exercise 6.4 leads you through an outline version of the proof.

In order to state the theorem properly we need to have a more precise idea of what we mean by a 'solution' of the initial value problem (6.2). The main point of the definition is that we allow for a solution to be defined only for some interval of t values, and do not require it to be defined for every $t \in \mathbb{R}$.

Definition 6.1 *Given an open interval I that contains t_0, a solution of the initial value problem*

$$\frac{dx}{dt}(t) = f(x, t) \qquad \text{with} \qquad x(t_0) = x_0 \qquad (6.4)$$

on I is a differentiable function $x(t)$ defined on I, with $x(t_0) = x_0$ and $\dot{x}(t) = f(x, t)$ for all $t \in I$.

The way that the definition specifies the interval on which the solution exists (rather than insisting that it be defined for every value of $t \in \mathbb{R}$) may seem pedantic at first, but we will soon see that this is necessary even for some very simple equations, since it is possible for the solution to 'blow up' in a finite time.

But for now, given our formal definition of a solution, we can state the existence and uniqueness theorem.[1]

Theorem 6.2 *If $f(x, t)$ and $\partial f/\partial x(x, t)$ are continuous for $a < x < b$ and for $c < t < d$ then for any $x_0 \in (a, b)$ and $t_0 \in (c, d)$ the initial value problem (6.4) has a unique solution on some open interval I containing t_0.*

Essentially the result says that if the function $f(x, t)$ is 'sufficiently nice' then the equation will have a unique solution, at least close to $t = t_0$ (see Figure 6.2). However, the result tells us nothing about how large the interval is on which the solution can be defined.

In almost all of the examples we meet, f will be 'sufficiently nice'; but we have already seen one simple example in (6.3) for which there is no uniqueness. This does not contradict Theorem 6.2, since the derivative of $x^{1/2}$ is infinite at $x = 0$: when $f(x) = x^{1/2}$, we have $f'(x) = \frac{1}{2}x^{-1/2}$, and this becomes infinite as $x \downarrow 0$, so f' is certainly not continuous at $x = 0$.

---

[1] In fact the conditions on f in the theorem are a little stronger than they need to be. It is only necessary that the function f is a *Lipschitz continuous* function of x, which means that

$$|f(x, t) - f(y, t)| \le L|x - y| \qquad (6.5)$$

for some constant L. Any function with continuous first derivative is Lipschitz continuous (see Exercise 6.2), but not every function that is Lipschitz continuous has continuous first derivative (e.g. $f(x) = |x|$).

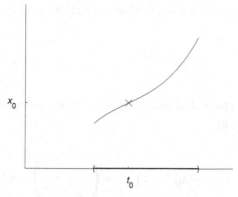

Fig. 6.2. Given an initial condition $x(t_0) = x_0$, the existence and uniqueness theorem only guarantees the existence of a solution defined on some open interval (marked by the bold line on the horizontal axis) containing the initial time t_0.

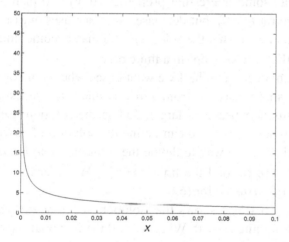

Fig. 6.3. The derivative of $x^{1/2}$, plotted here against x, is not continuous at zero, where it is infinite.

6.3 Maximal interval of existence

We now give an example showing that we need the freedom to specify the interval on which the solution of an equation exists if we want a result as general as Theorem 6.2:

$$\frac{\mathrm{d}x}{\mathrm{d}t} = x^2 \qquad x(0) = x_0. \tag{6.6}$$

Since x^2 and its derivative $2x$ are continuous, the equation certainly has a unique solution that exists in some open interval containing $t = 0$. We will see how to derive the solution of this equation in Chapter 8. For now observe that

$$x(t) = \frac{1}{x_0^{-1} - t} \tag{6.7}$$

satisfies the equation (provided that $x_0 \neq 0$): clearly when $t = 0$ we have $x(0) = x_0$, and differentiating gives

$$\frac{dx}{dt} = -1 \times -\left(x_0^{-1} - t\right)^{-2} = \left(\frac{1}{x_0^{-1} - t}\right)^2 = [x(t)]^2,$$

so that the equation is satisfied. Since we know that the solution of the equation is unique, we must have *the* solution.

Our solution has some interesting properties. If $x_0 > 0$ then the denominator is initially positive (at $t = 0$), but decreases as t increases until it reaches zero at time $t = x_0^{-1}$. This means that the solution, $x(t)$, has become infinite by the time $t = x_0^{-1}$; we say that it 'blows up' in a finite time.

Things are much nicer, though, if we want to see where our solution came from in the past. We can decrease t (from zero) as much as we like, since as t decreases the denominator becomes larger, and so the solution itself tends to zero as $t \to -\infty$. So when $x_0 > 0$ we can define the solution of (6.6) on the interval $(-\infty, x_0^{-1})$, but there is no way to define the solution *on an interval* that extends further into the future beyond the time $t = x_0^{-1}$. We refer to $(-\infty, x_0^{-1})$ as the *maximal interval of existence* for (6.6).

Note that Figure 6.4 also shows that solutions with $x_0 < 0$ tend to $-\infty$ as t decreases towards a finite $t^* < 0$. When $x_0 < 0$ the maximal interval of existence is $(x_0^{-1}, +\infty)$, and only for $x_0 = 0$ can we define a solution for all $t \in \mathbb{R}$ (and then the solution is $x(t) \equiv 0$).

The two ill-behaved equations in this chapter ($\dot{x} = x^{1/2}$ and $\dot{x} = x^2$) should serve as cautionary examples as to the limitations of the existence and uniqueness theorem. That said, almost all the examples we meet in what follows will have unique solutions that exist at least for all $t \geq 0$.

6.4 The Clay Mathematics Institute's $1 000 000 question

The questions of existence and uniqueness are somewhat dry, but there are still extremely important mathematical models for which these issues are not resolved. Outstandingly, it is still not known whether the Navier–Stokes equations that

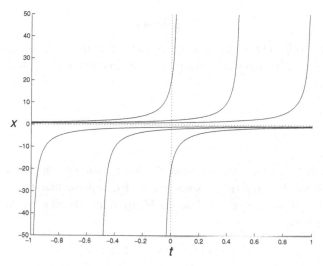

Fig. 6.4. For positive initial conditions the solutions of $\dot{x} = x^2$ blow up in a finite time, but exist for all negative values of t; while for negative initial conditions the solutions blow up for a finite value of $t < 0$ but exist for all $t > 0$.

model the flow of fluids,

$$\rho\left[\frac{\partial \mathbf{u}}{\partial t} + (\mathbf{u} \cdot \nabla)\mathbf{u}\right] - \mu \Delta \mathbf{u} + \nabla p = \mathbf{f} \qquad \nabla \cdot \mathbf{u} = 0$$

(cf. (3.3)), have unique solutions that exist for all positive times.

These equations are the basis of computational design of everything that involves fluid flow; given that the term 'fluid' includes both liquids (in particular water) and gases (in particular air), numerical methods based on these equations are extremely important commercially. Clearly given the financial investment involved, people are confident that these equations really can predict the behaviour of physical systems, but currently we have no guarantee. Most tellingly, you cannot prove that a numerical approximation is 'close' to the 'true solution' if you do not even know that such a solution exists.

For the year 2000 the Clay Mathematics Institute, based in America, announced seven Millennium Prize Problems; for the solution of any of these they will award a prize of one million dollars.[2] One of these problems is to determine whether or not the three-dimensional Navier–Stokes equations are indeed a good physical model, i.e. whether or not they have unique solutions valid for all positive times. There are of course, two ways to win this prize: either to invent some insightful new mathematics that will prove the existence of unique solutions; or to dream up a single initial condition for which the solution breaks down.

[2] See www.claymath.org/index.htm

Exercises

6.1 Which of the following differential equations have unique solutions (at least on some small time interval) for any non-negative initial condition $(x(0) \geq 0)$?

    (i) $\dot{x} = x(1 - x^2)$

    (ii) $\dot{x} = x^3$

    (iii) $\dot{x} = x^{1/3}$

    (iv) $\dot{x} = x^{1/2}(1 + x)^2$

    (v) $\dot{x} = (1 + x)^{3/2}$.

6.2 (T) The Mean Value Theorem says that if f is differentiable on an interval $[a, b]$ then $f(a) - f(b) = (b - a)f'(c)$ for some $c \in (a, b)$. Suppose that $f(x)$ is differentiable with $|f'(x)| \leq L$ for $a \leq x \leq b$. Use the Mean Value Theorem to show that for $a \leq x, y \leq b$ we have

$$|f(x) - f(y)| \leq L|x - y|.$$

6.3 (T) This exercise gives a simple proof of the uniqueness of solutions of

$$\dot{x} = f(x, t) \qquad x(t_0) = x_0, \tag{E6.1}$$

under the assumption that

$$|f(x, t) - f(y, t)| \leq L|x - y|. \tag{E6.2}$$

Suppose that $x(t)$ and $y(t)$ are two solutions of (E6.1). Write down the differential equation satisfied by $z(t) = x(t) - y(t)$, and hence show that

$$\frac{d}{dt}|z|^2 = 2z[f(x(t), t) - f(y(t), t)].$$

Now use (E6.2) to show that

$$\frac{d}{dt}|z|^2 \leq 2L|z|^2.$$

If $dZ/dt \leq cZ$ it follows that $Z(t) \leq Z(t_0)e^{c(t-t_0)}$ (see Exercise 9.7): use this to deduce that the solution of (E6.1) is unique. Hint: any two solutions of (E6.1) agree when $t = t_0$.

6.4 (T) The proof of existence of solutions is much more involved than the proof of their uniqueness. We will consider here the slightly simpler case

$$\dot{x} = f(x) \qquad \text{with} \qquad x(0) = x_0, \tag{E6.3}$$

assuming that

$$|f(x) - f(y)| \leq L|x - y|. \tag{E6.4}$$

The first step is to convert the differential equation into an integral equation that is easier to deal with: we integrate both sides of (E6.3) between times 0 and t to give

$$x(t) = x_0 + \int_0^t f(x(\tilde{t})) \, d\tilde{t}. \tag{E6.5}$$

This integral equation is equivalent to the original differential equation; any solution of (E6.5) will solve (E6.3), and vice versa.

The idea behind the method is to use the right-hand side of (E6.5) as a means of refining any 'guess' of the solution $x_n(t)$ by replacing it with

$$x_{n+1}(t) = x_0 + \int_0^t f(x_n(\tilde{t})) \, d\tilde{t}. \qquad (E6.6)$$

We start with $x_0(t) = x_0$ for all t, set

$$x_1(t) = x_0 + \int_0^t f(x_0) \, d\tilde{t},$$

and continue in this way using (E6.6). The hope is that $x_n(t)$ will converge to the solution of the differential equation as $n \to \infty$.

(i) Use (E6.4) to show that

$$|x_{n+1}(t) - x_n(t)| \le L \int_0^t |x_n(\tilde{t}) - x_{n-1}(\tilde{t})| \, d\tilde{t},$$

and deduce that

$$\max_{t \in [0, 1/2L]} |x_{n+1}(t) - x_n(t)| \le \frac{1}{2} \max_{t \in [0, 1/2L]} |x_n(t) - x_{n-1}(t)|. \qquad (E6.7)$$

(ii) Using (E6.7) show that

$$\max_{t \in [0, 1/2L]} |x_{n+1}(t) - x_n(t)| \le \frac{1}{2^{n-1}} \max_{t \in [0, 1/2L]} |x_1(t) - x_0(t)|.$$

(iii) By writing

$$x_n(t) = [x_n(t) - x_{n-1}(t)] + [x_{n-1}(t) - x_{n-2}(t)]$$
$$+ \cdots + [x_1(t) - x_0(t)] + x_0(t)$$

deduce that

$$\max_{t \in [0, 1/2L]} |x_n(t) - x_m(t)| \le \frac{1}{2^{N-2}} \max_{t \in [0, 1/2L]} |x_1(t) - x_0(t)| \qquad (E6.8)$$

for all $n, m \ge N$.

It follows that $x_n(t)$ converges to some function $x_\infty(t)$ as $n \to \infty$, and therefore taking limits in both sides of (E6.6) implies that

$$x_\infty(t) = x_0 + \int_0^t f(x_\infty(\tilde{t})) \, d\tilde{t}.$$

Thus $x_\infty(t)$ satisfies (E6.5), and so is a solution of the differential equation. The previous exercise shows that this solution is unique.

7

Scalar autonomous ODEs

For the most part when considering first order equations we will concentrate on finding explicit solutions. However, in this chapter we will see how, for the particular class of equations of the form

$$\frac{dx}{dt} = f(x),$$

we can understand the solutions 'qualitatively', even if we cannot (or do not) write down their solutions explicitly.

What this means is that instead of writing '$x(t) =$ something' we describe how the solutions behave, e.g. 'any solution starting with $x(0)$ between zero and one tends to $x = 1$ as $t \to \infty$' or 'the point $x = -1$ is stable'. There is a very simple way to represent all this information about solutions pictorially, and the method essentially reduces to sketching the graph of the function f (in fact we only need to know where f is positive and negative). Nevertheless, the qualitative results we will obtain are completely rigorous.

7.1 The qualitative approach

The key observation is that the existence and uniqueness result of Theorem 6.2 tells us that, provided f is 'nice', a solution of

$$\frac{dx}{dt} = f(x) \tag{7.1}$$

is completely determined by its value at any time t. The equation itself can then be used to determine whether $x(t)$ is increasing or decreasing, depending on the sign of f.

The easiest way to think of this kind of equation is to imagine that $x(t)$ represents the position of a particle moving on a line at time t. We can then talk about the 'velocity of the particle' rather than the more cumbersome 'rate of change of x'.

46

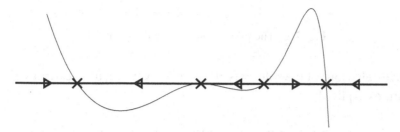

Fig. 7.1. A sketch of the function f against x, and the phase diagram for the equation $\dot{x} = f(x)$. On the phase diagram the stationary points are represented as crosses and the arrows indicate whether the solution is increasing or decreasing.

Whatever the equation really represents, we can use the particle idea while solving it, and then reinterpret our results for the original application when we have finished.

In order to understand how solutions behave we first find all the values of x at which the particle does not move; this happens at the points x^* where $f(x^*) = 0$. As is often the case with fundamental ideas, such points have many names; we will call them *stationary points*.[1]

In regions where $f(x) > 0$ the solution $x(t)$ is increasing, and so the particle is moving to the right; similarly wherever $f(x) < 0$ the solution $x(t)$ is decreasing, and the particle is moving to the left. Note that the particle cannot reverse the direction in which it is moving; if it were to do this then at some time t^* it would have to be instantaneously at rest, so that $\dot{x}(t^*) = f(x(t^*)) = 0$. But then $x(t^*)$ would be a stationary point, and so the particle would not be able to move.

The simplest way to present all this information is to draw a line representing the x coordinate. Stationary points are usually indicated by a cross ($\times$); we then draw arrows on the line indicating the direction in which $x(t)$ is changing: if $f(x) > 0$ then the particle will move to the right and if $f(x) < 0$ then the particle will move to the left. If we sketch the graph of f then it is easy to see the regions in which f is positive and negative. An example is shown in Figure 7.1.

The picture of the line, with the stationary points and the direction of travel of the solution indicated, is known as the 'phase diagram' or 'phase portrait', which is shown on its own in Figure 7.2. (Figure 7.1 has two components; the phase diagram, and the sketch of f that makes it easier to draw.) With this kind of picture

---

[1] Other common terms are *equilibrium points, fixed points* and *critical points*. 'Equilibrium point' has a more physical flavour than the general tone of this book, and we will reserve the term 'fixed point' for use with iterated maps in Chapter 23 (a fixed point will be a point for which $x^* = f(x^*)$). We will use the term 'critical point' for a point at which a function $F(\mathbf{x})$ has all its partial derivatives zero, see Appendix C.

Fig. 7.2. The phase diagram from Figure 7.1.

qualitative behaviour of the solutions at a glance, even when we cannot write down the solutions explicitly.

7.2 Stability, instability and bifurcation

Looking at the phase diagram for the above example, we can see that some stationary points are 'attracting' (nearby solutions approach), while some appear to be 'repelling' (nearby solutions move away). These ideas can be made mathematically precise and are extremely important in applications.

A stationary point is *stable* if when you start close enough to it you stay close to it. More precisely, a stationary point x^* is stable if given any $\epsilon > 0$ there exists a $\delta > 0$ such that

$$\underbrace{|x_0 - x^*| < \delta}_{\text{start close enough}} \qquad \Rightarrow \qquad \underbrace{|x(t) - x^*| < \epsilon \quad \text{for all} \quad t \geq 0}_{\text{stay close}}. \qquad (7.2)$$

The stationary points in the example above have the stronger property of being *attracting*, which means that if you start close enough you actually tend to the stationary point: there exists a $\delta > 0$ such that

$$\underbrace{|x_0 - x^*| < \delta}_{\text{start close enough}} \qquad \Rightarrow \qquad \underbrace{x(t) \to x^* \quad \text{as} \quad t \to +\infty}_{\text{tend to}}. \qquad (7.3)$$

In one-dimensional systems attracting points are stable (see Exercise 7.9) but this is not true in general. Conversely, an example with many stable stationary points that are not attracting is

$$\frac{\mathrm{d}x}{\mathrm{d}t} = \begin{cases} -x & x < 0 \\ 0 & 0 \leq x \leq 1 \\ 1 - x & x > 1. \end{cases}$$

Here, all the points in the interval $[0, 1]$ are stationary points, and they are all stable. However, none of them are attracting, since there are nearby points that move no closer. The phase diagram is shown in Figure 7.3.

A stationary point is *unstable* if it is not stable; this means that no matter how close (δ) you start you will always move some stationary distance (ϵ) away: there exists an $\epsilon > 0$ such that whatever $\delta > 0$ you take there is a point with $|x_0 - x^*| < \delta$ but $|x(t) - x^*| > \epsilon$ for some $t > 0$.

The stationary points from Figure 7.2 are labelled according to their stability or instability in Figure 7.4.

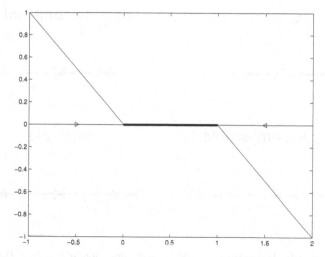

Fig. 7.3. The thick line consists entirely of stationary points, all of which are stable but none of which are attracting.

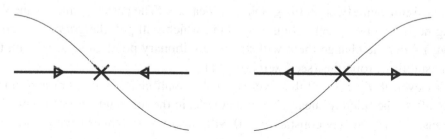

Fig. 7.4. The phase diagram from Figure 7.2, but with the stationary points labelled according to their stability type, S for stable and U for unstable.

Fig. 7.5. The stability or instability of a stationary point x^* can be determined from the value of $f'(x^*)$ provided that $f'(x^*) \neq 0$. If $f'(x^*) < 0$ then the stationary point is stable, and if $f'(x^*) > 0$ the stationary point is unstable.

7.3 Analytic conditions for stability and instability

There are very simple conditions on the derivative of f which will let us know whether a stationary point x^* is stable or unstable without having to sketch the graph of f.

If the graph of f near x^* looks as shown in the left-hand side of Figure 7.5, i.e. if $f'(x^*) < 0$, then the point will be stable, while if the graph of f looks like the right-hand side, i.e. if $f'(x^*) > 0$, then the point will be unstable. Only

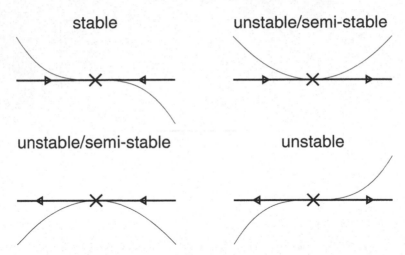

Fig. 7.6. The various stability possibilities when $f'(x^*) = 0$.

when $f'(x^*) = 0$ is there any ambiguity; there are four possibilities, pictured in Figure 7.6. The top right and bottom left cases are sometimes called 'semi-stable', since the stationary point is stable 'on one side' and unstable on the other.

7.4 Structural stability and bifurcations

Observe that if $f'(x^*) \neq 0$ then making small changes to the function f will not have a significant effect on the graph of f near x^*. (The easiest 'small change' to imagine is adding or subtracting a constant, which will pull the graph of f up or down.) After the change there will still be a stationary point close to x^* with the same stability properties (see Exercise 7.11).

However, if $f'(x^*) = 0$ then we can make small changes to $f(x)$ and drastically affect the 'picture' near x^*. For example, in the top right case of Figure 7.6, increasing $f(x)$ by any constant $c > 0$ will mean we no longer have a stationary point.

When we make a small change to f but the phase diagram changes drastically we say that the equation has undergone a *bifurcation*. In these simple examples, we cannot have a bifurcation near x^* unless $f'(x^*) = 0$. When small changes to $f(x)$ cannot affect the qualitative nature of the phase diagram we say that $\dot{x} = f(x)$ is *structurally stable*.

We will look at a particular example of a bifurcation in Section 7.6; watch for a stationary point with $f'(x^*) = 0$.

7.5 Some examples

We now consider various examples using this graphical method.

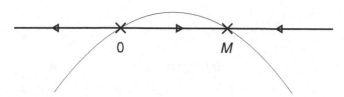

Fig. 7.7. The phase diagram for the population model (7.4). Note that this picture includes solutions with $p < 0$, which, while mathematically sensible, are irrelevant for this application.

7.5.1 A population model

The equation

$$\frac{\mathrm{d}p}{\mathrm{d}t} = kp \left(1 - \frac{p}{M}\right) \qquad \text{with} \qquad k, M > 0 \tag{7.4}$$

is a model for the change in the size of a population.[2] We will study this model in more detail in the next chapter, but for now we try to understand its qualitative behaviour.

The first step is to find the stationary points. These occur where the right-hand side is zero, i.e. when

$$kp \left(1 - \frac{p}{M}\right) = 0,$$

so they are $p = 0$ and $p = M$. If we sketch the graph of $f(p) = kp(1 - (p/M))$ then it is easy to draw the phase diagram, remembering that solutions move to the right whenever $f > 0$ and to the left whenever $f < 0$. The phase diagram is shown in Figure 7.7. It is easy to see from the diagram that provided that we start with a positive population then it will eventually settle down to the value at the stationary point $p = M$; smaller populations will tend to increase, while larger populations will shrink towards this value.

We can check the stability of the stationary points analytically by looking at the derivative of f,

$$f'(p) = k - (2kp/M).$$

At the origin $f'(0) = k > 0$, and the origin is unstable (as we expected), and $f'(M) = -k < 0$, confirming that the stationary point at $p = M$ is stable.

Note that this kind of solution can tell us what happens eventually, and how the population changes qualitatively. But it only allows us to make a very limited

---

[2] There are some very reasonable objections to this model; in particular, unlike the number of people in a population, the variable p does not have to be an integer. We can get round this to some extent by claiming that p is 'the population in millions', and then p can be a decimal. There are still values that do not correspond to whole numbers of people, but the equation will now be a good approximation.

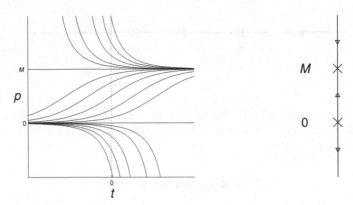

Fig. 7.8. The phase diagram (on the right) reflects the behaviour of solutions (a collection of which are shown on the left).

number of quantitative predictions. For example, if we know that equation (7.4) is the right model for our population, but do not know the values of k and M, there is no way that we can use our phase diagram to find k and M given a collection of data. However, this is possible using the explicit solution, as we will see in the next chapter.

Figure 7.8, shows how the phase diagram (rotated to be vertical in the figure) reflects the behaviour of the solutions, a collection of which are plotted against t.

7.5.2 Terminal velocity

Sometimes we do not need an explicit solution to find the quantitative information we require. Here we will use the phase diagram to find the terminal velocity of a falling object.

Suppose that a body of mass m is falling under gravity g and is subject to an air resistance proportional to the square of its velocity, kv^2. The equation for the downward velocity v is

$$m\frac{dv}{dt} = mg - kv|v|, \tag{7.5}$$

since gravity serves to accelerate the particle, and the air resistance acts in the opposite direction to v.

Provided that the particle is moving downwards, so that $v > 0$, equation (7.5) becomes

$$m\frac{dv}{dt} = mg - kv^2.$$

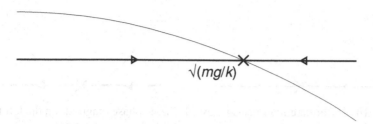

Fig. 7.9. You can simply read off the terminal velocity $v = \sqrt{mg/k}$.

We can rewrite this as

$$\frac{dv}{dt} = f(v) = g - \frac{k}{m}v^2.$$

For this equation there is only one stationary point, when

$$g - \frac{k}{m}v^2 = 0,$$

i.e. when $v = v^* = \sqrt{mg/k}$. This point is stable, as can be seen by looking at the derivative of f at v^*: since $f'(v) = -2kv/m$ we have $f'(v^*) = -2\sqrt{gk/m} < 0$.

The phase diagram is shown in Figure 7.9; it is clear that there is an attracting stationary point at $v = \sqrt{mg/k}$. This is the terminal velocity, since $v(t)$ approaches this value whatever the initial condition.

For a skydiver of mass 100 kg in freefall, we can take $k \approx 1/3$ kg/m and $g \approx 9.8$ m/s^2. It follows that the terminal velocity of the skydiver is

$$v = \sqrt{100 \times 9.8 \times 3} \approx 54.2 \, \text{m/s}.$$

7.5.3 What have we lost?

We will now see that we are missing some, at times vital, information if we only rely on the phase diagram. The phase diagrams for the two equations $\dot{x} = |x|$ and $\dot{x} = x^2$ are shown in Figure 7.10: although the equations are different their phase diagrams are the same.

In the next chapter we will see how to calculate the solutions of both of these equations. For now we will assume that we know what these solutions are; for an initial condition $x(0) = x_0 > 0$, the solution of $\dot{x} = |x|$ is

$$x(t) = x_0 e^t,$$

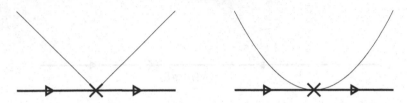

Fig. 7.10. Different equations can have the same phase diagram; on the left is the phase diagram for $\dot{x} = |x|$, and on the right the phase diagram for $\dot{x} = x^2$.

while the solution of $\dot{x} = x^2$ is

$$x(t) = \frac{1}{x_0^{-1} - t},$$

(cf. (6.7)).

The first solution increases to infinity, but is defined for all $t \geq 0$ (in fact for all $t \in \mathbb{R}$). However, we have already used the equation $\dot{x} = x^2$ to show that it is possible for solutions to blow up in finite time ($x(t) \to \infty$ as $t \to x_0^{-1}$). This finite time blowup behaviour is not captured in any way by our phase diagram.

So although the phase diagram gives the correct qualitative behaviour, we have lost all information on the rates at which things happen. When the equation exhibits blow up of solutions in a finite time, this is particularly unfortunate.

7.6 The pitchfork bifurcation

We now consider the equation

$$\dot{x} = x(k - x^2), \tag{7.6}$$

where k is a parameter. By varying k we can study a whole family of differential equations. We will see that the qualitative behaviour of the solutions of (7.6) changes drastically as k passes through zero.

When $k \leq 0$ there is only one stationary point, that at $x = 0$. If we write $f(x) = x(k - x^2)$ then $f'(x) = k - 3x^2$, and at the origin we have $f'(0) = k$. It follows that for $k < 0$ this stationary point is stable. The phase diagram is shown in Figure 7.11, together with the graph of f.

When $k = 0$ there is still only the one stationary point at $x = 0$, although now $f'(0) = 0$. In order to determine the stability of the origin when $k = 0$ we have to sketch the graph of $f(x) = -x^3$. It is then clear that the origin is still stable, and that the phase diagram is the same for $k = 0$ as it was for $k < 0$, see Figure 7.12.

Since $f'(0) = 0$ when $k = 0$, there is the possibility of a bifurcation as k changes from negative to positive (see Section 7.4). When $k > 0$ there are two

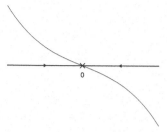

Fig. 7.11. The phase diagram when $k \leq 0$, and the graph of f.

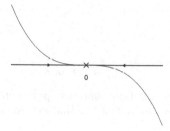

Fig. 7.12. The phase diagram when $k = 0$, and the graph of $f(x) = -x^3$.

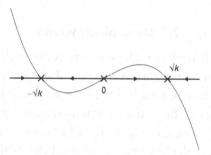

Fig. 7.13. The phase diagram when $k > 0$, and the graph of f.

new stationary points at $x = \pm\sqrt{k}$. While the origin is no longer stable, since $f'(0) = k > 0$, the new fixed points are both stable, since $f'(\pm\sqrt{k}) = -2k < 0$. The phase diagram for this case is that shown in Figure 7.13.

You can see that the phase diagram has changed drastically as k has gone from being negative to positive. We have gone from having one stable stationary point for $k < 0$ to having three stationary points when $k > 0$; two of these are stable, and the origin has become unstable.

We can draw a 'bifurcation diagram' to show these changes. The idea is to draw a graph where the horizontal axis represents the parameter k, and for each value of k we plot the location of the stationary points on the vertical axis, using a solid line when they are stable and a dashed line when they are unstable. This gives

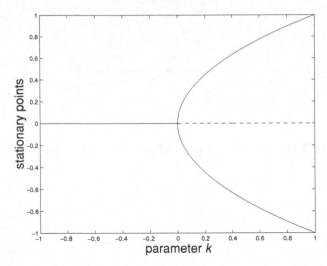

Fig. 7.14. The pitchfork bifurcation; stationary points are plotted against k. Solid lines indicate stable points, and the dashed line an unstable point.

the picture in Figure 7.14. For fairly obvious reasons this is known as a 'pitchfork bifurcation'.

7.7 Dynamical systems

The qualitative approach we have adopted here is the main viewpoint used in the general theory of *dynamical systems*. A dynamical system has two components: the *phase space* (or 'state space'), which consists of all possible 'states' of the system (for the scalar equations of this chapter this is the line $\mathbb{R}$, covering all possible values of x), and the 'dynamics' which describe how these states change in time (for us the dynamics were determined by the solutions of the differential equation $\dot{x} = f(x)$).

With the advent of more powerful computers there have been major advances in the theory of dynamical systems in recent years, and the subject received a lot of attention in the 1980s under the media-friendly 'chaos' banner. We will see more examples of dynamical systems later in the book.

Exercises

7.1 For each of the following differential equations draw the phase diagram, labelling the stationary points as stable or unstable.

    (i) $\dot{x} = -x + 1$
    (ii) $\dot{x} = x(2 - x)$
    (iii) $\dot{x} = (1 + x)(2 - x)\sin x$
    (iv) $\dot{x} = -x(1 - x)(2 - x)$
    (v) $\dot{x} = x^2 - x^4$

7.2 For the equations in Exercise 7.1 determine the stability of the stationary points analytically, by considering the sign of the derivative of the right-hand side.

7.3 For all positive values of c find all the stationary points of

$$\frac{dx}{dt} = \sin x + c,$$

and determine analytically which are stable and unstable. Draw the portion of the phase diagram between $-\pi$ and π. There are three different cases, $0 \le c < 1, c = 1,$ and $c > 1$. You will need to be more careful with the case $c = 1$.

7.4 A simple model of the spread of an infection in a population is

$$\dot{H} = -kIH$$
$$\dot{I} = kIH,$$

where $H(t)$ is the number of healthy people, $I(t)$ the number of infected people and k the rate of infection. Since $(d/dt)(H + I) = 0$, it follows that the size of the population is constant, $H + I = N$, say. Substitute $I = N - H$ in order to obtain a single equation for $H(t)$,

$$\frac{dH}{dt} = -kH(N - H).$$

Determine the stability of the stationary points for this equation, and draw its phase diagram. Deduce that eventually all the population becomes infected.

7.5 Consider the equation

$$\frac{dx}{dt} = f(x) \equiv x^2 - k.$$

Draw the phase diagram for the three cases $k < 0$, $k = 0$ and $k > 0$, labelling the stationary points as stable or unstable in each case. Find the stability of the stationary points using an analytic method when $k > 0$. Show that $f'(0) = 0$ when $k = 0$. Why is this significant?

Draw the bifurcation diagram, with k on the horizontal axis and the fixed points plotted against k, indicating stable fixed points by a solid line and unstable fixed points by a dashed line. (This is known as a *saddle node* bifurcation.)

7.6 Draw the phase diagram for the equation

$$\dot{x} = g(x) = kx - x^2$$

for $k < 0, k = 0$ and $k > 0$. Check the stability of the stationary points by considering $g'(x)$, and show that the two stationary points exchange stability as k passes through zero. Draw the bifurcation diagram for this *transcritical* bifurcation.

7.7 One equation can exhibit a number of bifurcations. Find, depending on the values of k, all the stationary points of the equation

$$\dot{x} = h(x) = -(1 + x)(x^2 - k)$$

and by considering $h'(x)$ determine their stability. At which points, and for which values of k, are there possible bifurcations?

Draw representative phase diagrams for the five distinct parameter ranges $k < 0$, $k = 0, 0 < k < 1, k = 1$ and $k > 1$, and then draw the bifurcation diagram. Identify the type of the two bifurcations.

In the remaining exercises assume that f is a C^1 function, i.e. that both f and df/dx are continuous functions. Note that such an f is smooth enough to guarantee that the equation $\dot{x} = f(x)$ with $x(t_0) = x_0$ has a unique solution. You may also assume that the solutions are defined for all $t \geq 0$.

7.8 (T) Let $x(t)$ be one solution of the differential equation

$$\dot{x} = f(x).$$

Show that
   (i) if $f(x(t^*)) = 0$ for some t^* then $x(t) = x(t^*)$ for all $t \in \mathbb{R}$ (the solution is con-
       stant, and $x(t^*)$ is a stationary point); and hence
   (ii) if $f(x(t^*)) > 0$ for some t^* then $f(x(t)) > 0$ for all $t \in \mathbb{R}$ (the solution can-
       not 'reverse direction'). Hint: Use the Intermediate Value Theorem: if g is
       a continuous function with $g(a) < 0$ and $g(b) > 0$ then there is a point c
       between a and b with $g(c) = 0$.
   Of course, a similar result to (ii) holds if $f(x(t^*)) < 0$ for some t^*.
7.9 (T) Show that for autonomous scalar equations, if x^* is attracting then it must also be
       stable. Hint: use (ii) above.
7.10 (T) Suppose that $x(t)$ is a solution of $\dot{x} = f(x)$ that is moving to the right. Show that
       either $x(t) \to +\infty$, or $x(t) \to x^*$, where x^* is a stationary point. (Hint: If $x(t)$ does
       not tend to infinity then it is increasing and bounded above, and so tends to a limit
       x^*. Show that in this case we must have $f(x^*) = 0$.) A similar result holds if $x(t)$ is
       moving to the left, with $+\infty$ replaced by $-\infty$.
7.11 (T) Suppose that $\dot{x} = f(x)$ has a stable stationary point at x_0, with $f'(x_0) < 0$. Let
       g be another C^1 function. Use the following scalar version of the Implicit Function
       Theorem to show that for ϵ sufficiently small the equation

$$\dot{x} = f(x) + \epsilon g(x)$$

has a unique stationary point near x_0 which is still stable.

Theorem. Suppose that $h(x, \epsilon)$, $\partial h/\partial x$, $\partial h/\partial \epsilon$ are all continuous functions of both x and ϵ. Suppose also that $h(x_0, 0) = 0$ and $\partial h/\partial x(x_0, 0) \neq 0$. Then there is an open interval I that contains x_0 such that for each ϵ sufficiently small there is a unique solution $y(\epsilon) \in I$ of

$$h(y(\epsilon), \epsilon) = 0,$$

and $y(\epsilon)$ depends continuously on ϵ.

8

Separable equations

We now begin our survey of the various different classes of equations that we can solve explicitly. Both the 'trivial' equations

$$\frac{\mathrm{d}x}{\mathrm{d}t}(t) = f(t)$$

of Chapter 5 and the autonomous equations

$$\frac{\mathrm{d}x}{\mathrm{d}t}(t) = f(x)$$

of the previous chapter are particular cases of the *separable equation*

$$\frac{\mathrm{d}x}{\mathrm{d}t} = f(x)g(t) \tag{8.1}$$

which we study in this chapter.

8.1 The solution 'recipe'

If you have already seen these equations, then you will probably be used to solving them in the following way. If these equations are new to you, take careful note; this is the practical way of finding a solution. However, there are steps here that should make you uneasy.

We start with the equation

$$\frac{\mathrm{d}x}{\mathrm{d}t} = f(x)g(t).$$

Now divide by $f(x)$ and 'multiply up by $\mathrm{d}t$' to obtain

$$\frac{1}{f(x)}\,\mathrm{d}x = g(t)\,\mathrm{d}t.$$

This is 'separating the variables', since we now have all the xs on one side and all the ts on the other. For the general solution we integrate both sides to get

$$\int \frac{1}{f(x)} \, dx = \int g(t) \, dt. \tag{8.2}$$

Alternatively, if we want to take into account an initial condition $x(t_0) = x_0$ then we integrate between the limits *that correspond to times t_0 and t*: for the left-hand side these are $x(t_0)$ and $x(t)$, while on the right-hand side they are just t_0 and t. This gives

$$\int_{x_0}^{x(t)} \frac{1}{f(x)} \, dx = \int_{t_0}^{t} g(\tilde{t}) \, d\tilde{t}. \tag{8.3}$$

We now use this recipe to find the solution of the equation we used in Chapter 6 to show that the solutions of a differential equation can blow up in a finite amount of time (i.e. $x(t) \to +\infty$ as $t \to t^* < \infty$). At the time we had no method for solving this equation, and just wrote down the solution, but now we can use the separation method to find it for ourselves.

Example 8.1 *Find the solution of the initial value problem*

$$\frac{dx}{dt} = x^2 \qquad x(0) = x_0.$$

If $x_0 = 0$ then $x(t) = 0$ for all t. Otherwise we can separate the variables to give

$$\frac{1}{x^2} \, dx = dt.$$

Integrating between limits corresponding to times 0 and t,

$$\int_{x_0}^{x(t)} \frac{1}{x^2} \, dx = \int_{0}^{t} d\tilde{t},$$

we obtain

$$\left[-\frac{1}{x} \right]_{x=x_0}^{x(t)} = t.$$

Therefore

$$-\frac{1}{x(t)} + \frac{1}{x_0} = t$$

which simplifies to give

$$x(t) = \frac{1}{x_0^{-1} - t},$$

as we claimed before (cf. (6.7)).   □

8.2 The linear equation $\dot{x} = \lambda x$

We now find the solution of the simplest possible linear differential equation,

$$\frac{\mathrm{d}x}{\mathrm{d}t} = \lambda x \tag{8.4}$$

with the initial condition $x(t_0) = x_0$.

This example is absolutely fundamental (the reasons for this will become apparent later) and you should really only have to solve this equation 'long-hand' once or twice before you are happy to write down the solution with no calculation.

First note that if $x_0 = 0$ then $x(t) = 0$ for all t. Otherwise, if $x \neq 0$ then we can divide by x and 'multiply up by dt' to give

$$\frac{\mathrm{d}x}{x} = \lambda \, \mathrm{d}t.$$

We now integrate both sides between the limits corresponding to the times t_0 and t; that is, x_0 and $x(t)$ on the left, and t_0 and t on the right; and get

$$\int_{x_0}^{x(t)} \frac{\mathrm{d}x}{x} = \int_{t_0}^{t} \lambda \, \mathrm{d}\tilde{t},$$

which gives[1]

$$\left[\ln |x| \right]_{x=x_0}^{x(t)} = \lambda(t - t_0).$$

So we have

$$\ln |x(t)| - \ln |x_0| = \lambda(t - t_0),$$

and taking exponentials (e to the power) of both sides gives

$$\frac{|x(t)|}{|x_0|} = \mathrm{e}^{\lambda(t - t_0)}.$$

To work out what to do about the modulus signs, the easiest thing is to draw the phase diagram. For the case $\lambda > 0$ this is shown in Figure 8.1, from which we can see that $x(t)$ and x_0 have the same sign. It follows that we can remove the modulus signs and multiply up to give

$$x(t) = x_0 \mathrm{e}^{\lambda(t - t_0)}.$$

---

[1] Many of the integrals in this chapter will involve logarithms, and the annoying modulus signs that come with them. We will have to take some care to work out how to remove them for our final answers; the most useful method is to use the phase diagram, as the examples show.

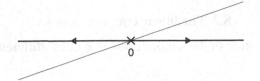

Fig. 8.1. The phase diagram for $\dot{x} = \lambda x$ when $\lambda > 0$ (taking $\lambda < 0$ would reverse the direction of the arrows).

Note that the *general solution* of (8.4) is

$$x(t) = Ae^{\lambda t},$$

see also Exercise 8.5.

8.2.1 Exponential decay and exponential growth

We looked at the solution of equation (8.4) with $\lambda < 0$ in Chapter 1, and applied it to the example of radioactive decay. We saw that the solutions decay to zero exponentially fast, and that the rate of decay could be characterised by the half-life; the solution halves in a fixed time.

When $\lambda > 0$ the solutions tend to infinity as $t \to \infty$, and increase exponentially fast. In this case the size of the solution will double after a fixed time, given by t_2, where

$$2 = e^{\lambda t_2},$$

i.e. $t_2 = \ln 2/\lambda$. In the following section we look at the use of this linear equation as a simple population model.

8.3 Malthus' population model

The simple linear equation

$$\frac{dp}{dt} = kp \qquad \text{with} \qquad k > 0 \tag{8.5}$$

was proposed in 1798 by the English economist Thomas Malthus as a basic model for population growth. Here the increase in the population is taken to be proportional to the total number of people, and k is a constant representing the rate of growth (the difference between the birthrate and the deathrate). This model predicts exponential growth of the population,

$$p(t) = p(t_0)e^{k(t-t_0)},$$

so that its size grows without bound and will double every d years, where $d = \ln 2/k$.

We will now compare the predictions of this model with census data gathered over the last two hundred years. The population of Great Britain and Ireland in 1801, 1851 and 1901 can be found in the results of the Census for each of those years:

$$
\begin{array}{ll}
year & population \\
1801 & 16\,345\,646 \\
1851 & 27\,533\,755 \\
1901 & 41\,609\,091
\end{array}
\tag{8.6}
$$

We can use the data from 1801 and 1851 to estimate k. Our solution predicts

$$p(1851) = p(1801)e^{50k},$$

and so

$$k = \frac{\ln p(1851) - \ln p(1801)}{50} \approx 0.010.$$

This implies that the population will double roughly every 69 years ($\ln 2/k \approx 69$).

Using this value of k, our solution, illustrated in Figure 8.2, gives a reasonable prediction for the population in 1901:

$$p(1901) = p(1801)e^{100k} \approx 46 \text{ million}.$$

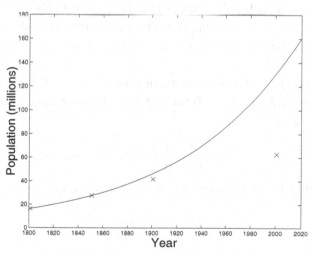

Fig. 8.2. The UK population as predicted by Malthus' linear model. The census values for 1801, 1851, 1901 and 2001 are indicated by crosses.

However, it vastly overestimates the population in 2001 as

$$p(2001) = p(1801)e^{200k} \approx 131 \text{ million},$$

whereas the 2001 census found just below 59 million (in fact[2] the figure is the delightfully precise 58 789 194). To be consistent we should include the figures for the Republic of Ireland, since the data in (8.6) dates from before the partition of Ireland in 1921. The census held there in 2002 found a population of around 4 million.[3] So the total figure for 2001 should be approximately 63 million. Malthus' model has predicted over twice this, so it turns out to be very unreliable when we try to extrapolate the population very far into the future. We will soon see another model that gives much more realistic results.

8.4 Justifying the method

We now give a careful justification of the 'recipe' we outlined in Section 8.1. In particular, you should have worried about the idea of 'multiplying up by dt', since this kind of manipulation of infinitesimal quantities is extremely dubious.

We start again with

$$\frac{dx}{dt} = f(x)g(t), \tag{8.7}$$

and assume that $f(x)$ is sufficiently smooth to ensure that Theorem 6.2 guarantees the existence of a unique solution for any specified initial condition.

First note that if $x(t)$ is a solution of (8.7) with $f(x(s)) = 0$ for some s then in fact $x(t) = x(s)$ for all $t \in \mathbb{R}$. This follows from the uniqueness of solutions; assuming that $x(t) = x(s)$ for all t implies that $f(x(t)) = f(x(s)) = 0$ for all $t \in \mathbb{R}$, and so $\dot{x}(t) = 0$ for all t, showing that this choice for $x(t)$ solves the equation. Since solutions of the IVP are unique, this $x(t)$ must be the only solution with the specified value of $x(s)$.

So either $f(x(t)) = 0$ for every value of t, or $f(x(t)) \neq 0$ for every value of t. We now treat the case $f(x(t)) \neq 0$ for all t, for which we can divide both sides of (8.7) by $f(x)$ to give

$$\frac{1}{f(x)}\frac{dx}{dt} = g(t). \tag{8.8}$$

Now, suppose that $H(x)$ is an anti-derivative of $1/f(x)$, i.e.

$$H'(x) = \frac{1}{f(x)}.$$

---

[2] See www.statistics.gov.uk/census2001/default.asp
[3] The exact figure was 3 917 336, see www.cso.ie/census/prelimimary_details.html#pop.

Then observe that by the chain rule (see Appendix C)

$$\frac{d}{dt}H(x(t)) = H'(x(t))\frac{dx}{dt} = \frac{1}{f(x)}\frac{dx}{dt},$$

and so (8.8) can be rewritten as

$$\frac{d}{dt}H(x(t)) = g(t).$$

To find the solution we can integrate both sides with respect to t to give

$$H(x(t)) = \int g(t)\,dt. \tag{8.9}$$

(To find $x(t)$ explicitly we have to be able to invert H, i.e. solve the equation $H(x) = z$ to obtain x in terms of z. In some cases the implicit form of (8.9) might be the best that we can do.) Since H is an anti-derivative of $1/f$, we could write this symbolically as

$$\int \frac{1}{f(x)}\,dx = \int g(t)\,dt, \tag{8.10}$$

which is precisely what we had before as equation (8.2).

We now see how to recover (8.3) (see equation (8.11) below). If G is an anti-derivative of g then (8.10) reads

$$H(x(t)) = G(t) + c,$$

and so when we want to take into account an initial condition $x(t_0) = x_0$ we need

$$H(x_0) = G(t_0) + c \quad \Rightarrow \quad c = H(x_0) - G(t_0),$$

and the solution is

$$H(x(t)) - H(x_0) = G(t) - G(t_0).$$

Using the method of evaluating an integral by anti-derivatives (which is formalised as (5.3) in the FTC) we can rewrite this as

$$\int_{x_0}^{x(t)} \frac{1}{f(x)}\,dx = \int_{t_0}^{t} g(\tilde{t})\,d\tilde{t}, \tag{8.11}$$

which agrees with the result of our more heuristic derivation above (equation (8.3)).

8.5 A more realistic population model

We now return to population modelling, but rather than allowing the unbounded exponential growth that resulted from Malthus' model

$$\frac{dp}{dt} = kp$$

we impose a maximum sustainable size for the population. The idea is that any species (including ours) is limited by the availability of natural resources. We will find that this new model gives a much better estimate of the current population, even extrapolated from the century-old data we used above.

The so-called 'logistic equation' is

$$\frac{dp}{dt} = kp\left(1 - \frac{p}{M}\right). \tag{8.12}$$

Interpreted as a population model, k is the growth rate of small populations; when p is small, p^2 is very small, so the equation is approximately $dp/dt = kp$, the model we had previously. The parameter $M > 0$ is the maximum sustainable population; when $p < M$ the population increases, and when $p > M$ the population decreases. We drew the phase diagram for this equation in the previous chapter as Figure 7.7, and it will be useful to recall it now (see Figure 8.3) for use below. The phase diagram predicts that eventually the population will settle to its maximum sustainable level, $p = M$.

We now solve the equation explicitly. Separating the variables gives

$$\frac{M}{kp(M - p)}\, dp = dt,$$

where we have multiplied top and bottom of the left-hand side by M. Using the method of partial fractions on the left-hand side this becomes

$$\frac{1}{k}\left[\frac{1}{p} + \frac{1}{M - p}\right]dp = dt$$

or

$$\left[\frac{1}{p} + \frac{1}{M - p}\right]dp = k\, dt. \tag{8.13}$$

Fig. 8.3. The phase diagram for the population model (8.12). In line with the interpretation of p as the size of a population, only the values $p \geq 0$ are shown.

Since

$$\int \frac{1}{p} + \frac{1}{M-p} \, dp = \ln|p| - \ln|M-p|$$

We can integrate both sides of (8.13) between the limits corresponding to times t_0 and t,

$$\int_{p(t_0)}^{p(t)} \frac{1}{p} + \frac{1}{M-p} \, dp = \int_{t_0}^{t} k \, d\tilde{t},$$

to give

$$\left[\ln|p| - \ln|M-p| \right]_{p=p(t_0)}^{p(t)} = \left[k\tilde{t} \right]_{\tilde{t}=t_0}^{t}.$$

Putting in the limits of integration,

$$\ln p(t) - \ln|M - p(t)| - \ln p(t_0) + \ln|M - p(t_0)| = k(t - t_0)$$

(since $p(t) > 0$ we do not need the modulus sign on $\ln|p(t)|$). Equivalently this is

$$\ln\left[\frac{p(t)|M - p(t_0)|}{|M - p(t)|p(t_0)} \right] = k(t - t_0).$$

From the phase diagram in Figure 8.3 it is clear that if $p(t_0) < M$ then $p(t) < M$ for all t, and similarly if $p(t_0) > M$ then $p(t) > M$ for all t. So the sign of $M - p(t)$ does not change for each solution. It follows that we can remove the modulus signs, and then exponentiating both sides we obtain

$$\frac{p(t)(M - p(t_0))}{(M - p(t))p(t_0)} = e^{k(t-t_0)}.$$

Finally, rearranging this gives

$$p(t) = M \left[\frac{p(t_0)e^{k(t-t_0)}}{M - p(t_0) + p(t_0)e^{k(t-t_0)}} \right].$$

With some thought we can read from this explicit solution the same qualitative behaviour we see in the phase diagram. In particular, since $e^{kt} \to \infty$ as $t \to \infty$, we can deduce once again that $p(t) \to M$ as $t \to \infty$.

Since we have an explicit solution we can now estimate the parameters M and k that occur in the equation using the census data quoted above in (8.6). Once we know M and k we can then see what the quantitative predictions of the model are for 2001. The calculations to find M and k are just simple algebra, but are not particularly instructive, so feel free to go straight to the values of M and k in equation (8.14).

Setting $\alpha = e^{50k}$, $p_0 = p(1801)$, $p_1 = p(1851)$, and $p_2 = p(1901)$, our solution requires

$$p_1 = \frac{M p_0 \alpha}{M - p_0 + p_0 \alpha}$$

$$p_2 = \frac{M p_0 \alpha^2}{M - p_0 + p_0 \alpha^2}.$$

(Since there are now two parameters in our equation we need all the data from (8.6) to estimate them.) Rearranging both equations to find M in terms of α and equating we have

$$\frac{p_0 p_1 (\alpha - 1)}{p_0 \alpha - p_1} = \frac{p_0 p_2 (\alpha^2 - 1)}{p_0 \alpha^2 - p_2},$$

and so

$$\alpha = \frac{p_2 (p_1 - p_0)}{p_0 (p_2 - p_1)} \quad \text{and} \quad M = \frac{p_1 (2 p_0 p_2 - p_1 p_2 - p_0 p_1)}{p_0 p_2 - p_1^2}.$$

Using the correct values for p_0, p_1 and p_2 gives

$$\alpha = e^{50k} = 2.0234 \quad \text{and} \quad M = 83.1 \text{ million} \tag{8.14}$$

which implies that $k \approx 0.014$, similar to the value ($k \approx 0.010$) we found for the simple exponential model (8.5).

We can now use these values of k and M to predict the population in 2001. The value we obtain is 66.8 million, surprisingly close to the true figure (which, remember, is about 63 million); the solution is illustrated in Figure 8.4. Of course, there are good reasons for the discrepancy – among them two world wars and the invention of the contraceptive pill in the 1960s.

Note that the constant M that arises in the model represents the maximum sustainable population; at 83 million this is still comfortably above its current level.

8.6 Further examples

We now treat some other examples. Note that often both the method of partial fractions and a quick sketch of the phase diagram are useful tools.

8.6.1 Partial fractions again

We drew the phase diagram for the equation $\dot{x} = x(k - x^2)$ as part of our study of the pitchfork bifurcation in Section 7.6. Here we will consider the case $k > 0$, and so replace k by κ^2, which will make the algebra that is to come a little simpler. We

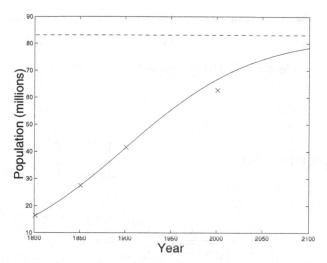

Fig. 8.4. Graph of the population of the UK and the Republic of Ireland, as predicted by the logistic model using the census data for 1801, 1851 and 1901. The curve is our theoretical prediction, and the crosses show the exact values (we have chosen our parameters to ensure that the first three crosses lie on this curve). The dashed line is the maximum sustainable population predicted by our model.

will solve the equation

$$\frac{\mathrm{d}x}{\mathrm{d}t} = x(\kappa^2 - x^2)$$

with a general initial condition $x(0) = x_0$. For the case $k < 0$ see Exercise 8.8.

Separating the variables we have

$$\frac{1}{x(\kappa^2 - x^2)}\, \mathrm{d}x = \mathrm{d}t.$$

We can use the method of partial fractions to rewrite the left-hand side as

$$\frac{1}{x(\kappa^2 - x^2)} = \frac{1}{x(\kappa - x)(\kappa + x)} = \frac{1}{\kappa^2}\left[\frac{1}{x} + \frac{1}{2(\kappa - x)} - \frac{1}{2(\kappa + x)}\right].$$

So we have

$$\frac{1}{x} + \frac{1}{2(\kappa - x)} - \frac{1}{2(\kappa + x)}\, \mathrm{d}x = \kappa^2\, \mathrm{d}t.$$

We can integrate this between the limits corresponding to times 0 and t to give

$$\int_{x_0}^{x(t)} \frac{1}{x} + \frac{1}{2(\kappa - x)} - \frac{1}{2(\kappa + x)}\, \mathrm{d}x = \int_0^t \kappa^2\, \mathrm{d}\tilde{t}$$

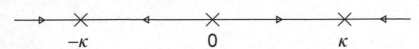

Fig. 8.5. The phase diagram for $\dot{x} = x(\kappa^2 - x^2)$.

which is

$$\left[\ln |x| - \tfrac{1}{2} \ln |\kappa - x| - \tfrac{1}{2} \ln |\kappa + x| \right]_{x=x_0}^{x(t)} = \left[\kappa^2 \tilde{t} \right]_{\tilde{t}=0}^{t}$$

or

$$\left[\ln \frac{|x|}{\sqrt{|\kappa^2 - x^2|}} \right]_{x=x_0}^{x(t)} = \kappa^2 t.$$

Putting in the limits this becomes

$$\ln \left[\frac{|x(t)|\sqrt{|\kappa^2 - x_0^2|}}{|x_0|\sqrt{|\kappa^2 - x(t)^2|}} \right] = \kappa^2 t,$$

and exponentiating both sides we have

$$\frac{|x(t)|\sqrt{|\kappa^2 - x_0^2|}}{|x_0|\sqrt{|\kappa^2 - x(t)^2|}} = e^{\kappa^2 t}.$$

Now if we square both sides and multiply up we have

$$x(t)^2 |\kappa^2 - x_0^2| = x_0^2 |\kappa^2 - x(t)^2| e^{2\kappa^2 t}. \tag{8.15}$$

We drew the phase diagram for this example in the previous chapter (see Figure 7.13), and it is reproduced here as Figure 8.5. It is easy to see from the phase diagram that if $x_0^2 < \kappa^2$ then $x(t)^2 < \kappa^2$ for all t and similarly if $x_0^2 > \kappa^2$ then $x(t)^2 > \kappa^2$ for all t. So we can remove the modulus signs and rearrange (8.15) to give

$$x(t)^2 = \frac{x_0^2 \kappa^2 e^{2\kappa^2 t}}{\kappa^2 + x_0^2 (e^{2\kappa^2 t} - 1)}$$

or

$$x(t) = \pm \sqrt{\frac{\kappa^2}{1 + e^{-2\kappa^2 t}(\kappa^2 x_0^{-2} - 1)}}. \tag{8.16}$$

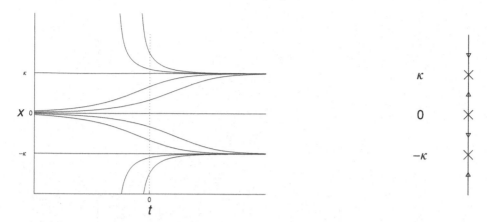

Fig. 8.6. Solutions of $\dot{x} = x(\kappa^2 - x^2)$ on the left, and the corresponding phase diagram (rotated through 90 degrees) on the right.

Whether we take the plus sign or the minus sign depends on the sign of the initial condition; for $t = 0$ we obtain

$$x(0) = \pm\sqrt{x_0^2} = \pm|x_0|$$

and we have to choose the sign so that $x(0) = x_0$.

From the phase diagram we can see that for any $x_0 > 0$ the solution tends to κ as $t \to \infty$, and for any $x_0 < 0$ the solution tends to $-\kappa$; we can also recover this behaviour from our explicit solution, since if $0 < x_0 < \kappa$ then $e^{-2\kappa^2 t}(\kappa^2 x_0^{-2} - 1)$ is always positive, and decreases from its initial value to zero as $t \to \infty$; it follows that $x(t)$ increases from x_0 to κ. Similarly if $x_0 > \kappa$ then $e^{-2\kappa^2 t}(\kappa^2 x_0^{-2} - 1)$ is always negative, and increases up to zero as $t \to \infty$, so that $x(t)$ decreases to κ as $t \to \infty$.

You can also see from the explicit solution that if $|x_0| > \kappa$ then the solution will blow up as $t \downarrow t^* < 0$, since then the expression $(\kappa^2 x_0^{-2} - 1)$ is negative (see Exercise 8.9 for more details).

Figure 8.6 shows the solutions of the equation, along with the corresponding phase diagram, rotated to illustrate how the behaviour of the solutions matches the predictions of the phase diagram.

8.6.2 Two competing species

Later we will look at some simple models of competing species, and will come across such equations as

$$\frac{dy}{dx} = \frac{y(5x - 2)}{x(1 - 3y)}.$$

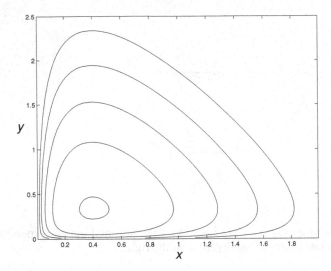

Fig. 8.7. Curves on which $\ln y + 2\ln x - 3y - 5x$ is constant.

This can be separated to give

$$\frac{1-3y}{y}\,\mathrm{d}y = \frac{5x-2}{x}\,\mathrm{d}x.$$

Integrating both sides we have

$$\int\left(\frac{1}{y}-3\right)\mathrm{d}y = \int\left(5-\frac{2}{x}\right)\mathrm{d}x;$$

taking x and y positive, because they represent the size of a population, we have no need of modulus signs in the logarithms arising from the integration,

$$\ln y - 3y = 5x - 2\ln x + c.$$

We can do no better than this implicit solution relating x and y. However, we can represent the curves defined by

$$F(x,y) = \ln y + 2\ln x - 3y - 5x = \text{constant}$$

graphically, and these are shown in Figure 8.7.

Exercises

8.1 Solve the following equations:
   (i) $\dot{x} = t^3(1-x)$ with $x(0) = 3$;
   (ii) $y' = (1+y^2)\tan x$ with $y(0) = 1$;
   (iii) $\dot{x} = t^2 x$ (general solution);
   (iv) $\dot{x} = -x^2$ (general solution);

(v) for $dy/dt = e^{-t^2} y^2$ give the solution in terms of an integral and describe the behaviour of the solution as $t \to +\infty$ depending on the initial condition $y(0)$. You may assume that $\int_0^\infty e^{-s^2}\,ds = \sqrt{\pi}/2$.

8.2 Solve the linear equation

$$\dot{x} + px = q$$

by separation of variables.

8.3 Find the general solution of the equation

$$xy' = ky$$

that is valid for $x > 0$.

8.4 Find the function $I(t)$ that satisfies

$$\frac{dI}{dt} = p(t)I.$$

(Your answer will involve an integral.)

8.5 Use the method of separation of variables to show that the general solution of the linear equation

$$\dot{x} = \lambda x$$

is $x(t) = Ae^{\lambda t}$ for any $A \in \mathbb{R}$.

8.6 In Exercise 5.7 we showed, neglecting air resistance, that an apple falling from a height h reaches the ground when $t = \sqrt{2h/g}$. If we include air resistance then provided that $v \le 0$ the equation becomes

$$m\frac{dv}{dt} = -mg + kv^2 \qquad v(0) = 0$$

with $k > 0$. Show that

$$v(t) = -\sqrt{\frac{mg}{k}}\ \tanh\left(\sqrt{\frac{gk}{m}}t\right),$$

and hence that the apple now takes a time

$$t^* = \sqrt{\frac{m}{kg}}\ \ln\left(e^{kh/m} + \sqrt{e^{2kh/m} - 1}\right)$$

to reach the ground. Check that this coincides with the answer with no air resistance ($t^* = \sqrt{2h/g}$) as $k \to 0$. Hint: for small x, $e^x \approx 1 + x$ and $\ln(1 + x) \approx x$.

8.7 Show that for $k \ne 0$ the solution of the differential equation

$$\frac{dx}{dt} = kx - x^2 \qquad \text{with} \qquad x(0) = x_0$$

is

$$x(t) = \frac{k\,e^{kt}x_0}{x_0(e^{kt} - 1) + k}.$$

Using this explicit solution describe the behaviour of $x(t)$ as $t \to \infty$ for $k < 0$ and $k > 0$. (Note that this is much easier to do using the phase diagram than using the explicit form of the solution.) For $k = 0$ see part (iv) of Exercise 8.1.

8.8 Show that the solution of the equation

$$\frac{dx}{dt} = -x(\kappa^2 + x^2)$$

with initial condition $x(0) = x_0$ is

$$x(t) = \pm \sqrt{\frac{\kappa^2}{(1 + \kappa^2 x_0^{-2})e^{2\kappa^2 t} - 1}},$$

where the $\pm$ is chosen according to the sign of the initial condition. Deduce that $x(t) \to 0$ as $t \to \infty$. As t decreases from zero the solution blows up as t approaches a finite value $t^* < 0$. When is this 'blow up time'?

8.9 We found the solution of the equation $\dot{x} = x(\kappa^2 - x^2)$ in Section 8.6.1,

$$x(t) = \pm \sqrt{\frac{\kappa^2}{1 + e^{-2\kappa^2 t}(\kappa^2 x_0^{-2} - 1)}}.$$

Show that if $|x_0| > |\kappa|$ the solution blows up as t decreases towards a finite negative value, and find this critical time.

8.10 Consider the equation

$$\dot{x} = x^\alpha \quad \text{with } x(0) \geq 0$$

for $\alpha > 0$. Show that the only value of α for which the equation has solutions that are both unique and exist for all time is $\alpha = 1$. You should be able to find an initial condition for which the solutions are not unique when $\alpha < 1$ (cf. (6.3)), and show that solutions with $x(0) > 0$ blow up in a finite time if $\alpha > 1$ (cf. (6.6)).

8.11 (T) Assuming that $f(x)$ and $f'(x)$ are continuous, show that if the solution of

$$\dot{x} = f(x) \quad \text{with } x(0) = x_0$$

blows up to $x = +\infty$ in finite time then

$$\int_{x_0}^{\infty} \frac{1}{f(x)} dx < \infty.$$

9

First order linear equations and the integrating factor

One type of first order equation that we can always solve, at least in theory, is a linear equation. The most general first-order linear equation (cf. (3.5)) is

$$a_1(t)\frac{dx}{dt} + a_0(t)x = f(t).$$

However, we will concentrate on equations that are 'always first order', so we assume that $a_1(t) \neq 0$ and divide through by $a_1(t)$ to obtain

$$\frac{dx}{dt} + p(t)x = q(t). \tag{9.1}$$

9.1 Constant coefficients

First we will consider the simplest case, when both p and q are constants,

$$\frac{dx}{dt} + px = q. \tag{9.2}$$

There are a number of ways to solve this equation. We have already met one, the method of separation of variables; we could write the equation as

$$\frac{dx}{q - px} = dt,$$

integrate both sides, and solve it this way (see Exercise 8.2).

However, we are going to solve it by another method. Although this way involves a trick, and may seem complicated in this simple case, it is also useful for the more general equation (9.1) where the coefficients do not have to be constants. The key point is to notice that

$$\frac{d}{dt}\left(x(t)e^{pt}\right) = \frac{dx}{dt}e^{pt} + px\,e^{pt} = e^{pt}\left(\frac{dx}{dt} + px\right) \tag{9.3}$$

(using the product rule). The right-hand side of (9.3) is the same as the left-hand side of our differential equation (9.2), except that it is multiplied by a factor e^{pt}. If we multiply both sides of (9.2) by e^{pt} we have

$$e^{pt}\left(\frac{dx}{dt} + px\right) = qe^{pt},$$

and using (9.3) this is simply

$$\frac{d}{dt}(xe^{pt}) = qe^{pt}. \tag{9.4}$$

For the general solution we integrate both sides to give

$$x(t)e^{pt} = \frac{q}{p}e^{pt} + C,$$

so that

$$x(t) = \frac{q}{p} + Ce^{-pt}. \tag{9.5}$$

(It follows that if $p > 0$ then $x(t) \to q/p$ as $t \to \infty$, independent of any initial condition.)

If we want the solution that has $x(a) = x_a$ then we need

$$x_a = \frac{q}{p} + Ce^{-pa} \quad\Rightarrow\quad C = \left(x_a - \frac{q}{p}\right)e^{pa},$$

and so this solution is

$$x(t) = \frac{q}{p} + \left(x_a - \frac{q}{p}\right)e^{-p(t-a)}. \tag{9.6}$$

9.2 Integrating factors

We now use the same sort of trick on the more general linear equation

$$\frac{dx}{dt} + p(t)x = q(t). \tag{9.7}$$

What we are doing is looking for an 'integrating factor' by which we can multiply both sides and so turn the left-hand side into something we can integrate easily. We will give a quick derivation of the form of the integrating factor that we need. However, in practice you should just write down the integrating factor; as we will now see it is

$$\exp\left(\int p(t)\,dt\right),$$

i.e. $e^{\int p(t)\,dt}$.

If we multiply both sides of (9.7) by a factor $I(t)$ then we get

$$I(t)\frac{dx}{dt} + I(t)p(t)x = I(t)q(t).$$

Concentrate on the left-hand side,

$$I(t)\frac{dx}{dt} + I(t)p(t)x;$$

we want this to be the derivative of something. The first term is part of the derivative of $I(t)x(t)$, so we will see whether we can find a function $I(t)$ such that

$$\frac{d}{dt}[I(t)x(t)] = I(t)\frac{dx}{dt} + I(t)p(t)x(t),$$

i.e. such that (differentiating the left-hand side using the product rule)

$$\frac{dI}{dt}x(t) + I(t)\frac{dx}{dt} = I(t)\frac{dx}{dt} + I(t)p(t)x(t).$$

For this we would need

$$x(t)\frac{dI}{dt} = I(t)p(t)x(t),$$

which is certainly true if

$$\frac{dI}{dt} = p(t)I.$$

This is a separable equation (see Exercise 8.4); we can divide by I and multiply up by dt to give,

$$\frac{1}{I}\,dI = p(t)\,dt,$$

and then by integration we get

$$\ln|I(t)| = \int p(t)\,dt.$$

Finally we exponentiate both sides and choose $I(t)$ to be positive to give

$$I(t) = \exp\left(\int p(t)\,dt\right).$$

Given this integrating factor we should now be able to solve our general linear equation

$$\frac{dx}{dt} + p(t)x = q(t). \tag{9.8}$$

If P is an anti-derivative of p (so that $\dot{P}(t) = p(t)$) then the integrating factor we need is $e^{P(t)}$. Multiplying both sides of (9.8) by this integrating factor the equation becomes

$$\frac{dx}{dt}e^{P(t)} + p(t)x(t)e^{P(t)} = q(t)e^{P(t)}.$$

The point of the integrating factor is that the left-hand side is now the derivative of $x(t)e^{P(t)}$, so we have

$$\frac{d}{dt}[x(t)e^{P(t)}] = q(t)e^{P(t)}.$$

In order to solve the problem completely we have to be able to integrate the right-hand side, and then the solution is

$$x(t)e^{P(t)} = \int q(t)e^{P(t)}\,dt$$

We now apply this method to some examples.

9.3 Examples

Example 9.1 *Solve the equation*

$$\frac{dx}{dt} + 3x = t \qquad \text{with} \qquad x(0) = 8/9.$$

The integrating factor is

$$I(t) = \exp\left(\int 3\,dt\right) = e^{3t}.$$

Multiplying both sides of the equation by e^{3t} we get

$$e^{3t}\frac{dx}{dt} + 3xe^{3t} = te^{3t}.$$

The whole point of the method is that we can now rewrite the left-hand side as a derivative:

$$\frac{d}{dt}(xe^{3t}) = te^{3t}.$$

Integrating this equation with respect to t between 0 and t – using integration by parts on the right-hand side – we obtain

$$x(t)e^{3t} - x(0) = \int_0^t \tilde{t}e^{3\tilde{t}}\,d\tilde{t}$$

$$= \left[\frac{\tilde{t}e^{3\tilde{t}}}{3} - \frac{e^{3\tilde{t}}}{9}\right]_{\tilde{t}=0}^t$$

$$= \frac{te^{3t}}{3} - \frac{e^{3t}}{9} + \frac{1}{9},$$

and so, since $x(0) = 8/9$,

$$x(t) = e^{-3t} + \frac{t}{3} - \frac{1}{9}.$$ □

Example 9.2 *Find the general solution of*

$$(x^2 + 1)\frac{dy}{dx} + 4xy = 12x.$$

This is a linear equation; if we divide both sides by $x^2 + 1$ then

$$\frac{dy}{dx} + \frac{4x}{x^2 + 1}y = \frac{12x}{x^2 + 1}$$

which is in the form (9.1). The integrating factor is

$$I(x) = \exp\left(\int \frac{4x}{x^2 + 1}\,dx\right)$$
$$= \exp(2\ln(x^2 + 1)) = (x^2 + 1)^2.$$

So, multiplying both sides by $(x^2 + 1)^2$ we get

$$(x^2 + 1)^2\frac{dy}{dx} + 4x(x^2 + 1)y = 12x(x^2 + 1),$$

which is

$$\frac{d}{dx}[y(x^2 + 1)^2] = 12x(x^2 + 1).$$

To find the general solution we integrate both sides to get

$$y(x)(x^2 + 1)^2 = 3(x^2 + 1)^2 + c$$

and so

$$y(x) = 3 + \frac{c}{(x^2 + 1)^2}.$$ □

9.4 Newton's law of cooling

An interesting example of a linear equation arises from Newton's law of cooling, which provides a mathematical model of the temperature $T(t)$ of an object in surroundings of temperature $A(t)$:

$$\frac{dT}{dt} = -k(T - A(t)), \tag{9.9}$$

where $k > 0$ measures the rate that heat is absorbed (or emitted) by the object.

9.4.1 Estimating the time of death

One forensic method for ascertaining the time of death of a body is based on Newton's law of cooling. The idea is to take the temperature of the body at two different times, in order to give an estimate of the constant k to be used in equation (9.9), and then to extrapolate back to find the time when T is the temperature of a living body, 37 °C.

To keep things simple we will suppose that a body is found in a room which is kept at a constant temperature of 24 °C. At 8 a.m. in the morning its temperature is 28 °C, while an hour later it is 26 °C.

With the time t measured in hours we need to find the solution of

$$\frac{dT}{dt} + kT = kA$$

(in fact we have done this already in Section 9.1). Multiplying both sides by the integrating factor e^{kt} we obtain

$$\frac{d}{dt}[T(t)e^{kt}] = kAe^{kt},$$

and then integrating both sides with respect to t between times t_1 and t_2 gives

$$T(t_2)e^{kt_2} - T(t_1)e^{kt_1} = A(e^{kt_2} - e^{kt_1});$$

rearranging this gives the temperature at time t_2 in terms of the temperature at time t_1,

$$T(t_2) = A + [T(t_1) - A]e^{-k(t_2 - t_1)}.$$

This implies that

$$T(9) = A + [T(8) - A]e^{-k}.$$

To find k we set $T(8) = 28$, $T(9) = 26$ and $A = 24$, so we then have

$$26 = 24 + [28 - 24]e^{-k},$$

which implies that

$$e^{-k} = 0.5$$

giving $k = \ln 2$.

If the time of death was t_0 then our solution gives

$$T(8) = A + [T(t_0) - A]e^{-k(8 - t_0)}.$$

Since $T(t_0) = 37$, $T(8) = 28$ and $A = 24$, we want to find the value of t_0 such that

$$28 = 24 + [37 - 24]e^{-k(8-t_0)},$$

i.e.

$$4 = 13e^{-k(8-t_0)}.$$

Taking logarithms gives

$$\ln 4 = \ln 13 - k(8 - t_0),$$

and using $k = \ln 2$ we have

$$(8 - t_0) = \frac{\ln 4 - \ln 13}{-\ln 2}.$$

Solving this for t_0 we obtain $t_0 \approx 1.7$, putting the time of death at approximately 1:42 a.m.

9.4.2 The temperature in an unheated building

We now look at a case where the ambient temperature is not taken to be constant. To make things more definite, and a little less gruesome, we suppose that $T(t)$ represents the temperature inside an unheated church.

Once again we want to use the integrating factor method, so we rewrite equation (9.9) as

$$\frac{dT}{dt} + kT = kA(t);$$

as before, the integrating factor is e^{kt}. Multiplying both sides by e^{kt} gives

$$\frac{d}{dt}[Te^{kt}] = kA(t)e^{kt},$$

and then integrating between 0 and t we get

$$T(t)e^{kt} - T(0) = k \int_0^t A(\tilde{t})e^{k\tilde{t}} \, d\tilde{t}. \tag{9.10}$$

Rearranging this gives

$$T(t) = T(0)e^{-kt} + ke^{-kt} \int_0^t A(\tilde{t})e^{k\tilde{t}} \, d\tilde{t}.$$

Now we model the outside temperature as a regular oscillation about an average temperature μ, setting

$$A(t) = \mu + a \cos \omega t.$$

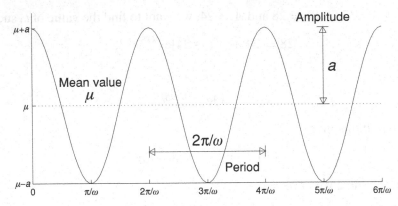

Fig. 9.1. The graph of $A(t) = \mu + a\cos\omega t$ against t, showing the period and amplitude.

We say that $A(t)$ oscillates about a mean value μ; the size of the oscillations, a, is called the amplitude, and the frequency of the oscillations is how many occur for every unit of time, in this case $\omega/2\pi$. The reciprocal of the frequency, $2\pi/\omega$, is known as the period: this is the time between any two successive maxima of $A(t)$, i.e. the time for one 'complete oscillation'. These concepts are illustrated in Figure 9.1.

Over a British year we can take the mean temperature as 9 °C, with the amplitude of oscillations 10 °C. With t measured in years we then want $\omega = 2\pi$ so that the oscillations have a period of one year.

With this particular form for $A(t)$ equation (9.10) becomes

$$T(t)e^{kt} - T(0) = \mu k \int_0^t e^{k\tilde{t}}\, d\tilde{t} + ak \int_0^t e^{k\tilde{t}} \cos\omega\tilde{t}\, d\tilde{t}.$$

An anti-derivative of $e^{kt}\cos\omega t$ is[1]

$$\frac{k}{k^2 + \omega^2}e^{kt}\cos\omega t + \frac{\omega}{k^2 + \omega^2}e^{kt}\sin\omega t, \qquad (9.11)$$

and so we have

$$e^{kt}T(t) - T(0) = \mu\left[e^{k\tilde{t}}\right]_{\tilde{t}=0}^{t} + ak\left[\frac{k}{k^2 + \omega^2}e^{k\tilde{t}}\cos\omega\tilde{t} + \frac{\omega}{k^2 + \omega^2}e^{k\tilde{t}}\sin\omega\tilde{t}\right]_{\tilde{t}=0}^{t}$$

$$= \mu(e^{kt} - 1) + ak\left[\frac{k}{k^2 + \omega^2}e^{kt}\cos\omega t + \frac{\omega}{k^2 + \omega^2}e^{kt}\sin\omega t - \frac{k}{k^2 + \omega^2}\right].$$

---

[1] You might expect the anti-derivative to be of the form $Ae^{kt}\cos\omega t + Be^{kt}\sin\omega t$, since differentiating the first of these two terms gives one term involving $e^{kt}\cos\omega t$ (which we want) but also an $e^{kt}\sin\omega t$ term. Differentiating this guess gives an equation for A and B which is straightforward to solve. There is also a more systematic way of obtaining this result using the complex form for $\cos\omega t$ and $\sin\omega t$, see Exercise 9.8.

This is looking complicated, but we can rearrange it, in particular multiplying through by e^{-kt}, to give

$$T(t) = \mu + \left[T(0) - \mu - \frac{ak^2}{k^2 + \omega^2}\right]e^{-kt} + ak\left[\frac{k}{k^2 + \omega^2}\cos\omega t + \frac{\omega}{k^2 + \omega^2}\sin\omega t\right].$$

This still looks complicated,[2] but if we consider each term individually then we should be able to understand what this solution is actually saying about the temperature. The first term is the average outside temperature, which we would expect to form the main contribution to the temperature inside the church; the second term decays exponentially, so will have very little effect after some time has passed; and the last two terms both oscillate with the same frequency as the ambient temperature. Before we write the equation in its final form we will see in the next section how to combine these two oscillating terms in order to make it clear that they result in just one oscillation.

9.4.3 Combining two oscillating terms

We are now going to show that it is possible to combine two oscillating terms and rewrite them as one:

$$A\cos\omega t + B\sin\omega t = M\cos(\omega t - \phi). \tag{9.12}$$

(In our case the constants A and B are given by $A = k/(k^2 + \omega^2)$ and $B = \omega/(k^2 + \omega^2)$, but the argument is much easier to follow with the more general A and B.)

The idea is to use the double angle formula[3]

$$\cos\alpha\cos\beta + \sin\alpha\sin\beta = \cos(\beta - \alpha) \tag{9.13}$$

in an appropriate way. Note that the left-hand side of (9.12) looks slightly like the left-hand side of (9.13) if we choose $\beta = \omega t$; it would look just like the left-hand side of (9.13) if we could find a ϕ such that

$$\cos\phi = A \qquad \text{and} \qquad \sin\phi = B. \tag{9.14}$$

In general we are unable to do this, because we know that, whatever the value of θ, $\cos^2\theta + \sin^2\theta = 1$, and there is no reason why $A^2 + B^2$ should be equal to

---

[2] If you solve an equation and end up with a long expression like this then there are various ways that you can check your answer. You always have the option of differentiating and substituting back into the original equation to check that it works; sometimes this itself might be daunting. One thing that you can do quite quickly here is to check that at $t = 0$ the right-hand side reduces to $T(0)$, as it should.

[3] The right-hand side could also be written as $\cos(\alpha - \beta)$, since $\cos\theta = \cos(-\theta)$.

one. The way to circumvent this is to take out an appropriate factor from (9.12) and write

$$A\cos\omega t + B\sin\omega t = \sqrt{A^2 + B^2}\left[\frac{A}{\sqrt{A^2 + B^2}}\cos\omega t + \frac{B}{\sqrt{A^2 + B^2}}\sin\omega t\right].$$

The coefficients of $\cos\omega t$ and $\sin\omega t$ within the square brackets now satisfy

$$\left(\frac{A}{\sqrt{A^2 + B^2}}\right)^2 + \left(\frac{B}{\sqrt{A^2 + B^2}}\right)^2 = 1,$$

and so we can find a ϕ with

$$\cos\phi = \frac{A}{\sqrt{A^2 + B^2}} \quad \text{and} \quad \sin\phi = \frac{B}{\sqrt{A^2 + B^2}}, \qquad (9.15)$$

which is just given by

$$\phi = \tan^{-1}(B/A).$$

(One way to think about this is in terms of the right-angled triangle pictured in Figure 9.2.)

With this choice of ϕ we now have

$$\begin{aligned} A\cos\omega t + B\sin\omega t &= \sqrt{A^2 + B^2}\left(\frac{A}{\sqrt{A^2 + B^2}}\cos\omega t + \frac{B}{\sqrt{A^2 + B^2}}\sin\omega t\right) \\ &= \sqrt{A^2 + B^2}\,(\cos\phi\cos\omega t + \sin\phi\sin\omega t) \\ &= \sqrt{A^2 + B^2}\,\cos(\omega t - \phi). \end{aligned}$$

The sum of the two oscillations has amplitude $M = \sqrt{A^2 + B^2}$ and oscillates at frequency $\omega/2\pi$, with a time lag of ϕ/ω.

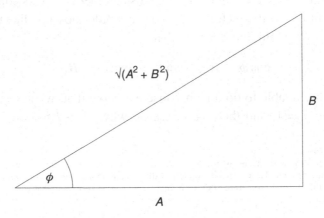

Fig. 9.2. A, B, ϕ, and a right-angled triangle.

9.4.4 Back to our example

In our particular example, we want to combine

$$\frac{k}{k^2 + \omega^2} \cos \omega t + \frac{\omega}{k^2 + \omega^2} \sin \omega t; \qquad (9.16)$$

so if we take out the factor

$$\sqrt{\frac{k^2}{(k^2 + \omega^2)^2} + \frac{\omega^2}{(k^2 + \omega^2)^2}} = \sqrt{\frac{k^2 + \omega^2}{(k^2 + \omega^2)^2}} = \sqrt{\frac{1}{k^2 + \omega^2}},$$

and choose ϕ with

$$\phi = \tan^{-1}(\omega/k),$$

then the two terms in (9.16) become

$$\sqrt{\frac{1}{k^2 + \omega^2}} \cos(\omega t - \phi),$$

and the full solution is

$$T(t) = \mu + \left[T(0) - \mu - \frac{ak}{k^2 + \omega^2} \right] e^{-kt} + ak \sqrt{\frac{1}{k^2 + \omega^2}} \cos(\omega t - \phi). \quad (9.17)$$

For a particular choice of parameters this solution is shown in Figure 9.3, along with the ambient temperature.

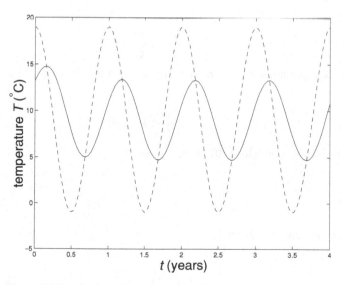

Fig. 9.3. The solid line is the solution (9.17) when $A(t) = 9 + 10 \cos 2\pi t$, $k = 3$ and $T(0) = 12$. The ambient temperature $A(t)$ is shown as a dashed line.

Now we can see clearly what the solution describes. The underlying temperature is the average ambient temperature μ; this is adjusted by a term that decays to zero as $t \to \infty$, and an oscillating term. When t is large the decaying term will be small, and the solution will be approximately

$$T_{\text{approx}}(t) = \mu + a\sqrt{\frac{k^2}{k^2 + \omega^2}}\cos(\omega t - \alpha);$$

eventually the temperature oscillates at the same frequency as the ambient temperature, but its amplitude is a multiplied by the factor

$$\sqrt{\frac{k^2}{k^2 + \omega^2}}. \tag{9.18}$$

Since $\cos(\omega t - \alpha) = \cos \omega[t - (\alpha/\omega)]$ there is a time lag between the oscillations of the ambient temperature and the oscillations of $T(t)$ which is

$$\frac{\alpha}{\omega} = \frac{1}{\omega}\tan^{-1}\frac{\omega}{k}. \tag{9.19}$$

Note that if the constant k is large, so that heat is easily conducted from the surroundings to the building, the factor in (9.18) will be close to 1, while the time lag will be close to zero; in such a case $T(t)$ follows $A(t)$ very closely.

Exercises

9.1 Use an integrating factor to solve the following differential equations:

(i)

$$\frac{dy}{dx} + \frac{y}{x} = x^2$$

(find the general solution and the only solution that is finite when $x = 0$),

(ii)

$$\frac{dx}{dt} + tx = 4t$$

(find the solution with $x(0) = 2$),

(iii)

$$\frac{dz}{dy} = z \tan y + \sin y$$

(find the general solution),

(iv)

$$y' + e^{-x}y = 1$$

(find the solution when $y(0) = e$, leaving your answer as an integral),

(v)

$$\dot{x} + x \tanh t = 3$$

(find the general solution, and compare it to that for $\dot{x} + x = 3$),

(vi)

$$y' + 2y \cot x = 5$$

(find the solution with $y(\pi/2) = 1$),

(vii)

$$\frac{\mathrm{d}x}{\mathrm{d}t} + 5x = t$$

(find the general solution),

(viii) with $a > 0$ find the solution of the equation

$$\frac{\mathrm{d}x}{\mathrm{d}t} + \left[a + \frac{1}{t}\right] x = b$$

for a general initial condition $x(1) = x_0$, and show that $x(t) \to b/a$ as $t \to \infty$ (you would get the same result if you replaced $a + t^{-1}$ by a).

9.2 A body is found in a cold room (temperature 5 °C) at 3 p.m. and its temperature then is 19 °C. An hour later its temperature has dropped to 15 °C. Use Newton's law of cooling to estimate the time of death, assuming that body temperature is 37 °C.

9.3 At 7 a.m. in the morning I make my wife a cup of tea using boiling water; after adding some milk it is about 90 °C. When we leave for the station at 7:30 a.m. the tea is still drinkable at about 45 °C. When I get back home at 8 a.m. the neglected tea has cooled to about 30 °C. What is the temperature of our house?

9.4 Use the integrating factor method to find $T(t_2)$ in terms of $T(t_1)$ when

$$\frac{\mathrm{d}T}{\mathrm{d}t} = -k(T(t) - A(t))$$

and

$$A(t) = \mu + a \cos \omega(t - \phi).$$

9.5 A dead body is found outside on a winter's morning at 7 a.m.; its temperature is measured as 20 °C. Measured an hour later it has dropped to 15 °C. The air temperature $A(t)$ fluctuates on a daily cycle about a mean of 3 °C with $A(t) = 3 - 5 \cos \omega(t - 2)$, where t is measured in hours with $t = 0$ corresponding to midnight, and $\omega = \pi/12$.

  (i) Use the solution from Exercise 9.4 and the temperature observations at 7 a.m. and 8 a.m. to show that

$$k = -\ln \left\{ \frac{12(k^2 + \omega^2) - 5k(k \cos 6\omega + \omega \sin 6\omega)}{17(k^2 + \omega^2) - 5k(k \cos 5\omega + \omega \sin 5\omega)} \right\}. \tag{E9.1}$$

  (ii) (C) This is a MATLAB exercise. Choose an initial guess for k, and then substitute this into the right-hand side of (E9.1) to obtain a new guess. Continue doing this until your 'guess' stabilises. Once this happens you have actually obtained the required solution of (E9.1). Can you see why? (You should find that $k \approx 0.3640$.)

(iii) If the time of death was t_0, use the fact that body temperature is $37\,°C$ (so $T(t_0) = 37$) and $T(7) = 20$ to show that

$$t_0 = 7 + \frac{1}{k} \ln \left[\frac{17(k^2 + \omega^2) - 5k(k \cos 5\omega + \omega \sin 5\omega)}{34(k^2 + \omega^2) - 5k(k \cos \omega(t_0 - 2) + \omega \sin \omega(t_0 - 2))} \right].$$

(iv) (C) Use MATLAB again to refine an initial guess for the time of death as in part (ii). You should find that $t_0 \approx 4.8803$, or 4:53 a.m.

9.6 Show that if y_1 and y_2 are any two solutions of

$$\frac{dy}{dx} + p(x)y = 0$$

then $y_1(x)/y_2(x)$ is constant. (You do not need to solve the equation!)

9.7 (T) Suppose that

$$\frac{dx}{dt} \leq ax$$

(this is known as a differential *inequality*). Use an appropriate integrating factor to show that

$$\frac{d}{dt}[e^{-at}x] \leq 0,$$

and then integrate both sides between appropriate limits to deduce that

$$x(t) \leq x(s)e^{a(t-s)}$$

for any t and s. Hint: it is a fundamental property of integration that if $f(x) \leq g(x)$ then

$$\int_a^b f(x)\,dx \leq \int_a^b g(x)\,dx.$$

9.8 (T) The function $\sin \omega t$ can be written as a combination of complex exponentials,

$$\sin \omega t = \frac{e^{i\omega t} - e^{-i\omega t}}{2i}.$$

Using this form for $\sin \omega t$, and assuming that the usual rules of integration apply to such complex exponentials, find

$$\int e^{kt} \sin \omega t\, dt.$$

You may also need to use the identity

$$\cos \omega t = \frac{e^{i\omega t} + e^{-i\omega t}}{2}.$$

See Appendix A for more on these complex exponentials.

10

Two 'tricks' for nonlinear equations

This chapter deals with two tricks that can be used to solve certain nonlinear equations. Since these techniques can only be used for equations of particular kinds, it is important to be able to spot them.

10.1 Exact equations

Suppose that x and y are related implicitly by

$$F(x, y) = c, \qquad (10.1)$$

so that x and y form a 'curve of constant F'. Then if we take the derivative of (10.1) with respect to x we get, using the chain rule (see Appendix C),

$$\frac{\partial F}{\partial x}(x, y) + \frac{\partial F}{\partial y}(x, y)\frac{dy}{dx} = 0. \qquad (10.2)$$

This is a differential equation for $y(x)$, whose solution is the implicit equation (10.1) that we started with.

The nice thing is that there is an easy way to check whether or not a differential equation

$$f(x, y) + g(x, y)\frac{dy}{dx} = 0 \qquad (10.3)$$

is one of these 'exact equations' (so called since the equation is exactly the derivative of the function $F(x, y)$ with respect to x). Notice that if we compare (10.3) with (10.2) then we would need

$$f(x, y) = \frac{\partial F}{\partial x} \qquad \text{and} \qquad g(x, y) = \frac{\partial F}{\partial y}. \qquad (10.4)$$

Since the order of taking two partial derivatives does not matter (see Appendix C again), if we have (10.4) then we will have

$$\frac{\partial f}{\partial y} = \frac{\partial^2 F}{\partial x\, \partial y} = \frac{\partial^2 F}{\partial y\, \partial x} = \frac{\partial g}{\partial x},$$

i.e.

$$\frac{\partial f}{\partial y} = \frac{\partial g}{\partial x}. \tag{10.5}$$

This equation (10.5) is in fact a necessary and sufficient condition for the original equation to be 'exact', i.e. for there to be a function $F(x, y)$ such that (10.4) holds. We prove this by showing how to find such a function F under condition (10.5). First we want to make sure that the first equation in (10.4) holds, i.e.

$$\frac{\partial F}{\partial x} = f(x, y).$$

We want to reverse the *partial* differentiation with respect to x. Remember that when we perform a partial differentiation with respect to x we have to keep y constant; so any function of y alone, $C(y)$, behaves as a constant would if we were carrying out an ordinary differentiation (i.e. it disappears). It follows that when we integrate this equation with respect to x, the '$+c$' term could depend on y:

$$F(x, y) = \int f(x, y)\, dx + C(y). \tag{10.6}$$

In order to fix $C(y)$ we partially differentiate (10.6) with respect to y,

$$\frac{\partial F}{\partial y} = \frac{\partial}{\partial y} \int f(x, y)\, dx + \frac{dC}{dy}$$

(we have an ordinary derivative in the last term since C only depends on y) and now we can use the second equation in (10.4) to get

$$\frac{dC}{dy} = g(x, y) - \frac{\partial}{\partial y} \int f(x, y)\, dx. \tag{10.7}$$

The condition in (10.5) means that this expression for dC/dy only depends on y, since

$$\frac{\partial}{\partial x}\left(g(x, y) - \frac{\partial}{\partial y} \int f(x, y)\, dx \right) = \frac{\partial g}{\partial x} - \frac{\partial}{\partial x}\frac{\partial}{\partial y} \int f(x, y)\, dx$$

$$= \frac{\partial g}{\partial x} - \frac{\partial}{\partial y}\frac{\partial}{\partial x} \int f(x, y)\, dx$$

$$= \frac{\partial g}{\partial x} - \frac{\partial f}{\partial y} = 0.$$

By integrating the right-hand side of (10.7) we will find $C(y)$ up to an arbitrary additive constant and so obtain the solution.

Example 10.1 *Check that the equation*

$$\underbrace{x^3 + \frac{y}{x}}_{f(x,y)} + \underbrace{(y^2 + \ln x)}_{g(x,y)} \frac{dy}{dx} = 0$$

(valid for $x > 0$) is exact and hence find its solution.

First we calculate

$$\frac{\partial f}{\partial y} = \frac{1}{x} \quad \text{and} \quad \frac{\partial g}{\partial x} = \frac{1}{x},$$

and so the equation is exact and $F(x, y) = c$ for some F with

$$\frac{\partial F}{\partial x} = x^3 + \frac{y}{x} \quad \text{and} \quad \frac{\partial F}{\partial y} = y^2 + \ln x.$$

Integrating $\partial F / \partial x = x^3 + (y/x)$ with respect to x we get

$$F(x, y) = \frac{x^4}{4} + y \ln x + C(y).$$

To find C, we differentiate this partially with respect to y,

$$\ln x + \frac{dC}{dy} = y^2 + \ln x,$$

and so $C'(y) = y^2$ which implies that $C(y) = y^3/3$; we can omit the constant of integration since this can be absorbed into the 'c' that occurs in the resulting solution:

$$F(x, y) = \frac{x^4}{4} + y \ln x + \frac{y^3}{3} = c.$$

You cannot rearrange this to solve for y as a function of x; the best you can do is to have the solution in this implicit form. □

10.1.1 Integrating factors

It may be the case that an equation is not exact, but can be turned into an exact equation if it is multiplied by the correct integrating factor.

We have already done this for the simple case of the linear equation

$$\frac{dy}{dx} + p(x)y = 0.$$

Clearly this equation as it stands is not exact, since

$$\frac{\partial}{\partial y}[p(x)y] = p(x) \neq 0 = \frac{\partial}{\partial x}[1].$$

However, if we multiply by the integrating factor $e^{P(x)}$, where P is an anti-derivative of p, then the resulting equation

$$p(x)e^{P(x)}y + e^{P(x)}\frac{dy}{dx} = 0$$

is exact, since

$$\frac{\partial}{\partial y}[p(x)e^{P(x)}y] = p(x)e^{P(x)} = \frac{\partial}{\partial x}e^{P(x)}.$$

Of course, linear equations are a very special case, and in general there is no simple way to find an integrating factor. Suppose that we start with the more general equation

$$f(x, y) + g(x, y)\frac{dy}{dx} = 0, \tag{10.8}$$

and try to turn this into an exact equation by multiplying both sides by $I(x, y)$. Then we obtain

$$f(x, y)I(x, y) + g(x, y)I(x, y)\frac{dy}{dx} = 0,$$

and for this equation to be exact we need

$$\frac{\partial}{\partial y}[f(x, y)I(x, y)] = \frac{\partial}{\partial x}[g(x, y)I(x, y)],$$

or

$$\left(\frac{\partial f}{\partial y} - \frac{\partial g}{\partial x}\right)I = g\frac{\partial I}{\partial x} - f\frac{\partial I}{\partial y}. \tag{10.9}$$

This is a partial differential equation for I, and is certainly no easier to solve than the original equation (10.8).

However, there are situations in which it is possible to simplify (10.9). For example, suppose that we assume that there is an integrating factor I that depends only on x. Then (10.9) becomes

$$\frac{dI}{dx} = \frac{1}{g}\left(\frac{\partial f}{\partial y} - \frac{\partial g}{\partial x}\right)I. \tag{10.10}$$

This equation will have a solution that depends only on x provided that

$$\frac{1}{g}\left(\frac{\partial f}{\partial y} - \frac{\partial g}{\partial x}\right) \tag{10.11}$$

depends only on x.

Example 10.2 *Find an integrating factor depending only on x that will make*

$$\underbrace{3\sin y + 5ye^{5x} + \frac{2ye^{5x}}{x}}_{f(x)} + \underbrace{(x\cos y + e^{5x})}_{g(x)}\frac{dy}{dx} = 0$$

an exact equation, and hence find its solution.

There will be an integrating factor that depends only on x if (10.11) holds. This gives

$$\frac{1}{g}\left(\frac{\partial f}{\partial y} - \frac{\partial g}{\partial x}\right) = \frac{1}{x\cos y + e^{5x}}\left[3\cos y + 5e^{5x} + \frac{2e^{5x}}{x} - (\cos y + 5e^{5x})\right]$$

$$= \frac{1}{x\cos y + e^{5x}}\left(2\cos y + \frac{2e^{5x}}{x}\right)$$

$$= \frac{2}{x}.$$

So, from (10.10), we need

$$\frac{dI}{dx} = \frac{2I}{x}.$$

Separating variables gives

$$\frac{1}{I}\frac{dI}{dx} = \frac{2}{x},$$

and so

$$\ln I = 2\ln x,$$

which implies that $I(x) = x^2$. Multiplying the equation by x^2 yields

$$\underbrace{3x^2\sin y + 5x^2ye^{5x} + 2xye^{5x}}_{\tilde{f}(x)} + \underbrace{(x^3\cos y + x^2e^{5x})}_{\tilde{g}(x)}\frac{dy}{dx} = 0,$$

which is now exact, since

$$\frac{\partial\tilde{f}}{\partial y} = 3x^2\cos y + 5x^2e^{5x} + 2xe^{5x} = \frac{\partial\tilde{g}}{\partial x}.$$

To find $F(x, y)$, we first integrate

$$\frac{\partial F}{\partial x} = \tilde{f}(x, y) = 3x^2 \sin y + 5x^2 y e^{5x} + 2xy e^{5x}$$

partially with respect to x to give

$$F(x, y) = x^3 \sin y + x^2 y e^{5x} + C(y).$$

To fix $C(y)$ we differentiate F partially with respect to y,

$$\frac{\partial F}{\partial y} = x^3 \cos y + x^2 e^{5x} + C'(y).$$

We therefore have $\partial F/\partial y = \tilde{g}(x)$ if $C'(y) = 0$. So we finally have our solution,

$$F(x, y) = x^3 \sin y + x^2 y e^{5x} + C = 0. \qquad \square$$

10.2 Substitution methods

In some cases it is possible to simplify an equation considerably by making an appropriate substitution. Just as with integration, knowing what this 'appropriate substitution' might be is not always clear, and given a general differential equation it is probably *not* the case that it can be solved by a clever substitution. However, in this section we cover two types of first order equation that can be easily identified and then solved in this way.

10.2.1 Homogeneous equations

A first order differential equation is said to be *homogeneous*[1] if it can be written in the form

$$\frac{dy}{dx} = F\left(\frac{y}{x}\right).$$

In this case we can make the substitution $u = y/x$. So then $y = ux$ and using the product rule gives

$$\frac{dy}{dx} = u + x\frac{du}{dx},$$

so that

$$x\frac{du}{dx} = F(u) - u,$$

which is a separable equation.

---

[1] We also use this word in a different, but related, sense to describe a linear equation of the form $a_n d^n y/dx^n + \cdots + a_0 y = 0$, see Section 3.3.

Example 10.3 *By means of an appropriate substitution solve the equation*

$$xy\frac{dy}{dx} = 2x^2 + 3y^2. \tag{10.12}$$

Dividing both sides of (10.12) by xy gives

$$\frac{dy}{dx} = 2\frac{x}{y} + 3\frac{y}{x},$$

and the right-hand side is a function of $u = y/x$,

$$F(u) = \frac{2}{u} + 3u.$$

We substitute $u = y/x$; therefore $y = xu$, and so $y' = u + xu'$. Thus

$$u + xu' = \frac{2}{u} + 3u,$$

which gives

$$x\frac{du}{dx} = \frac{2}{u} + 2u.$$

Separating the variables we get

$$\frac{u}{1 + u^2}\,du = \frac{2}{x}\,dx,$$

and so

$$\int \frac{u}{1 + u^2}\,du = \int \frac{2}{x}\,dx$$

which gives

$$\tfrac{1}{2}\ln(1 + u^2) = 2\ln|x| + c,$$

or

$$\sqrt{1 + u^2} = Ax^2,$$

i.e. $u(x) = \pm\sqrt{A^2x^4 - 1}$. Since $y(x) = xu(x)$ the final answer is

$$y(x) = \pm x\sqrt{\alpha x^4 - 1},$$

where we have replaced A^2 by $\alpha > 0$. You might like to check that this really is the solution of (10.12). $\qquad\square$

10.2.2 Bernoulli equations

Another type of equation that can be solved by substitution is the so-called Bernoulli equation,

$$\frac{dy}{dx} + p(x)y = q(x)y^n. \tag{10.13}$$

When $n = 0$ or 1 this is just a linear equation. For n taking other (perhaps negative) values this falls into none of the classes we have considered so far.

However, the substitution $u = y^{1-n}$ turns (10.13) into a linear equation:

$$\begin{aligned}
\frac{du}{dx} &= (1-n)y^{-n}\frac{dy}{dx} \\
&= (1-n)y^{-n}[-p(x)y + q(x)y^n] \\
&= (1-n)[-p(x)y^{1-n} + q(x)] \\
&= (1-n)[-p(x)u + q(x)].
\end{aligned}$$

The resulting equation for u is

$$\frac{du}{dx} + (1-n)p(x)u = (1-n)q(x),$$

a linear equation that we can solve using the integrating factor method of Chapter 9.

Example 10.4 *Use an appropriate substitution to find the general solution of*

$$\frac{dy}{dx} - 6xy = 2xy^2. \tag{10.14}$$

This is clearly of the form in (10.13) with $n = 2$, so we set $u = y^{-1}$. Then we have

$$\begin{aligned}
\frac{du}{dx} &= -\frac{1}{y^2}\frac{dy}{dx} \\
&= -\frac{6x}{y} - 2x \\
&= -6xu - 2x,
\end{aligned}$$

or

$$\frac{du}{dx} + 6xu = -2x.$$

The integrating factor for this equation is

$$I(x) = \exp\left(\int 6x\, dx\right) = \exp(3x^2).$$

Multiplying both sides by e^{3x^2} we have

$$\frac{d}{dx}\left[ue^{3x^2}\right] = -2xe^{3x^2}.$$

Integrating both sides with respect to x gives

$$u(x)e^{3x^2} = -\frac{1}{3}e^{3x^2} + c,$$

and so

$$u(x) = -\frac{1}{3} + ce^{-3x^2}.$$

Since $y = 1/u$ we have

$$y(x) = \frac{3}{Ce^{-3x^2} - 1}.$$

Again, you might like to check that this really is a solution of the original differential equation.　□

Exercises

10.1 Check that the following equations are exact and hence solve them.
   (i)

$$(2xy - \sec^2 x) + (x^2 + 2y)\frac{dy}{dx} = 0,$$

   (ii)

$$(1 + e^x y + xe^x y) + (xe^x + 2)\frac{dy}{dx} = 0,$$

   (iii)

$$(x\cos y + \cos x)\frac{dy}{dx} + \sin y - y\sin x = 0,$$

   and
   (iv)

$$e^x \sin y + y + (e^x \cos y + x + e^y)\frac{dy}{dx} = 0.$$

10.2 Find an integrating factor depending only on x that makes the equation

$$e^{-y}\sec x + 2\cot x - e^{-y}\frac{dy}{dx} = 0$$

exact, and hence find its solution. Hint: $\int \csc x \, dx = \ln|\csc x - \cot x|$.

10.3 Show that any equation that can be written in the form

$$f(x) + g(y)\frac{dy}{dx} = 0$$

is exact, and find its solution in terms of integrals of f and g. Hence find the solutions of

(i)

$$V'(x) + 2y\frac{dy}{dx} = 0$$

and

(ii)

$$\left(\frac{1}{y} - a\right)\frac{dy}{dx} + \frac{2}{x} - b = 0,$$

for $x, y > 0$.

10.4 By substituting $u = y/x$ solve the following homogeneous equations:

(i)

$$xy + y^2 + x^2 - x^2\frac{dy}{dx} = 0$$

(the solution is $y = x\tan(\ln|x| + c)$).

(ii)

$$\frac{dx}{dt} = \frac{x^2 + t\sqrt{t^2 + x^2}}{tx}$$

(the solution is $x(t) = \pm t\sqrt{(\ln|t| + c)^2 - 1}$).

10.5 You could solve

$$\frac{dx}{dt} = kx - x^2.$$

by separating variables (see Exercise 8.7). Instead, substitute $u = x^{-1}$ and show that u satisfies the linear equation

$$\frac{du}{dt} = 1 - ku.$$

Solve this equation for $u(t)$, and hence find the solution $x(t)$.

10.6 Use an appropriate substitution to solve the equation

$$\dot{x} = x(\kappa^2 - x^2).$$

You should recover the solution (8.16) found by separating variables.

Part II

Second order linear equations with constant coefficients

11

Second order linear equations: general theory

We will now turn to second order differential equations,

$$\frac{d^2x}{dt^2} = f(\dot{x}, x, t). \tag{11.1}$$

In this chapter we address the kinds of theoretical question that we covered for first order equations in Chapter 6. As such there are few examples, but we will return to more concrete problems and solution methods in the next chapter.

11.1 Existence and uniqueness

First we discuss the existence and uniqueness of solutions. Before we do this formally, we give an indication of why we will need to specify both x and $\dot{x}$ in our 'initial condition'.

Consider the simplest type of second order equation,

$$\frac{d^2x}{dt^2} = f(t), \tag{11.2}$$

the second order equivalent of the 'trivial' equations we considered in Chapter 5. We can solve (11.2) by integrating twice: if F is any anti-derivative of f then

$$\frac{dx}{dt}(t) = F(t) + c_1 \tag{11.3}$$

and then if $\mathcal{F}$ is any anti-derivative of F

$$x(t) = \mathcal{F}(t) + c_1 t + c_2.$$

The two integrations result in two arbitrary constants (c_1 and c_2); specifying $x(t_0)$ alone will not be enough to tie down the solution, but we need to specify $\dot{x}(t_0)$ (to fix c_1) and then $x(t_0)$ (to determine c_2).

101

To put this in a physical context, equation (11.2) is the equation of motion for a particle moving under the influence of a force $f(t)$ per unit mass. In order to predict the motion of the particle in the future we need to know both its current position *and* its current velocity (e.g. the difference between dropping and throwing a piece of chalk).

We now state the existence and uniqueness theorem. As with our previous existence and uniqueness theorem (Theorem 6.2) it can be paraphrased as 'unique solutions exist provided that f is sufficiently nice'.

Theorem 11.1 *Given a function $f(x_2, x_1, t)$, suppose that f, $\partial f/\partial x_1$ and $\partial f/\partial x_2$ are continuous functions for $a_1 < x_1 < a_2$, $b_1 < x_2 < b_2$ and $t_1 < t < t_2$. Then for all initial conditions*

$$x(t_0) = x_0 \qquad \text{and} \qquad \dot{x}(t_0) = y_0 \qquad (11.4)$$

with $a_1 < x_0 < a_2$, $b_1 < y_0 < b_2$ and $t_1 < t_0 < t_2$ there exists a unique solution of

$$\ddot{x} = f(\dot{x}, x, t) \qquad (11.5)$$

on some interval I containing t_0, i.e. a continuous function with two continuous derivatives that satisfies (11.4) and the equation (11.5) on I.

11.2 Linearity

In the following chapters (Chapters 12–20) we will concentrate on linear second order equations, the most general form of which is

$$a_2(t)\frac{d^2x}{dt^2} + a_1(t)\frac{dx}{dt} + a_0(t)x = g(t).$$

Most of the time during any general treatment we will assume that $a_2(t) \neq 0$, divide by a_2, and rewrite the equation as

$$\frac{d^2x}{dt^2} + p(t)\frac{dx}{dt} + q(t)x = f(t) \qquad (11.6)$$

(cf. (3.5)), since this saves a little algebra.

When we have not specified the initial conditions for (11.6) the equation will not have a unique solution, and we would expect its general solution to have two arbitrary constants so that we can fit any pair of initial conditions, $x(t_0) = x_0$ and $\dot{x}(t_0) = y_0$. We now investigate this a little further, starting with the homogeneous problem

$$\frac{d^2x}{dt^2} + p(t)\frac{dx}{dt} + q(t)x = 0 \qquad (11.7)$$

(recall that (11.6) is called homogeneous when $f(t) = 0$).

Our first observation is crucial: if $x_1(t)$ and $x_2(t)$ are two solutions of (11.7) then so is

$$x(t) = \alpha x_1(t) + \beta x_2(t)$$

for any choice of two real numbers α and β. This is known as the principle of superposition of solutions. To see this, first note that

$$\frac{dx}{dt} = \alpha\frac{dx_1}{dt} + \beta\frac{dx_2}{dt} \qquad \text{and} \qquad \frac{d^2x}{dt^2} = \alpha\frac{d^2x_1}{dt^2} + \beta\frac{d^2x_2}{dt^2};$$

therefore

$$\frac{d^2x}{dt^2} + p(t)\frac{dx}{dt} + q(t)x = \alpha\left[\frac{d^2x_1}{dt^2} + p(t)\frac{dx_1}{dt} + q(t)x_1\right]$$

$$+ \beta\left[\frac{d^2x_2}{dt^2} + p(t)\frac{dx_2}{dt} + q(t)x_2\right] = 0. \quad (11.8)$$

We can express this in a more elegant way if we are prepared to define some extra notation. Although this might appear complicated, all that we are going to do is to define a shorthand so that instead of always having to talk about

$$\frac{d^2x}{dt^2} + p(t)\frac{dx}{dt} + q(t)x \qquad (11.9)$$

we can refer simply to $L[x]$.

Given a function $x(t)$ that has two derivatives we simply define $L[x]$ to be (11.9), i.e. the left-hand side of our equation:

$$L[x](t) = \frac{d^2x}{dt^2}(t) + p(t)\frac{dx}{dt}(t) + q(t)x(t). \qquad (11.10)$$

The argument '(t)' has been included here to emphasise that $L[x]$ is a function of t; starting with $x(t)$, which is itself a function of t, $L[x](t)$ gives another function of t. This L is referred to as a *differential operator*, since it 'operates' on $x(t)$ by performing various differentiations. We will be somewhat more rigorous about the definition of L in Section 11.5.

Now we can express our observation about adding multiples of solutions by saying that *the operator L is linear*; this simply means that

$$L[\alpha x_1 + \beta x_2] = \alpha L[x_1] + \beta L[x_2] \qquad (11.11)$$

for all choices of $\alpha, \beta \in \mathbb{R}$ and functions x_1 and x_2 that have two derivatives (so that $L[x_1]$ and $L[x_2]$ make sense). If we write equation (11.11) in full using the definition of L then it gives precisely the first equality in (11.8). It is this property of *linearity* which makes linear equations so much easier to solve than nonlinear ones.

11.3 Linearly independent solutions

We will now see that in order to find the general solution of (11.7) we have to find two 'different' solutions $x_1(t)$ and $x_2(t)$; from these we will be able to construct any solution $x(t)$ as a linear combination

$$x(t) = \alpha x_1(t) + \beta x_2(t). \tag{11.12}$$

11.3.1 Linear independence of functions

At the moment we have no well-formed idea of what it means for two solutions to be 'different'. The appropriate notion is that the two solutions are linearly independent, an idea borrowed from vector algebra. The n vectors $v_1, \ldots, v_n$ are said to be *linearly independent* if the only solution of

$$\alpha_1 v_1 + \cdots + \alpha_n v_n = 0$$

is $\alpha_1 = \alpha_2 = \cdots = \alpha_n = 0$, i.e. if none of them can be written as a linear combination of the others. We now make the same definition replacing the vectors $v_1, \ldots, v_n$ with functions $x_1(t), \ldots, x_n(t)$.

Definition 11.2 *The functions $x_1(t), \ldots, x_n(t)$ are* linearly independent *on an interval I if the only solution of*

$$\alpha_1 x_1(t) + \cdots + \alpha_n x_n(t) = 0 \qquad \text{for all} \qquad t \in I$$

is $\alpha_1 = \cdots = \alpha_n = 0$.

When we have only two functions their linear independence simply says that they are not proportional on I. If $x_1(t)$ and $x_2(t)$ are proportional on I then for some constant c

$$x_2(t) = cx_1(t);$$

it follows that $cx_1(t) - x_2(t) = 0$ on I, and so x_1 and x_2 are linearly dependent. Conversely, if $x_1(t)$ and $x_2(t)$ are linearly dependent on I then for some non-zero α_1 and α_2

$$\alpha_1 x_1(t) + \alpha_2 x_2(t) = 0 \qquad \text{for all} \qquad t \in I$$

which implies that

$$x_1(t) = -\frac{\alpha_2}{\alpha_1} x_2(t),$$

i.e. x_1 and x_2 are proportional.

11.3.2 Two linearly independent solutions are necessary and sufficient

First of all, we will see that it is not possible to obtain all possible solutions of

$$\frac{d^2x}{dt^2} + p(t)\frac{dx}{dt} + q(t)x = 0 \tag{11.13}$$

(equation (11.7)) as multiples of a single special solution. Suppose that $x_1(t)$ is the solution of (11.13) satisfying

$$x_1(t_0) = 1 \quad \text{and} \quad \dot{x}_1(t_0) = 0,$$

while $x_2(t)$ is the solution that satisfies

$$x_2(t_0) = 0 \quad \text{and} \quad \dot{x}_2(t_0) = 1.$$

Using Theorem 11.1 both these solutions exist and are unique; but it is clear that one cannot be a multiple of the other. This shows that at least two linearly independent solutions are necessary.

We now see that given two solutions $x_1(t)$ and $x_2(t)$ that are not proportional (like the two just defined above) we can find α and β such that the linear combination

$$x(t) = \alpha x_1(t) + \beta x_2(t)$$

(which must also solve (11.13)) satisfies any given initial condition

$$x(t_0) = x_0 \quad \text{and} \quad \dot{x}(t_0) = v_0. \tag{11.14}$$

The correct values of α and β can be obtained by solving the simultaneous equations

$$\alpha x_1(t_0) + \beta x_2(t_0) = x_0$$
$$\alpha \dot{x}_1(t_0) + \beta \dot{x}_2(t_0) = v_0.$$

Writing these as a matrix equation

$$\begin{pmatrix} x_1(t_0) & x_2(t_0) \\ \dot{x}_1(t_0) & \dot{x}_2(t_0) \end{pmatrix} \begin{pmatrix} \alpha \\ \beta \end{pmatrix} = \begin{pmatrix} x_0 \\ v_0 \end{pmatrix} \tag{11.15}$$

it is easy to see that we can solve for α and β provided that the matrix on the left-hand side is non-singular. This happens whenever its determinant is non-zero (see Appendix B).

We will assume that the determinant is zero,

$$x_1(t_0)\dot{x}_2(t_0) - x_2(t_0)\dot{x}_1(t_0) = 0, \tag{11.16}$$

and deduce a contradiction. It follows from this assumption that[1]

$$\frac{x_1(t_0)}{x_2(t_0)} = \frac{\dot{x}_1(t_0)}{\dot{x}_2(t_0)} = c, \text{ say,}$$

and so

$$x_1(t_0) = cx_2(t_0) \quad \text{and} \quad \dot{x}_1(t_0) = c\dot{x}_2(t_0). \tag{11.17}$$

Because the equation is linear this implies that $x_2(t)$ and $x_1(t)$ are proportional; since $x_2(t)$ is a solution so is $y(t) = cx_2(t)$, and clearly $y(t)$ satisfies the initial conditions

$$y(t_0) = cx_2(t_0) = x_1(t_0) \quad \text{and} \quad \dot{y}(t_0) = c\dot{x}_2(t_0) = \dot{x}_1(t_0).$$

Since solutions are unique, it follows that $x_1(t) = y(t) = cx_2(t)$ for all t. However, we chose $x_1(t)$ and $x_2(t)$ to be two linearly independent solutions, so we know that they are *not* proportional.

Since our assumption that the matrix in (11.15) is singular has led us to a contradiction, the matrix must be non-singular. Hence it is possible to solve this equation and find values of α and β such that $x(t) = \alpha x_1(t) + \beta x_2(t)$ satisfies the required initial conditions.

Thus, as claimed, two linearly independent solutions are necessary (we need at least two) and also sufficient (two will do) to form any solution as a linear combination

$$x(t) = \alpha x_1(t) + \beta x_2(t).$$

11.4 *The Wronskian

We have just seen that the determinant of the matrix in (11.15) is closely related to the linear independence of the functions $x_1(t)$ and $x_2(t)$. We now investigate this a little further. If x_1 and x_2 are linearly independent on an interval I then the only

---

[1] The following line assumes implicitly that $x_2(t_0) \neq 0$ and that $\dot{x}_2(t_0) \neq 0$. It is certainly not possible that $x_2(t_0) = \dot{x}_2(t_0) = 0$, since then the uniqueness of solutions would imply that $x_2(t) = 0$ for all t. If $x_2(t_0) = 0$ (and $\dot{x}_2(t_0) \neq 0$) then (11.16) implies that $x_1(t_0) = 0$, and then equation (11.17) follows once more. A similar argument yields (11.17) if $\dot{x}_2(t_0) = 0$ and $x_2(t_0) \neq 0$.

solution of

$$\alpha x_1(t) + \beta x_2(t) = 0 \qquad \text{for all} \qquad t \in I \tag{11.18}$$

should be $\alpha = \beta = 0$. If (11.18) holds for all $t \in I$ then we can differentiate and obtain a second equation

$$\alpha \dot{x}_1(t) + \beta \dot{x}_2(t) = 0 \qquad \text{for all} \qquad t \in I.$$

Putting these two equations together we obtain the matrix equation

$$\begin{pmatrix} x_1(t) & x_2(t) \\ \dot{x}_1(t) & \dot{x}_2(t) \end{pmatrix} \begin{pmatrix} \alpha \\ \beta \end{pmatrix} = \begin{pmatrix} 0 \\ 0 \end{pmatrix}.$$

If the matrix in this equation is non-singular for some $t_0 \in I$ then we can find the solution (α, β) by multiplying by the inverse of the matrix. This will give $\alpha = \beta = 0$, implying that $x_1(t)$ and $x_2(t)$ are linearly independent on I.

There is a special name for the determinant of this matrix, the *Wronskian* of x_1 and x_2, written as $W[x_1, x_2](t)$ (note that the Wronskian is a function of t):

$$W[x_1, x_2](t) = \begin{vmatrix} x_1(t) & x_2(t) \\ \dot{x}_1(t) & \dot{x}_2(t) \end{vmatrix} = x_1(t)\dot{x}_2(t) - x_2(t)\dot{x}_1(t).$$

We can re-express what we said above by saying that if $W[x_1, x_2](t) \not\equiv 0$ on I (is not identically equal to zero on I) then the functions x_1 and x_2 are linearly independent on I.

Conversely, in Section 11.3.2 we showed that (11.16) implies that $x_1(t) = cx_2(t)$ for all $t \in I$. Since (11.16) is just $W[x_1, x_2](t_0) = 0$, we have already shown that if $W[x_1, x_2](t) = 0$ anywhere on the interval I then x_1 and x_2 are linearly dependent.

Therefore two solutions of a linear second order equation are linearly independent if and only if their Wronskian is non-zero. For some other properties of the Wronskian see Exercises 11.2 and 11.3.

11.5 *Linear algebra

Linear algebra is the abstract study of the properties of linear spaces, and linear maps between such spaces. Because the subject is abstract it can be daunting, but it is its very abstraction that makes it widely applicable. Here we will see that the above results about linear differential equations can be very naturally recast within the linear algebra framework. Those unfamiliar with the ideas of linear algebra should feel free to move on to the next chapter.

The fundamental concept in the theory is the notion of a vector space; a (real) vector space is a collection V of elements along with notions of addition and multiplication, such that if v_1 and v_2 are elements of V then

$$\alpha v_1 + \beta v_2 \in V \qquad \text{for all} \qquad \alpha, \beta \in \mathbb{R}.$$

The prime example is the collection of all vectors in $\mathbb{R}^n$.

In order to set our results in this context, the first thing we need to do is to be a little more careful about our definition of the linear operator L. In order to do this we have to be more precise about the 'functions x with two derivatives' for which $L[x]$ is sensible.

We will denote by $C^0(I)$ the collection of all continuous functions that are defined on the interval I. This is a vector space, since if $f, g \in C^0(I)$ then

$$\alpha f + \beta g \in C^0(I) \qquad \text{for all} \qquad \alpha, \beta \in \mathbb{R}.$$

Similarly, the space $C^2(I)$, consisting of all continuous functions on I with continuous first and second derivatives, is also a vector space.

Given a function $x \in C^2(I)$, the linear operator L defined as in (11.10) by

$$L[x] = \frac{d^2x}{dt^2} + p(t)\frac{dx}{dt} + q(t)x$$

is certainly sensible, since x has two derivatives. Assuming that $p(t)$ and $q(t)$ are continuous functions, it follows that $L[x] \in C^0(I)$, since x, $\dot{x}$, and $\ddot{x}$ are continuous.

Thus L is a map from $C^2(I)$ into $C^0(I)$. Furthermore, as we have already seen, it is a linear map, i.e.

$$L[\alpha x_1 + \beta x_2] = \alpha L[x_1] + \beta L[x_2]$$

for any $\alpha, \beta \in \mathbb{R}$.

The kernel of a linear operator $L : E \to V$ (where E and V are vector spaces) consists of all those elements of E that are mapped to zero by L,

$$\ker(L) = \{x \in E : L[x] = 0\}.$$

The kernel of our differential operator L consists of all the elements of $C^2(I)$ for which $L[x] = 0$, i.e. precisely the set of all solutions of the homogeneous equation.

Now, it is a general result that the kernel of a linear operator is itself a vector space, in other words if $x_1 \in \ker(L)$ and $x_2 \in \ker(L)$ then

$$x_1 \in \ker(L) \quad \text{and} \quad x_2 \in \ker(L) \qquad \Rightarrow \qquad \alpha x_1 + \beta x_2 \in \ker(L).$$

We have seen this already for our operator L, since it is just another way of writing

$$L[x_1] = 0 \quad \text{and} \quad L[x_2] = 0 \qquad \Rightarrow \qquad L[\alpha x_1 + \beta x_2] = 0$$

(cf. (11.11)), i.e. we can add multiples of solutions of homogeneous linear equations and still have a solution (the superposition principle).

In order to construct the general solution of a homogeneous second order linear ODE we have just seen that we need two linearly independent solutions $x_1(t)$ and $x_2(t)$; using these we can construct any solution as

$$\alpha x_1(t) + \beta x_2(t).$$

Put another way, any element of ker(L) can be written as a linear combination of the two linearly independent elements x_1 and x_2. This says precisely that x_1 and x_2 form a basis of ker(L), i.e. a linearly independent spanning set.

Since the number of basis elements for a vector space is exactly what we mean by its dimension, it follows that the dimension of ker(L) is two. Given our definition of L, the statement

$$\dim \ker(L) = 2$$

is an elegant (albeit abstract) way of saying 'we can construct any solution of a homogeneous second order linear ODE given two linearly independent solutions (and two are needed)'.

Exercises

11.1 By finding the Wronskian of the following pairs of functions, show that they are linearly independent:
   (i) $x_1(t) = e^{k_1 t}$ and $x_2(t) = e^{k_2 t}$ with $k_1 \neq k_2$,
   (ii) $x_1(t) = e^{kt}$ and $x_2(t) = te^{kt}$, and
   (iii) $x_1(t) = e^{\rho t} \sin \omega t$ and $x_2(t) = e^{\rho t} \cos \omega t$.

11.2 Show that the Wronskian for two solutions $x_1(t)$ and $x_2(t)$ of the second order differential equation

$$\frac{d^2 x}{dt^2} + p_1(t)\frac{dx}{dt} + p_2(t)x = 0 \tag{E11.1}$$

   satisfies

$$\dot{W}(t) = -p_1(t)W(t).$$

   (Write $W(t) = x_1(t)\dot{x}_2(t) - x_2(t)\dot{x}_1(t)$, differentiate, and use the fact that $x_1(t)$ and $x_2(t)$ satisfy the equation (E11.1).) Deduce either that $W(t) = 0$ for all t, or that $W(t) \neq 0$ for all t.

11.3 We have seen that if x_1 and x_2 *are two solutions of a linear differential equation*, then they are linearly independent if and only if their Wronskian is non-zero. The simple example of this question shows that this is not true for general functions that are not the solutions of some differential equation.

(i) Check carefully that if $f(t) = t^2|t|$ then $df/dt = 3t|t|$ (this is easy when $t \neq 0$; you will have to use the formal definition of the derivative at $t = 0$).

(ii) Let

$$f_1(t) = t^2|t| \qquad \text{and} \qquad f_2(t) = t^3.$$

Show that although these two functions are linearly independent on $\mathbb{R}$, their Wronskian is identically zero.

12
Homogeneous second order linear equations with constant coefficients

In this chapter we will find the general solution of the homogeneous linear equation

$$a\frac{d^2x}{dt^2} + b\frac{dx}{dt} + cx = 0. \tag{12.1}$$

From our analysis in the previous chapter we would expect this general solution to be of the form

$$x(t) = Ax_1(t) + Bx_2(t)$$

where $x_1(t)$ and $x_2(t)$ are two linearly independent solutions of (12.1).

In order to find these two solutions, we 'guess' that they are of the form

$$x(t) = e^{kt}$$

and substitute this into (12.1). To see that this is reasonable, first, remember that e^{kt} is a solution of the constant coefficient first order linear equation

$$\frac{dx}{dt} = kx,$$

so that we have already seen exponential functions in the context of linear equations. More tellingly, we know that taking derivatives of $x(t) = e^{kt}$ only multiplies $x(t)$ by k. So $\ddot{x}$, $\dot{x}$, and x will all be just some constant times e^{kt}; we should be able to cancel the e^{kt}s and, all being well, end up with an equation that we can solve.

This is exactly what happens. Since

$$\frac{d}{dt}(e^{kt}) = ke^{kt} \qquad \text{and} \qquad \frac{d^2}{dt^2}(e^{kt}) = k^2e^{kt},$$

substituting $x(t) = e^{kt}$ in (12.1) gives

$$ak^2e^{kt} + bke^{kt} + ce^{kt} = 0.$$

111

Since e^{kt} is never zero we can divide by e^{kt} and find that we are left with a quadratic equation for k, known as the auxiliary equation,

$$ak^2 + bk + c = 0. \tag{12.2}$$

Our substitution has reduced the original differential equation for $x(t)$ into a simple algebraic equation for k. Since (12.2) is a quadratic equation we stand a good chance of finding two roots, which would provide us with the two independent solutions we need to form our general solution. However, although we might have two distinct real roots, it is also possible to have only one repeated real root, or even two complex conjugate roots (and therefore no real roots). These three cases lead to different types of solutions for (12.1) and we consider each in turn.

12.1 Two distinct real roots

Using the quadratic formula, the solutions of (12.2) are given by

$$k = \frac{-b \pm \sqrt{b^2 - 4ac}}{2a}.$$

Provided that $b^2 - 4ac > 0$ we can solve the equation to give two distinct real roots, which we will call k_1 and k_2. We have found two different values of k that will make our guess e^{kt} a solution of equation (12.1), so we have already obtained the two solutions we require; it is clear that if $k_1 \neq k_2$ then $e^{k_1 t}$ and $e^{k_2 t}$ are linearly independent on any interval (we could also check this using the Wronskian, see Exercise 11.1).

The general solution of (12.1) is therefore given by a linear combination of these two solutions,

$$x(t) = Ae^{k_1 t} + Be^{k_2 t}.$$

Note that if one of k_1 and k_2 is positive (or both of them) then there are solutions that tend exponentially fast to $+\infty$, and solutions that tend exponentially fast to $-\infty$ (depending on the initial condition), both as $t \to +\infty$. It is only if both k_1 and k_2 are negative that all the solutions decay exponentially to zero. See Exercise 12.2 for more on this.

Example 12.1 *Find the general solution of the equation*

$$\ddot{x} + \dot{x} - 6x = 0,$$

and the solution that satisfies the initial conditions

$$x(0) = 1 \quad \text{and} \quad \dot{x}(0) = 2.$$

The equation has $x(t) = e^{kt}$ as a solution when k satisfies

$$k^2 + k - 6 = 0.$$

The roots of this are $k = 2$ and $k = -3$, so the general solution is

$$x(t) = Ae^{2t} + Be^{-3t}. \tag{12.3}$$

To fit the initial conditions we substitute in to (12.3) and in to $\dot{x}(t) = 2Ae^{2t} - 3Be^{-3t}$ and then solve the resulting simultaneous equations

$$A + B = 1 \qquad 2A - 3B = 2;$$

so $A = 1$ and $B = 0$ which gives the solution

$$x(t) = e^{2t}. \qquad \square$$

If one of the roots of the auxiliary equation is zero then one of the two linearly independent solutions will be $e^{0 \times t} = e^0 = 1$, i.e. a constant.

Example 12.2 *Find the general solution of*

$$\ddot{x} - 2\dot{x} = 0.$$

If we try $x(t) = e^{kt}$ for this example then

$$k^2 - 2k = 0,$$

with roots $k = 0$ and $k = 2$, giving the two solutions $e^{0 \times t} = e^0 = 1$ and e^{2t}. It follows that the general solution is

$$x(t) = A + Be^{2t}. \tag{12.4}$$

$\square$

12.2 A repeated real root

When $b^2 = 4ac$ in the auxiliary equation ($ak^2 + bk + c = 0$) we get a repeated root k, and so we only obtain a single solution $x(t) = e^{kt}$. However, we know that in order to write down the general solution we need to have two linearly independent solutions.

In this case the second solution[1] includes an extra factor of t: it is te^{kt}. We will see in Chapter 17 that it is possible to derive this solution systematically, but for now we will just check that it really is a solution.

When the auxiliary equation has a repeated root we must have $b^2 = 4ac$ and we can use this to put the linear equation $a\ddot{x} + b\dot{x} + cx = 0$ into a standard form.

---

[1] For the use of the Wronskian to make sure that these two solutions are linearly independent see Exercise 11.1.

Dividing through by a gives

$$\ddot{x} + \frac{b}{a}\dot{x} + \frac{c}{a}x = 0.$$

Since $b^2 = 4ac$ this is the same as

$$\ddot{x} + \frac{b}{a}\dot{x} + \frac{b^2}{4a^2}x = 0,$$

and writing $\lambda = -b/2a$ this becomes

$$\ddot{x} - 2\lambda\dot{x} + \lambda^2 x = 0. \tag{12.5}$$

If we try $x(t) = e^{kt}$ in this equation then k must solve

$$k^2 - 2\lambda k + \lambda^2 = (k - \lambda)^2 = 0,$$

and there is indeed just one repeated root $k = \lambda$.

We now try the second solution given above, $x(t) = te^{\lambda t}$. For this we have

$$\dot{x} = e^{\lambda t} + \lambda te^{\lambda t} \qquad \ddot{x} = 2\lambda e^{\lambda t} + \lambda^2 te^{\lambda t},$$

and so substituting into the left-hand side of (12.5) we get

$$(2\lambda e^{\lambda t} + \lambda^2 te^{\lambda t}) - 2\lambda(e^{\lambda t} + \lambda te^{\lambda t}) + \lambda^2(te^{\lambda t})$$

which is zero.

So the two linearly independent solutions that we get in the case of a repeated root k are e^{kt} and te^{kt}, and the general solution is

$$x(t) = (A + Bt)e^{kt}. \tag{12.6}$$

Clearly if $k > 0$ then solutions tend to infinity ($+\infty$ or $-\infty$ depending on the initial condition), while if $k < 0$ the solutions decay to zero; te^{kt} tends to zero if $k < 0$ (see Figure 12.1).

Example 12.3 *Find the solution of*

$$\ddot{x} + 2\dot{x} + x = 0$$

that satisfies

$$x(0) = 0 \qquad \text{and} \qquad \dot{x}(0) = 1.$$

The equation has a solution $x(t) = e^{kt}$ if k solves

$$k^2 + 2k + 1 = 0.$$

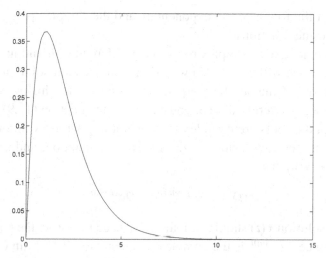

Fig. 12.1. The graph of te^{-t} against t: despite an initial increase the function decays to zero as $t \to \infty$.

This equation is $(k+1)^2 = 0$, so there is a repeated root $k = -1$. Two independent solutions of the equation are e^{-t} and te^{-t}, so the general solution is therefore

$$x(t) = (A + Bt)e^{-t}.$$

In order to satisfy the initial conditions we substitute in to find A and B, so we need, since $\dot{x}(t) = (B - A)e^{-t} - Bte^{-t}$,

$$A = 0 \qquad B - A = 1,$$

and the solution in this case is

$$x(t) = te^{-t},$$

whose graph is shown in Figure 12.1. □

12.3 No real roots

When $b^2 < 4ac$ the expression within the square root in the quadratic formula is negative, and so there are no real roots. Instead we obtain a pair of complex roots,

$$k = -\frac{b}{2a} \pm i\frac{\sqrt{4ac - b^2}}{2a}.$$

For brevity we will write these as $k = \rho \pm i\omega$ (so $\rho = -b/2a$ and $\omega = \sqrt{4ac - b^2}/2a$). The general solution corresponding to these roots is

$$x(t) = e^{\rho t}(A \cos \omega t + B \sin \omega t). \tag{12.7}$$

The real part of the root gives an exponential, and the complex part gives oscillating sine and cosine functions.

When you come across complex roots in a problem you should just write down (12.7). However, we will now see how to derive the general solution (12.7) using the complex roots of the auxiliary equation. If you are not happy with complex numbers, or just not interested, then you can move to equation (12.9)

Even though the roots are complex there is nothing wrong with our argument. With $k = \rho \pm i\omega$ the expression e^{kt} does solve the equation, and so the general solution can be written as

$$x(t) = Ce^{(\rho+i\omega)t} + De^{(\rho-i\omega)t}.$$

However, our solution $x(t)$ should be real, so we need to restrict the possible values of C and D; since $e^{(\rho-i\omega)t}$ is the complex conjugate of $e^{(\rho+i\omega)t}$, in order to make the whole expression real we want D to be the complex conjugate of C (we write $D = C^*$). Real solutions of the equation are given by

$$x(t) = Ce^{(\rho+i\omega)t} + C^*e^{(\rho-i\omega)t} \tag{12.8}$$

for an arbitrary complex number $C = \alpha + i\beta$. Note that the solution still has the two arbitrary constants (now α and β) that we would expect. With a little work we can rewrite this expression in a form that involves no complex numbers and thereby recover (12.7).

To do this we need to use the fact that

$$e^{i\theta} = \cos\theta + i\sin\theta$$

and that for any complex number z, $z + z^* = 2\,\text{Re}(z)$ (see Appendix A). Going back to (12.8) and using these two facts we get

$$\begin{aligned}
x(t) &= 2\,\text{Re}[Ce^{(\rho+i\omega)t}] \\
&= 2e^{\rho t}\,\text{Re}[Ce^{i\omega t}] \\
&= 2e^{\rho t}\,\text{Re}[(\alpha+i\beta)(\cos\omega t + i\sin\omega t)] \\
&= 2e^{\rho t}(\alpha\cos\omega t - \beta\sin\omega t) \\
&= e^{\rho t}(A\cos\omega t + B\sin\omega t),
\end{aligned}$$

if we set $A = 2\alpha$ and $B = -2\beta$. Since α and β were entirely arbitrary, so are A and B.

It is worth emphasising again that when you are trying to solve an equation and come across a complex conjugate pair of roots $\rho \pm i\omega$, you should immediately write down the general solution (12.9), rather than going through the above analysis.

The solution

$$x(t) = e^{\rho t}(A \cos \omega t + B \sin \omega t) \qquad (12.9)$$

naturally splits into two parts: a multiplying factor of $e^{\rho t}$, and an oscillating expression inside the brackets (remember that we have already seen in Section 9.4.3 how to combine $A \cos \omega t + B \sin \omega t$ to give one oscillating term $M \cos(\omega t - \phi)$).

All such solutions oscillate, and whether solutions decay or grow depends on the sign of ρ. If $\rho < 0$ then all the solutions decay exponentially to zero; if $\rho = 0$ then the amplitude is constant as we have pure oscillations; and if $\rho > 0$ then the solution oscillates with an amplitude that grows exponentially fast. These possibilities are illustrated in Figure 12.2.

The simplest example involving complex roots is when $k = \pm i\omega$, known as simple harmonic motion; we consider this in more detail in the next chapter.

Example 12.4 *Find the solution of*

$$\ddot{x} + 2\dot{x} + 5x = 0 \qquad x(0) = 1 \qquad \dot{x}(0) = 0. \qquad (12.10)$$

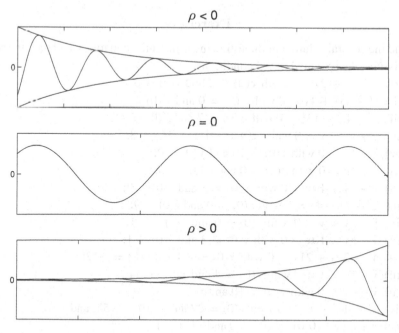

Fig. 12.2. Sample solutions of the form $x(t) = e^{\rho t}(A \cos \omega t + B \sin \omega t)$ for (from top to bottom) $\rho < 0$, $\rho = 0$ and $\rho > 0$. Each graph shows $x(t)$ against t.

For this equation $x(t) = e^{kt}$ is a solution if k satisfies

$$k^2 + 2k + 5 = 0$$

which has roots

$$k = \frac{-2 \pm \sqrt{4 - 20}}{2} = -1 \pm \sqrt{-4} = -1 \pm 2i.$$

So the general solution of (12.10) is

$$x(t) = e^{-t}(A \cos 2t + B \sin 2t),$$

showing that the origin is stable. Since

$$\dot{x}(t) = e^{-t}((2B - A) \cos 2t - (2A + B) \sin 2t)$$

the initial conditions pick out the solution with

$$A = 1 \qquad 2B - A = 0,$$

i.e. $A = 1$ and $B = \frac{1}{2}$, so that

$$x(t) = e^{-t}\left(\cos 2t + \tfrac{1}{2} \sin 2t\right). \qquad \square$$

Exercises

12.1 Find the general solution of the following differential equations, and then the solution satisfying the specified initial conditions.

    (i) $\ddot{x} - 3\dot{x} + 2x = 0$ with $x(0) = 2$ and $\dot{x}(0) = 6$;

    (ii) $y'' - 4y' + 4y = 0$ with $y(0) = 0$ and $y'(0) = 3$;

    (iii) $z'' - 4z' + 13z = 0$ with $z(0) = 7$ and $z'(0) = 42$;

    (iv) $\ddot{y} + \dot{y} - 6y = 0$ with $y(0) = -1$ and $\dot{y}(0) = 8$;

    (v) $\ddot{y} - 4\dot{y} = 0$ with $y(0) = 13$ and $\dot{y}(0) = 0$;

    (vi) $\ddot{\theta} + 4\theta = 0$ with $\theta(0) = 0$ and $\dot{\theta}(0) = 10$;

    (vii) $\ddot{y} + 2\dot{y} + 10y = 0$ with $y(0) = 3$ and $\dot{y}(0) = 0$;

    (viii) $2\ddot{z} + 7\dot{z} - 4z = 0$ with $z(0) = 0$ and $\dot{z}(0) = 9$;

    (ix) $\ddot{y} + 2\dot{y} + y = 0$ with $y(0) = 0$ and $\dot{y}(0) = -1$;

    (x) $\ddot{x} + 6\dot{x} + 10x = 0$ with $x(0) = 3$ and $\dot{x}(0) = 1$;

    (xi) $4\ddot{x} - 20\dot{x} + 21x = 0$ with $x(0) = -4$ and $\dot{x}(0) = -12$;

    (xii) $\ddot{y} + \dot{y} - 2y = 0$ with $y(0) = 4$ and $\dot{y}(0) = -4$;

    (xiii) $\ddot{y} - 4y = 0$ with $y(0) = 10$ and $\dot{y}(0) = 0$;

    (xiv) $y'' + 4y' + 4y = 0$ with $y(0) = 27$ and $y'(0) = -54$; and

    (xv) $\ddot{y} + \omega^2 y = 0$ with $y(0) = 0$ and $\dot{y}(0) = 1$.

12.2 If the roots of the auxiliary equation are $k_1 > 0$ and $-k_2 < 0$ then the solution is

$$x(t) = Ae^{k_1 t} + Be^{-k_2 t}.$$

For most choices of initial conditions

$$x(0) = x_0 \qquad \dot{x}(0) = y_0$$

we will have $x(t) \to \pm\infty$ as $t \to \infty$. However, there are some special initial conditions for which $x(t) \to 0$ as $t \to \infty$. Find the relationship between x_0 and y_0 that ensures this.

12.3 (T) Solutions of linear equations with constant coefficients cannot blow up in finite time; it follows that their solutions exist for all $t \in \mathbb{R}$. To see this, we will consider

$$\ddot{x} + p\dot{x} + qx = 0 \qquad \text{with} \qquad x(0) = x_0 \qquad \text{and} \qquad \dot{x}(0) = y_0$$

for $t \geq 0$ (a similar argument applies for $t \leq 0$). By setting $y = \dot{x}$, we can rewrite this as a coupled pair of first order equations

$$\dot{x} = y$$
$$\dot{y} = -py - qx.$$

Show that

$$\frac{1}{2}\frac{d}{dt}(x^2 + y^2) = (1 - q)xy - py^2,$$

and hence that

$$\frac{d}{dt}(x^2 + y^2) \leq (1 + |q| + 2|p|)(x^2 + y^2).$$

Using the result of Exercise 9.7 deduce that for $t \geq 0$

$$x(t)^2 + y(t)^2 \leq (x(0)^2 + y(0)^2)e^{(1+|q|+2|p|)t},$$

showing that finite-time blowup is impossible. Hint: $xy \leq \frac{1}{2}(x^2 + y^2)$. (The same argument works, essentially unchanged, for

$$\ddot{x} + p(t)\dot{x} + q(t)x = 0$$

provided that $|p(t)| \leq p$ and $|q(t)| \leq q$ for all $t \in \mathbb{R}$).

13

Oscillations

In this chapter we look at oscillating mechanical systems, which form one very natural class of examples of second order linear equations.

13.1 The spring

The simplest system that gives rise to oscillations is a mass on a spring. If the spring has a natural length l, then Hooke's law says the force exerted by the spring when it is extended an additional length x is proportional to this extension x. So the equation of motion for a mass (with mass m) on the end of spring (see Figure 13.1) is

$$m\ddot{x} = -\sigma x. \tag{13.0}$$

where $\sigma > 0$ is the 'spring constant'.

If we divide by m and set $\omega^2 = \sigma/m$ then this reads

$$\ddot{x} = -\omega^2 x. \tag{13.1}$$

Equation (13.1) is a linear equation. Trying $x(t) = e^{kt}$ we get

$$k^2 = -\omega^2$$

as the auxiliary equation that determines k. This equation has complex roots $k = \pm i\omega$, and it follows that the general solution is of the form

$$x(t) = A \cos \omega t + B \sin \omega t. \tag{13.2}$$

We can use the same trick we used in Section 9.4 to rewrite our solution to make it clear that it is one oscillation. If we write

$$x(t) = \sqrt{A^2 + B^2} \left[\frac{A}{\sqrt{A^2 + B^2}} \cos \omega t + \frac{B}{\sqrt{A^2 + B^2}} \sin \omega t \right] \tag{13.3}$$

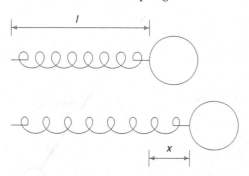

Fig. 13.1. A spring: the top picture shows the spring at its natural length l, unextended, while the lower picture shows the spring extended by a length x.

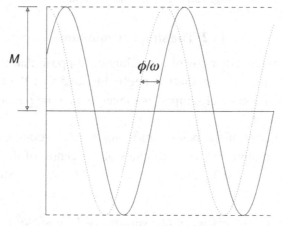

Fig. 13.2. The solution $x(t) = M \cos(\omega t - \phi)$ plotted against t is shown here as a solid line. For comparison the dotted line shows the graph of $x(t) = M \cos \omega t$; the delay between the two solutions is ϕ/ω, as indicated.

then the squares of the coefficients of cos and sin within the $[\cdots]$ now sum to one, so we can find ϕ with

$$\cos \phi = \frac{A}{\sqrt{A^2 + B^2}} \quad \text{and} \quad \sin \phi = \frac{B}{\sqrt{A^2 + B^2}}$$

(i.e. with $\tan \phi = B/A$). Writing $M = \sqrt{A^2 + B^2}$ then (13.3) becomes

$$\begin{aligned} x(t) &= M[\cos \phi \cos \omega t + \sin \phi \sin \omega t] \\ &= M \cos(\omega t - \phi), \end{aligned}$$

using the double angle formula $\cos a \cos b + \sin a \sin b = \cos(a - b)$. This solution is illustrated in Figure 13.2.

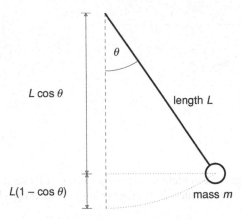

Fig. 13.3. The simple pendulum.

13.2 The simple pendulum

We now consider the oscillations of a pendulum. Suppose that the pendulum has length L, that the mass of the shaft is negligible, and that the bob has mass m. Figure 13.3 shows the general setup; θ is the angle that the pendulum makes with the downward vertical.

The potential energy of the pendulum is $mg \times L(1 - \cos\theta)$. If the pivot is at the origin then at an angle θ the coordinates of the centre of the mass m are $\mathbf{x} = (x, y) = (L\sin\theta, L(-\cos\theta))$. Then $\dot{\mathbf{x}} = (\dot{x}, \dot{y}) = (L\cos\theta\dot{\theta}, L\sin\theta\dot{\theta})$, and so the kinetic energy is

$$\tfrac{1}{2}m|\dot{\mathbf{x}}|^2 = \tfrac{1}{2}m(\dot{x}^2 + \dot{y}^2) = \tfrac{1}{2}m(L^2\sin^2\theta\dot{\theta}^2 + L^2\cos^2\theta\dot{\theta}^2) = \tfrac{1}{2}mL^2\dot{\theta}^2.$$

Since the total energy

$$E = \tfrac{1}{2}mL^2\dot{\theta}^2 + mgL(1 - \cos\theta)$$

is constant, if we differentiate we obtain

$$0 = mL^2\dot{\theta}\ddot{\theta} + mgL\sin\theta\dot{\theta},$$

which, on dividing by $\dot{\theta}$, yields the equation of motion[1]

$$m\frac{d^2\theta}{dt^2} = -\frac{mg}{L}\sin\theta. \qquad (13.4)$$

We will investigate equation (13.4) itself in Chapter 35, but for now we approximate it by a linear equation so that we can use the theory of the preceding chapter.

---

[1] With some care, assuming the continuity of $\theta(t)$ and its derivatives, it is possible to show that this equation is also valid when $\dot{\theta} = 0$.

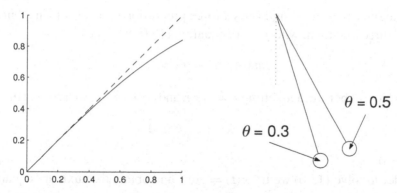

Fig. 13.4. The left-hand picture is the graph of $\sin\theta$ (solid) and θ (dashed) plotted against θ for $0 \leq \theta \leq 1$. Note that the curves are fairly close for $\theta \leq 0.5$ and almost indistinguishable when $\theta \leq 0.3$. The right-hand picture shows the pendulum at angles $\theta = 0.3$ and $\theta = 0.5$ to the downward vertical. Clearly requiring $-0.3 \leq \theta \leq 0.3$ is little restriction for a calculation involving the pendulum in a clock.

When the oscillations are small, i.e. when θ is small, we can approximate $\sin\theta$ by θ (see Figure 13.4) and so obtain the linear equation

$$m\frac{d^2\theta}{dt^2} = -\frac{mg}{L}\theta.$$

Defining $\omega^2 = g/L$ we can rewrite this as

$$\frac{d^2\theta}{dt^2} = -\omega^2\theta. \tag{13.5}$$

Apart from the change of dependent variable from x to θ, this is the same equation as we had before in (13.1). The solution we found in the previous section remains valid here, and so we have

$$\theta(t) = M\cos(\omega t - \phi).$$

Thus the pendulum oscillates about $\theta = 0$, the downward vertical, with period $2\pi/\omega = 2\pi\sqrt{L/g}$. Note that in this linear approximation the amplitude M of the motion (how wide the swing is) has no effect on the period.

13.3 Damped oscillations

The simple equations above neglect any effects of friction or air resistance, or what we might more generally call 'damping'.

If we assume that damping exerts a force proportional to the velocity, but in the opposite direction, then, with $\mu > 0$, equation (13.0) becomes

$$m\ddot{x} + \mu\dot{x} + \sigma x = 0.$$

Dividing by m as before, and setting $\lambda = \mu/m$ and $\omega^2 = \sigma/m$, we arrive at the model

$$\ddot{x} + \lambda\dot{x} + \omega^2 x = 0 \qquad (13.6)$$

with $\lambda > 0$.

In order to solve (13.6) we try $x(t) = Ae^{kt}$ and obtain the auxiliary equation

$$k^2 + \lambda k + \omega^2 = 0.$$

The roots of this equation are

$$k = \frac{-\lambda \pm \sqrt{\lambda^2 - 4\omega^2}}{2},$$

which gives rise to three possibilities depending on the nature of the roots, all illustrated in Figure 13.5.

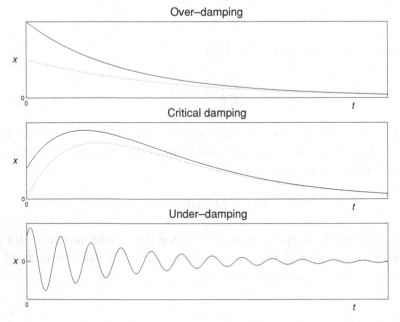

Fig. 13.5. From top to bottom: over-damping $x(t) = Ae^{-k_1 t} + Be^{-k_2 t}$ (for large t, $x(t) \approx Ae^{-k_1 t}$ (the dotted curve), where $0 < k_1 < k_2$); critical damping $x(t) = Ate^{-kt} + Be^{-kt}$ (for large t, $x(t) \approx Ate^{-kt}$, shown in the dotted curve); under-damping.

Over-damping

When $\lambda^2 > 4\omega^2$ there are two distinct real roots $-k_1$ and $-k_2$, and both are negative (since $0 < \lambda^2 - 4\omega^2 < \lambda^2$). So the general solution is

$$x(t) = Ae^{-k_1 t} + Be^{-k_2 t},$$

and all the solutions are exponentially decaying and approach zero as $t \to \infty$. Furthermore there are no oscillations of the system on its way to equilibrium.

Critical damping

When $\lambda^2 = 4\omega^2$ we have $k = -\lambda/2$ 'twice', so that the general solution is a combination of $e^{-\lambda t/2}$ and $te^{-\lambda t/2}$,

$$x(t) = (A + Bt)e^{-\lambda t/2}.$$

Again, the system settles down to its equilibrium without any oscillations. However, this is the critical case, in that any further reduction in the damping allows for oscillations, as we will soon see. It is also possible for $x(t)$ to increase for a short time interval (see Figure 13.5), see Exercise 13.9 for more details.

Under-damping

When $\lambda^2 < 4\omega^2$ we have a complex conjugate pair for k,

$$k = -\frac{\lambda}{2} \pm i\sigma$$

(with $\sigma = \frac{1}{2}\sqrt{4\omega^2 - \lambda^2}$). This gives the general solution

$$x(t) = e^{-\lambda t/2}(A\cos\sigma t + B\sin\sigma t). \qquad (13.7)$$

The system is always oscillating, but the amplitude of the oscillations decays to zero exponentially fast.

If this solution represents the oscillations of a pendulum damped by frictional forces (air resistance, friction at the pivot) then although the amplitude decays the period of the oscillations remains constant. It is this effect that makes the pendulum an effective mechanism to run a clock.

To see that this is the case, we rewrite $x(t)$ from (13.7) in the compact form

$$x(t) = Me^{-\lambda t/2}\cos(\sigma t - \phi),$$

and differentiate with respect to t to give

$$\frac{dx}{dt} = -M\frac{\lambda}{2}e^{-\lambda t/2}\cos(\sigma t - \phi) - \sigma Me^{-\lambda t/2}\sin(\sigma t - \phi).$$

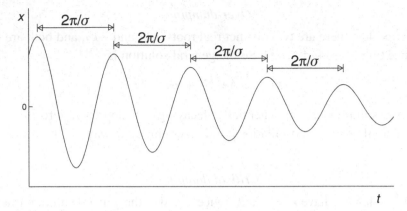

Fig. 13.6. Even though the oscillations are decaying, the period of the oscillations remains constant.

Thus maxima and minima occur when

$$\tan(\sigma t - \phi) = -\frac{\lambda}{2\sigma},$$

i.e. when

$$\sigma t - \phi = \tan^{-1}\left(-\frac{\lambda}{2\sigma}\right) + n\pi$$

for any integer n. The time between successive maxima is therefore $2\pi/\sigma$, independent of the amplitude, see Figure 13.6.

Exercises

13.1 A spring of natural length l and spring constant k is suspended vertically from a fixed point, and a weight of mass m attached. If the system is at rest ($\ddot{x} = \dot{x} = 0$) how far has the spring extended? If the mass is pulled down slightly from this rest position and then released, show that it then oscillates about its equilibrium position with period $2\pi/\omega$, where $\omega^2 = k/m$.

13.2 The acceleration due to gravity in fact depends on the distance R from the centre of the Earth: $g = GM/R^2$, where M is the mass of the Earth and G Newton's gravitational constant. Show that the period of oscillation of a pendulum will increase as it is taken higher.

13.3 The Earth bulges at the equator; at a latitude θ, the distance to the centre of the Earth (measured in kilometres) is approximately

$$R(\theta) = \sqrt{R_e^2 \cos^2\theta + R_p^2 \sin^2\theta},$$

where $R_e = 6378$ and $R_p = 6357$.

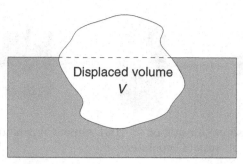

Displaced volume
V

Fig. 13.7. The buoyancy force on an object is equal to the weight of water that it displaces.

I decide to move from Leamington Spa, at a latitude of 52°, to Seville, which is lies at a latitude of 37°. My grandfather clock, which keeps perfect time, has a pendulum of length 75 cm. How long would the pendulum need to be to keep perfect time in Seville?

13.4 The buoyancy force on an object is equal to the weight of water that it displaces. If an object has mass M and displaces a volume V of water then the downward forces on it are $Mg - Vg$, in units for which the density of water is 1; see Figure 13.7.

A bird of mass m is sitting on a cylindrical buoy of density ρ, radius R, and height h, which is floating at rest. How much of the buoy lies below the surface?

The bird flies away. Show that the buoy now bobs up and down, with the amount below the surface oscillating about ρh with period $2\pi\sqrt{\rho h/g}$ and amplitude $m/\pi R^2$.

13.5 An open tin can, half full of water, is floating in a canal. The can is 11 cm tall, has a diameter of 7.5 cm, and has a mass of 50 g. Show that at rest the can is submerged a distance of approximately 6.63 cm below the surface of the canal. If the can is pushed down further it will then perform oscillations about its equilibrium position. Show that the can bobs up and down every 0.21 seconds (a little under five times per second). The acceleration due to gravity is approximately 9.8 m/s² $= 980$ cm/s²; the density of water is 1 g/cm³. You can check your answers in a sink with a baked bean can.

13.6 A right circular cone, of height h, density ρ, and with base radius R, is placed point downward in a lake. Assuming that the apex remains point vertically downwards, show that if the cone is submerged to a depth x then

$$\ddot{x} = g - \left(\frac{x}{h}\right)^3 \frac{g}{\rho}.$$

(You need not solve this equation.) At equilibrium how far is the cone submerged?

13.7 A dashpot is a device designed to add damping to a system, consisting essentially of a plunger in a cylinder of liquid or gas, see Figure 13.8.

Fig. 13.8. A dashpot. Illustration © 2001 Airpot Corporation. Airpot is a registered trademark of Airpot Corporation.

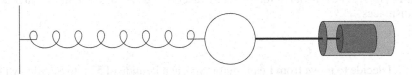

Fig. 13.9. A mass-spring-dashpot system.

It produces a resisting force proportional to the velocity, precisely the kind of 'damping' that we used in our model

$$m\ddot{x} + \mu\dot{x} + kx = 0, \tag{E13.1}$$

with μ indicating the 'strength' of the dashpot. Dashpots are used in a variety of applications, for example, cushioning the opening mechanism on a tape recorder, or in car shock absorbers.

A mass-spring-dashpot system consists of a mass attached to a spring and a dashpot, and is shown in Figure 13.9. A weight of mass 10 kg is attached to a spring with spring constant 5, and to a dashpot of strength μ. How strong should the dashpot be to ensure that the system is over-damped? What would the period of oscillations be if $\mu = 14$?

13.8 When first opened, the Millennium Bridge in London (see Figure 13.10) wobbled from side to side as people crossed; you can see this on video at www.arup.com/MillenniumBridge. Footfalls created small side-to-side movements of the bridge, which were then enhanced by the tendency of people to adjust their steps to compensate for the wobbling. With more than a critical number of pedestrians (around 160) the bridge began to wobble violently.[2]

Without any pedestrians, the displacement x of a representative point on the bridge away from its normal position would satisfy

$$M\ddot{x} + k\dot{x} + \lambda x = 0,$$

---

[2] A detailed analysis is given in P. Dallard *et al.*, The London Millennium Footbridge, *The Structural Engineer* **79** (2001), 17–33.

Fig. 13.10. The Millennium Bridge in London (courtesy of Arup).

where

$$M \approx 4 \times 10^5 \, \text{kg}, \quad k \approx 5 \times 10^4 \, \text{kg/s}, \quad \text{and} \quad \lambda \approx 10^7 \, \text{kg/s}^2.$$

Show that the level of damping here is only around 1% of the critical level.

The effective forcing from each pedestrian was found by experiment (which involved varying numbers of people walking across the bridge) to be proportional to $\dot{x}$, with

$$F \approx 300\dot{x}.$$

If there are N pedestrians, the displacement of the bridge satisfies

$$M\ddot{x} + k\dot{x} + \lambda x = 300N\dot{x}.$$

Find the critical number N_0 of pedestrians, such that if there are more than N_0 pedestrians the bridge is no longer damped. Show that if there are 200 pedestrians then there will be oscillations with a frequency of approximately 0.8 hertz (oscillations per second) the amplitude of oscillation of which grows as $e^{t/80}$.

The problem was corrected by adding additional damping, a large part of which was essentially a collection of dashpots, in order to bring the damping up to 20% of the critical level. What would this do to the value of k, and how many people can now walk across the bridge without counteracting all the damping?

13.9 In the case of critical damping (see Section 13.3), the general solution of (13.6) is of the form

$$x(t) + (A \quad Bt)e^{-\lambda t/2}.$$

Show that if $B > 0$ and $\lambda A < 2B$ then $x(t)$ increases initially, reaching its maximum value at

$$t = \frac{2}{\lambda} - \frac{A}{B}.$$

14

Inhomogeneous second order linear equations

We now investigate how to obtain solutions of the inhomogeneous equation

$$a\frac{d^2x}{dt^2} + b\frac{dx}{dt} + cx = f(t). \tag{14.1}$$

As a convenient shorthand we define an operator L, cf. (11.10), by

$$L[x] = a\frac{d^2x}{dt^2} + b\frac{dx}{dt} + cx.$$

Recall that $L[x](t)$ is a function of t, as is $x(t)$. Using this notation we can rewrite (14.1) more compactly as

$$L[x] = f(t).$$

Since (14.1) is a second order equation, we expect that the general solution will have two arbitrary constants in order to fit any choice of initial condition $x(t_0) = x_0$ and $\dot{x}(t_0) = y_0$.

14.1 Complementary function and particular integral

Because of the linearity of the equation, we can split the problem of finding its solution into two parts. First we try to find the general solution of the corresponding homogeneous problem

$$L[y] = 0.$$

We have already seen how to solve this problem in Chapter 12; its general solution will be a linear combination of two independent solutions $y_1(t)$ and $y_2(t)$,

$$y(t) = Ay_1(t) + By_2(t).$$

This is called the *complementary function*, and will provide us with the two arbitrary constants that we require in our final result.

The second part of solving the problem is to find one particular function $x_p(t)$ that satisfies the equation, i.e. for which

$$L[x_p] = f(t).$$

The standard method for finding such a *particular integral* requires a mixture of experience and inspired guesswork, but we will see later (Chapter 18) that there is also a systematic way of solving this problem.

Given our complementary function $y(t)$ and a particular integral $x_p(t)$, the general solution of the original problem will be given by $x(t) = y(t) + x_p(t)$, since

$$L[y(t) + x_p(t)] = L[y(t)] + L[x_p(t)] = 0 + f(t) = f(t),$$

using the linearity of L. Since $y(t) = Ay_1(t) + By_2(t)$, our solution

$$x(t) = Ay_1(t) + By_2(t) + x_p(t)$$

contains the two arbitrary constants that we need in order to fit any choice of initial conditions.

There is nothing special about applying this technique to second order equations. The essential point is the linearity, and it is possible to solve first order equations this way, as we now see in the following simple example.

Example 14.1 *Use the technique of the complementary function and particular integral to solve the first order equation*

$$\frac{dx}{dt} + px = q.$$

The 'complementary function' is the general solution of the homogeneous problem

$$\dot{y} + py = 0,$$

which is $y(t) = Ae^{-pt}$. A particular integral is any solution $x_p(t)$ that solves

$$\dot{x}_p + px_p = q;$$

one such solution is $x_p(t) = q/p$. So the general solution is given by the complementary function plus the particular integral,

$$x(t) = Ae^{-pt} + \frac{q}{p}$$

as we found before in (9.5).   □

We now look at certain choices of right-hand side $f(t)$ for which we can find a 'particular integral' for

$$a\frac{d^2x}{dt^2} + b\frac{dx}{dt} + cx = f(t), \tag{14.2}$$

i.e. one particular solution $x_p(t)$ that when plugged into the left-hand side of (14.2) gives the correct function $f(t)$ on the right-hand side.

We can do this when $f(t)$ is a combination of powers of t, exponentials (like e^{kt}), and sines and cosines. Essentially we try a solution $x(t)$ that looks very much like the original $f(t)$, with some adjustments when $f(t)$ is a part of the complementary function. This is known as the 'method of undetermined coefficients'; essentially we guess the form of the solution, and include some coefficients in our guess, which are then determined by substituting into the equation.

14.2 When $f(t)$ is a polynomial

When $f(t)$ is a polynomial in t then our 'guess' for the particular integral $x_p(t)$ is a general polynomial of the same order as f, i.e. if the highest power in $f(t)$ is t^n then our guess is

$$x_p(t) = c_n t^n + c_{n-1} t^{n-1} + \cdots + c_0.$$

However, we will see that if $x(t) = c$ satisfies the homogeneous equation then we have to multiply our guess by t. These ideas are most clearly illustrated by finding the particular integral for a number of examples.

Example 14.2 *Find the general solution of the equation*

$$\ddot{x} + \dot{x} - 6x = 12.$$

We have already found the complementary function, i.e. the solution of

$$\ddot{y} + \dot{y} - 6y = 0,$$

which was $y(t) = Ae^{2t} + Be^{-3t}$, see (12.3). Since the right-hand side is a constant, we try $x_p(t) = C$. All the derivative terms in the differential equation vanish and we are left with

$$-6C = 12,$$

so we need $C = -2$: the particular integral is $x_p(t) = -2$. The general solution is made up of the complementary function plus the particular integral,

$$x(t) = Ae^{2t} + Be^{-3t} - 2. \qquad \square$$

Example 14.3 *Find the general solution of the equation*

$$\ddot{x} + \dot{x} - 6x = 36t.$$

In line with our policy above we try $x_p(t) = Ct + D$ for a particular integral. (Note that if we were to try $x_p(t) = Ct$ without the constant term then although we can choose C to give $36t$ $(C = -6)$ the $\dot{x}_p$ term would give an extra factor of -6.) Then we have $\dot{x}_p = C$, and so we require

$$C - 6(Ct + D) = 36t,$$

which gives $C = -6$ and $D = -1$. So the particular integral is $x_p(t) = -6t - 1$ and the general solution is

$$Ae^{2t} + Be^{-3t} - 6t - 1.$$ □

Example 14.4 *Find the general solution of*

$$\ddot{x} + \dot{x} - 6x = 216t^3$$

For the particular integral we have to try $x_p(t) = Ct^3 + Dt^2 + Et + F$. Then

$$\dot{x}_p = 3Ct^2 + 2Dt + E \quad \text{and} \quad \ddot{x}_p = 6Ct + 2D,$$

and so we need

$$\underbrace{6Ct + 2D}_{\ddot{x}_p} + \underbrace{3Ct^2 + 2Dt + E}_{\dot{x}_p} - \underbrace{6(Ct^3 + Dt^2 + Et + F)}_{x_p} = 216t^3,$$

i.e.

$$-6Ct^3 + (3C - 6D)t^2 + (6C + 2D - 6E)t + (2D + E - 6F) = 216t^3.$$

We want

$$-6C = 216, \quad 3C - 6D = 0, \quad 6C + 2D - 6E = 0, \quad \text{and} \quad 2D + E - 6F = 0,$$

which yields $C = -36$, $D = -18$, $E = -42$, $F = -13$, and the particular integral is

$$x_p(t) = -36t^3 - 18t^2 - 42t - 13.$$

Thus the general solution is

$$x(t) = Ae^{2t} + Be^{-3t} - 36t^3 - 18t^2 - 42t - 13.$$ □

We now see that there is a possible catch with this technique.

Example 14.5 *Find the general solution of*

$$\ddot{x} - 2\dot{x} = 4.$$

We saw earlier, in (12.4), that the complementary function is $A + Be^{2t}$. If we try $x_p(t) = C$ for our particular integral then $\ddot{x}_p - 2\dot{x}_p$ will be zero, since we have a constant as part of the complementary function. To deal with this we multiply our 'guess' by an extra factor of t, as we did to obtain the second independent solution of the homogeneous equation when the auxiliary equation had a repeated root.

Thus instead of $x_p(t) = C$ we try $x_p(t) = Ct$ and then we have $\dot{x}_p = C$ and $\ddot{x}_p = 0$, therefore we want

$$-2C = 4$$

which gives a particular integral when $x_p(t) = -2t$, and so the general solution is

$$x(t) = A + Be^{2t} - 2t. \qquad \qquad \square$$

Example 14.6 *Find the general solution of the equation*

$$\ddot{x} = 4.$$

This is an extreme example, and it would be easy to solve by integrating twice. However, we can also apply our general method for linear equations with constant coefficients. The general solution of the homogeneous equation $\ddot{y} = 0$ is $y(t) = At + B$ (there is a repeated root $k = 0$ of the auxiliary equation), so we cannot try $x_p(t) = C$, and nor can we try $x_p(t) = Ct$; we have to try $x_p(t) = Ct^2$ in this case. Since $\ddot{x}_p = 2C$ we need $C = 4/2 = 2$ and the particular integral is $2t^2$. The general solution is therefore $x(t) = 2t^2 + At + B$. $\qquad \square$

14.3 When $f(t)$ is an exponential

The second kind of right-hand side for which we can find a particular integral is an exponential, $f(t) = ce^{kt}$. Essentially we try a multiple of the same exponential for our particular integral, $x_p(t) = Ce^{kt}$. However, if the exponential e^{kt} on the right-hand side is a solution of the homogeneous equation then Ce^{kt} cannot be a particular integral, since substituting this into the left-hand side will just give zero ($L[Ce^{kt}] = 0$). So in this case we need to try t times the exponential, $x_p(t) = Cte^{kt}$. In the extreme case when k is a repeated real root of the auxiliary equation we have to try $x_p(t) = Ct^2e^{kt}$.

We look at the simplest case first.

Example 14.7 *Find the general solution of*

$$\ddot{x} + \dot{x} - 6x = 4e^{-2t}.$$

We saw above (in Example 12.1) that the complementary function, i.e. the general solution of the homogeneous equation $\ddot{y} + \dot{y} - 6y = 0$, is $y(t) = Ae^{2t} + Be^{-3t}$, and so e^{-2t} is not a solution of the homogeneous problem. This means that we can try $x_p(t) = Ce^{-2t}$ as a particular integral. Because

$$\dot{x}_p = -2Ce^{-2t} \qquad \text{and} \qquad \ddot{x}_p = 4Ce^{-2t}$$

we need

$$4C - 2C - 6C = 4,$$

i.e. $C = -1$, so the particular integral is

$$x_p(t) = -e^{-2t}$$

and the general solution is

$$x(t) = Ae^{2t} + Be^{-3t} - e^{-2t}. \qquad \square$$

Example 14.8 *Find the general solution of*

$$\ddot{x} + \dot{x} - 6x = 5e^{-3t}. \tag{14.3}$$

The complementary function is the same as in the previous example, $y(t) = Ae^{2t} + Be^{-3t}$. Because e^{-3t} is a solution of the homogeneous equation we have to try $x_p(t) = Cte^{-3t}$ for a particular integral for this example. With this guess for $x_p(t)$ we have

$$\dot{x}_p = Ce^{-3t} + -3Cte^{-3t} \qquad \text{and} \qquad \ddot{x}_p = -6Ce^{3t} + 9Cte^{3t},$$

and so we need

$$-6Ce^{-3t} + 9Cte^{-3t} + Ce^{-3t} - 3Cte^{-3t} - 6Cte^{-3t} = 5e^{-3t}.$$

The te^{-3t} terms cancel, and so we want $C = -1$. The particular integral turns out to be

$$x_p(t) = -te^{-3t}$$

giving the general solution

$$x(t) = Ae^{2t} + Be^{-3t} - te^{-3t}. \qquad \square$$

Example 14.9 *Find the general solution of*

$$\ddot{x} + 2\dot{x} + x = 6e^{-t}.$$

The complementary function $y(t)$ is the solution of

$$\ddot{y} + 2\dot{y} + y = 0.$$

Trying $y(t) = e^{kt}$ yields the auxiliary equation $k^2 + 2k + 1 = 0$, and so $k = -1$ 'twice'. It follows that $y(t) = Ae^{-t} + Bte^{-t}$, which means that both our 'standard guess' $x_p(t) = Ce^{-t}$ and our adjusted guess $x_p(t) = Cte^{-t}$ solve the homogeneous equation. Because of this we have to try $x_p(t) = Ct^2e^{-t}$; once again we deal with the problem by including an extra factor of t. We have

$$\dot{x}_p = 2Cte^{-t} - Ct^2e^{-t} \quad \text{and} \quad \ddot{x}_p = 2Ce^{-t} - 4Cte^{-t} + Ct^2e^{-t}.$$

Substituting into the left-hand side we get

$$[2Ce^{-t} - 4Cte^{-t} + Ct^2e^{-t}] + 2[2Cte^{-t} - Ct^2e^{-t}] + Ct^2e^{-t} = 2Ce^{-t}.$$

So we need $C = 3$; the particular integral is $x_p(t) = 3t^2e^{-t}$ and the general solution is

$$x(t) = (3t^2 + At + B)e^{-t}. \qquad \square$$

14.4 When $f(t)$ is a sine or cosine

If $f(t) = \alpha \sin \sigma t + \beta \cos \sigma t$ (including the cases $\alpha = 0$ or $\beta = 0$) then we need to try a combination of $\sin \sigma t$ and $\cos \sigma t$ for the particular integral,

$$x_p(t) = C \sin \sigma t + D \cos \sigma t. \tag{14.4}$$

Note that we need a combination of sine and cosine even if $f(t)$ only involves one of these two functions, since trying $x_p(t) = C \sin \sigma t$ means that $\dot{x}_p(t) = C\sigma \cos \sigma t$. If $\sin \sigma t$ and $\cos \sigma t$ satisfy the homogeneous equation then instead we need to try

$$x_p(t) = Ct \sin \sigma t + Dt \cos \sigma t \tag{14.5}$$

with an extra factor of t.

Example 14.10 *Find the general solution of*

$$\ddot{x} + 2\dot{x} + x = 100 \cos 2t.$$

We found the complementary function $y(t) = (A + Bt)e^{-t}$ in Example 14.9. Since $\sin 2t$ is not a part of the complementary function, we can try

$$x_p(t) = C \sin 2t + D \cos 2t$$

as a potential particular integral, cf. (14.4). Then

$$\dot{x}_p(t) = 2C \cos 2t - 2D \sin 2t \quad \text{and} \quad \ddot{x}_p(t) = -4C \sin 2t - 4D \cos 2t,$$

and so

$$\ddot{x}_p + 2\dot{x}_p + x_p = (4C - 3D)\cos 2t - (3C + 4D)\sin 2t.$$

Therefore we need

$$4C - 3D = 100 \quad \text{and} \quad 3C + 4D = 0,$$

which gives $C = 16$ and $D = -12$. Thus the particular integral is

$$x_p(t) = 16\sin 2t - 12\cos 2t$$

and the general solution is

$$x(t) = (A + Bt)e^{-t} + 16\sin 2t - 12\cos 2t. \qquad \square$$

Example 14.11 *Find the general solution of*

$$\ddot{x} + x = 8\cos t. \tag{14.6}$$

This example is not quite so simple, since $\sin t$ and $\cos t$ are the solutions of the homogeneous problem (the complementary function is $A\sin t + B\cos t$). Using the same remedy as that for Example 14.5, namely multiplying our original guess by t, we now try a combination of $t\sin t$ and $t\cos t$,

$$x_p(t) = Ct\sin t + Dt\cos t,$$

cf. (14.5). Then we have

$$\dot{x}_p(t) = C\sin t + Ct\cos t + D\cos t - Dt\sin t$$
$$\ddot{x}_p(t) = 2C\cos t - Ct\sin t - 2D\sin t - Dt\cos t,$$

so that

$$\ddot{x}_p + x_p = 2C\cos t - 2D\sin t.$$

We need $C = 8/2 = 4$ and $D = 0$, giving the particular integral

$$x_p(t) = 4t\sin t.$$

The general solution is therefore

$$x(t) = A\sin t + B\cos t + 4t\sin t.$$

Note that although the complementary function has a fixed amplitude, the amplitude of the oscillations produced by having the $4t\sin t$ term on the right-hand side grows linearly in t. This is the phenomenon of resonance, and is discussed in more detail in the next chapter. $\qquad \square$

14.5 Rule of thumb

As a rule of thumb for finding particular integrals for second order equations:

- The 'standard guess' is a general version of what you are aiming for, e.g. given an nth order polynomial on the right-hand side, try a general nth order polynomial; but
- If the standard guess contains terms that satisfy the homogeneous equation, multiply by t, repeating this step until the guess no longer contains any terms that solve the homogeneous equation.

14.6 More complicated functions $f(t)$

Similar methods will also work for more complicated choices of $f(t)$ that are products and sums of those that we have already considered. To find a particular integral $x_p(t)$ for the equation

$$L[x] = \alpha f_1(t) + \beta f_2(t) \qquad (14.7)$$

(from the definition of L this is simply $a\ddot{x} + b\dot{x} + cx = \alpha f_1(t) + \beta f_2(t)$) we can use the linearity of L to reduce this to finding particular integrals x_1 and x_2 satisfying

$$L[x_1] = f_1(t) \qquad \text{and} \qquad L[x_2] = f_2(t).$$

A particular integral for (14.7) is then $x_p(t) = \alpha x_1(t) + \beta x_2(t)$, since

$$L[x_p] = L[\alpha x_1 + \beta x_2] = \alpha L[x_1] + \beta L[x_2] = \alpha f_1(t) + \beta f_2(t).$$

We can also find particular integrals for products; to do this we can try the product of the guesses for each individual factor. For example, to find a particular integral for

$$\ddot{x} + \dot{x} - 6x = te^{-t}\cos 2t$$

we would try

$$x_p(t) = (At + B)Ce^{-t}(D\cos 2t + E\sin 2t).$$

Multiplying this out and simplifying the arbitrary constants, we try a particular integral in the form

$$x_p(t) = (At + B)e^{-t}\cos 2t + (Ct + D)e^{-t}\sin 2t.$$

For this choice of $x_p(t)$,

$$\dot{x}_p(t) = ((-A + 2C)t + (A - B + 2D))e^{-t}\cos 2t$$
$$+ ((-2A - C)t + (-2B + C - D))e^{-t}\sin 2t$$

and

$$\ddot{x}_p(t) = ((-3A - 4C)t + (-2A - 3B + 4C - 4D))e^{-t}\cos 2t$$
$$+((4A - 3C)t + (-4A + 4B - 2C - 3D))e^{-t}\sin 2t$$

Substituting in gives

$$\ddot{x}_p + \dot{x}_p - 6x_p = ((-10A - 2C)t + (-A - 10B + 4C - 2D))e^{-t}\cos 2t$$
$$+((2A - 10C)t + (-4A + 2B - C - 10D))e^{-t}\sin 2t$$

Solving for A, B, C and D to ensure that this equals $te^{-t}\cos 2t$ gives

$$x_p(t) = \frac{(-8 - 130t)e^{-t}\cos 2t + (53 - 26t)e^{-t}\sin 2t}{1352}.$$

Exercises

14.1 Find the general solution to the following differential equations (the homogeneous parts of the equations are all treated in Exercise 12.1) In part (n) also find the one solution that has $x(0) = n$ and $\dot{x}(0) = 0$.

   (i) $\ddot{x} - 4x = t^2$,
   (ii) $\ddot{x} - 4\dot{x} = t^2$,
   (iii) $\ddot{x} + \dot{x} - 2x = 3e^{-t}$,
   (iv) $\ddot{x} + \dot{x} - 2x = e^t$,
   (v) $\ddot{x} + 2\dot{x} + x = e^{-t}$,
   (vi) for $\alpha \neq \omega$: $\ddot{x} + \omega^2 x = \sin \alpha t$,
   (vii) for $\alpha = \omega$: $\ddot{x} + \omega^2 x = \sin \alpha t$,
   (viii) $\ddot{x} + 2\dot{x} + 10x = e^{-t}$,
   (ix) $\ddot{x} + 2\dot{x} + 10x = e^{-t}\cos 3t$,
   (x) $\ddot{x} + 6\dot{x} + 10x = e^{-3t}\cos t$, and
   (xi) $\ddot{x} + 4\dot{x} + 4x = e^{2t}$.

14.2 Find a particular integral for

$$\ddot{x} + \dot{x} - 2x = 12e^{-t} - 6e^t.$$

(You might find parts (iii) and (iv) of the previous exercise useful.)

14.3 If you are feeling strong, find a particular integral for

$$\ddot{x} + 4x = 289te^t\sin 2t.$$

15

Resonance

We now consider in more detail the resonance phenomenon mentioned briefly in the previous chapter while we were dealing with Example 14.11.

15.1 Periodic forcing

Suppose that $x(t)$ denotes the distance of some system from its equilibrium position, and that without any external forcing the system would oscillate about this position, with $x(t)$ satisfying

$$\ddot{x} = -\omega^2 x. \tag{15.1}$$

We know from Chapter 13 that this equation has the general solution

$$x(t) = A \cos \omega t + B \cos \omega t \equiv M \cos(\omega t - \phi), \tag{15.2}$$

where in the more compact form $M = \sqrt{A^2 + B^2}$ and $\phi = \tan^{-1}(B/A)$. We refer to $2\pi/\omega$ as the 'natural frequency' of the system; it is how the system 'likes' to oscillate if left to itself.

We now consider what happens if we apply an external forcing to the system that is also oscillating,

$$\ddot{x} + \omega^2 x = a \cos \alpha t. \tag{15.3}$$

Note that the forcing function on the right-hand side has amplitude a and oscillates with frequency $2\pi/\alpha$, cf. Figure 9.1.

Such equations arise in many physical situations, and we now provide a physical motivation by considering a simple model of the spinning drum of a washing machine. We suppose that the cylindrical drum has mass M, and that it is loaded unevenly with clothes of mass m, such that the centre of mass of the clothes lies a distance r from the axle. In order to control the oscillations of the drum, it

141

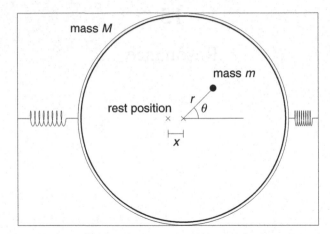

Fig. 15.1. A simple-minded model of a washing machine.

rotates within another cylinder that is attached to the sides of the housing with large springs, each with spring constant $k/2$, see Figure 15.1.

For simplicity we will assume that the drum can only move from side-to-side, and we denote by x the horizontal displacement of the axle from its rest position. The horizontal coordinate of the centre of mass of the drum and clothes is given by

$$X = \frac{mr\cos\theta + Mx}{m + M}.$$

Assuming that the drum spins with constant angular velocity $\dot\theta = \alpha$, so that $\theta(t) = \alpha t$, this gives

$$X(t) = \frac{mr\cos\alpha t + Mx}{m + M}.$$

The springs provide a restoring force of magnitude kx for some constant k, and so Newton's second law,

$$(M + m)\frac{d^2 X}{dt^2} = -kx,$$

gives

$$(M + m)\left[\frac{-mr\alpha^2\cos\alpha t + M\ddot x}{m + M}\right] = -kx.$$

Simplifying this equation we obtain

$$M\ddot x + kx = mr\alpha^2\cos\alpha t.$$

With no load ($m = 0$) or with a perfectly balanced load ($r = 0$) the displacement from the rest position satisfies

$$\ddot{x} + \omega^2 x = 0,$$

where we have defined $\omega^2 = k/M$. With the off-centred load the equation is

$$\ddot{x} + \omega^2 x = a \cos \alpha t,$$

precisely (15.3), where the forcing term on the right-hand side has amplitude $a = mr\alpha^2/M$.

15.1.1 No resonance: bounded response

First we consider what happens when the system is forced at a frequency that differs from its own natural frequency, i.e. when $\alpha \neq \omega$. The complementary function (given in (15.2)) is $y(t) = A \cos \omega t + B \sin \omega t$, and so we can try $x_p(t) = C \cos \alpha t$ for the particular integral (there is no need for the $\sin \alpha t$ term since there is no $\dot{x}$ in the equation). Substituting in we get

$$-C\alpha^2 \cos \alpha t + \omega^2 C \cos \alpha t = a \cos \alpha t,$$

and so we want

$$C = \frac{a}{\omega^2 - \alpha^2}.$$

The general solution is therefore

$$x(t) = M \cos(\omega t - \phi) + \frac{a}{\omega^2 - \alpha^2} \cos \alpha t, \tag{15.4}$$

and we can see that the motion of $x(t)$ combines oscillations at two frequencies: its 'natural frequency' ω, and the forcing frequency α. However, notice that as α gets closer to ω, the amplitude of the second term increases (although it is bounded for each fixed choice of α).

15.1.2 'Ideal' resonance: unbounded response

When $\alpha = \omega$ our usual 'guess' ($x_p(t) = C \sin \omega t + D \cos \omega t$) solves the homogeneous equation, and therefore we now have to try $x_p(t) = Ct \sin \omega t + Dt \cos \omega t$ as the particular integral. For this we have

$$\dot{x}_p = C \sin \omega t + C\omega t \cos \omega t + D \cos \omega t - D\omega \sin \omega t$$

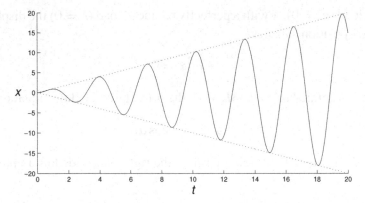

Fig. 15.2. A graph of the function $x(t) = t \sin 2t$. The amplitude of the response to a forcing at an object's natural frequency grows linearly in time.

and so

$$\ddot{x}_p = 2C\omega \cos \omega t - 2D\omega \sin \omega t - \omega^2 [Ct \sin \omega t + Dt \cos \omega t].$$

Therefore

$$\ddot{x}_p + \omega^2 x_p = 2C\omega \cos \omega t - 2D\omega \sin \omega t;$$

since we require the right-hand side to equal $a \cos \omega t$, for our particular integral we need

$$x_p(t) = \frac{a}{2\omega} t \sin \omega t.$$

You can see from a graph of $x_p(t)$ in Figure 15.2 that the amplitude of the resulting oscillations grows linearly. After a time this will, of course, become the main component of the solution

$$x(t) = C \sin \omega t + D \cos \omega t + \frac{a}{2\omega} t \sin \omega t,$$

since the first two terms represent an oscillation of fixed amplitude.

Forcing a structure at its natural frequency can have disastrous consequences; on 14 April 1831, the Broughton suspension bridge over the River Irwell collapsed when the 60th rifle corps marched over it in step, thereby forcing it at one of its natural frequencies. Figure 15.3 shows the report from *The Times* published on 15 April 1831. Armies on the march now break step over bridges to prevent this occurring (see Figure 15.4).

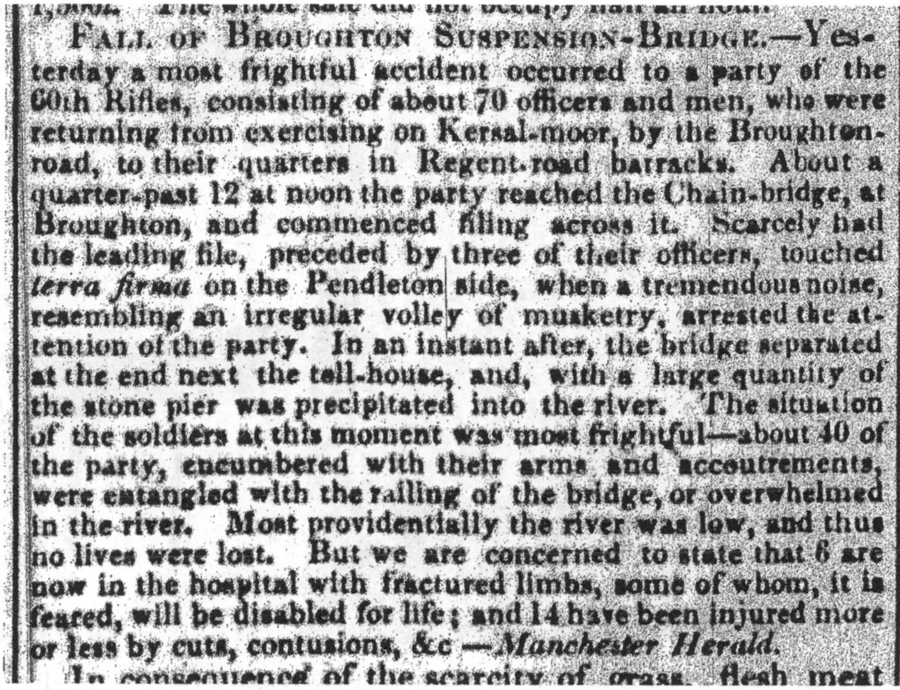

FALL OF BROUGHTON SUSPENSION-BRIDGE.—Yesterday a most frightful accident occurred to a party of the 60th Rifles, consisting of about 70 officers and men, who were returning from exercising on Kersal-moor, by the Broughton-road, to their quarters in Regent-road barracks. About a quarter-past 12 at noon the party reached the Chain-bridge, at Broughton, and commenced filing across it. Scarcely had the leading file, preceded by three of their officers, touched *terra firma* on the Pendleton side, when a tremendous noise, resembling an irregular volley of musketry, arrested the attention of the party. In an instant after, the bridge separated at the end next the toll-house, and, with a large quantity of the stone pier was precipitated into the river. The situation of the soldiers at this moment was most frightful—about 40 of the party, encumbered with their arms and accoutrements, were entangled with the railing of the bridge, or overwhelmed in the river. Most providentially the river was low, and thus no lives were lost. But we are concerned to state that 6 are now in the hospital with fractured limbs, some of whom, it is feared, will be disabled for life; and 14 have been injured more or less by cuts, contusions, &c.—*Manchester Herald.*

Fig. 15.3. *The Times*, 15 April 1831 'FALL OF BROUGHTON SUSPENSION-BRIDGE'.

15.2 Pseudo resonance in physical systems

There is almost invariably some damping in physical systems, and when there is damping you will not see this 'ideal' resonance (meaning linear growth of the amplitude). However, there will still be a frequency at which the amplitude of the resulting oscillations is significantly larger than the amplitude of the forcing.

We will consider the equation

$$\ddot{x} + \lambda\dot{x} + \omega^2 x = a\cos\alpha t$$

where the damping coefficient λ is strictly positive but not too large (we will soon be precise about what is 'not too large'). In Chapter 13 we discussed the homogeneous equation ($\ddot{y} + \lambda\dot{y} + \omega^2 y = 0$) in detail, and for $\lambda^2 < 4\omega^2$ we found the complementary function

$$y(t) = \mathrm{e}^{-\lambda t/2}(A\cos\sigma t + B\sin\sigma t),$$

where $\sigma = \frac{1}{2}\sqrt{4\omega^2 - \lambda^2}$. Since $\lambda > 0$ this oscillates and decays to zero exponentially (in the language of Chapter 13 the system is 'under-damped').

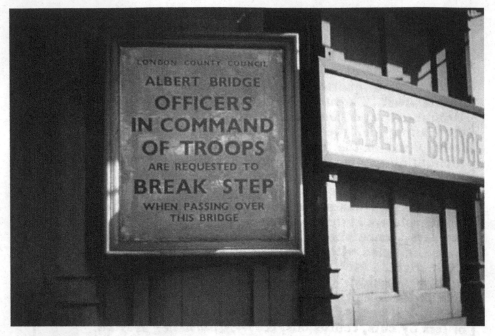

Fig. 15.4. A sign on the Albert Bridge in London, instructing troops to break step to prevent destructive resonance effects. (Courtesy of the National Information Service for Earthquake Engineering, University of California, Berkeley.)

The particular integral will be some combination of $\sin \alpha t$ and $\cos \alpha t$ (these are not part of the complementary function),

$$x_p(t) = C \sin \alpha t + D \cos \alpha t.$$

Substituting this guess into the equation, we want

$$-C\alpha^2 \sin \alpha t - D\alpha^2 \cos \alpha t + \lambda[C\alpha \cos \alpha t - D\alpha \sin \alpha t] +$$
$$\omega^2[C \sin \alpha t + D \cos \alpha t] = a \cos \alpha t.$$

Collecting coefficients of $\sin \alpha t$ gives

$$C(\omega^2 - \alpha^2) = D\lambda\alpha,$$

while by equating the coefficients of $\cos \alpha t$ we obtain

$$D(\omega^2 - \alpha^2) = a - C\lambda\alpha.$$

Solving these simultaneous equations gives

$$C = \frac{a\lambda\alpha}{(\omega^2 - \alpha^2)^2 + (\lambda\alpha)^2} \quad \text{and} \quad D = \frac{a(\omega^2 - \alpha^2)}{(\omega^2 - \alpha^2)^2 + (\lambda\alpha)^2},$$

and so the particular integral is

$$x_p(t) = \frac{a\lambda\alpha}{(\omega^2 - \alpha^2)^2 + (\lambda\alpha)^2} \sin\alpha t + \frac{a(\omega^2 - \alpha^2)}{(\omega^2 - \alpha^2)^2 + (\lambda\alpha)^2} \cos\alpha t. \quad (15.5)$$

We have already seen (in Section 9.4.3) that we can combine the two terms in an expression like $A\cos\alpha t + B\sin\alpha t$ to give one oscillating term; in particular the amplitude of the resulting oscillation is given by the square root of the sum of the squares of the two coefficients, $\sqrt{A^2 + B^2}$. So the amplitude of the oscillations that arise in response to the forcing $a\cos\alpha t$ is

$$R(\alpha) = a\sqrt{\frac{(\lambda\alpha)^2 + (\omega^2 - \alpha^2)^2}{[(\omega^2 - \alpha^2)^2 + (\lambda\alpha)^2]^2}}$$

$$= \frac{a}{\sqrt{(\omega^2 - \alpha^2)^2 + (\lambda\alpha)^2}}.$$

The ratio of the amplitude of the response to that of the forcing is therefore

$$F(\alpha) = \frac{1}{\sqrt{(\omega^2 - \alpha^2)^2 + (\lambda\alpha)^2}}, \quad (15.6)$$

which depends on the frequency $2\pi/\alpha$ of the forcing. Graphs of $F(\alpha)$ for various values of λ when $\omega = 1$ are shown in Figure 15.5. Note that the maximum value

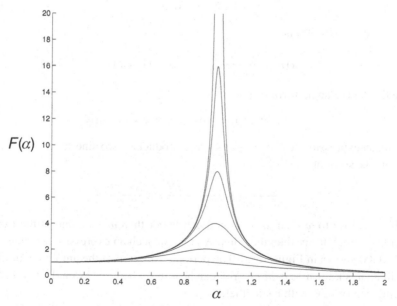

Fig. 15.5. Pseudo resonance when $\omega = 1$: graphs of $F(\alpha)$ for $\lambda = 1, 1/2, 1/4$, $1/8, 1/16$ and 0, increasing as λ decreases.

of $F(\alpha)$ occurs near $\alpha = 1$ (i.e. where $\alpha = \omega$), and increases as the damping level λ decreases.

The maximum value of the response occurs when the denominator in $F(\alpha)$ is a minimum, so when

$$\frac{d}{d\alpha}[(\omega^2 - \alpha^2)^2 + (\lambda\alpha)^2] = 0,$$

i.e. when

$$-4\alpha(\omega^2 - \alpha^2) + 2\lambda^2\alpha = 0,$$

which gives $\alpha^2 = \omega^2 - (1/2)\lambda^2$. It is therefore clear that as the damping becomes ever smaller the value of α for which the response is maximum becomes increasingly close to $\alpha = \omega$.

The maximum value of the amplitude is

$$R_{\max}(\lambda) = \frac{2a}{\lambda\sqrt{4\omega^2 - \lambda^2}},$$

and of course increases in magnitude as λ decreases. As $\lambda \to 0$, $R_{\max} \sim a(\lambda\omega)^{-1}$.

Exercises

15.1 For $\alpha \neq \omega$ show that the solution of the equation

$$\ddot{x} + \omega^2 x = \cos\alpha t \tag{E15.1}$$

with $x(0) = \dot{x}(0) = 0$ is

$$x(t) = \frac{1}{\omega^2 - \alpha^2}(\cos\alpha t - \cos\omega t). \tag{E15.2}$$

15.2 Use the double angle formulae

$$\cos(\theta \pm \phi) = \cos\theta\cos\phi \mp \sin\theta\sin\phi$$

to find an expression for $\cos x - \cos y$ as a product of two sine functions, and hence rewrite the solution in (E15.2) as

$$\frac{2}{\omega^2 - \alpha^2}\sin\frac{(\omega + \alpha)t}{2}\sin\frac{(\omega - \alpha)t}{2}.$$

If α is close to ω then $|\alpha + \omega|$ is much larger than $\omega - \alpha$; one of the two terms oscillates much faster than the other. A graph of such an expression when $\omega = 1$ and $\alpha = 0.8$ is shown in Figure 15.6. The periodic variation of the amplitude of the basic oscillation is known as *beating*. You can hear this when, for example, two flutes play slightly out of tune with each other.

15.3 When $\alpha = \omega$ show that the solution of (E15.1) with $x(0) = \dot{x}(0) = 0$ is $x(t) = t\sin\omega t/2\omega$. Recover this solution from that for $\alpha \neq \omega$ by letting $\alpha \to \omega$ in (E15.2) and using L'Hôpital's rule.

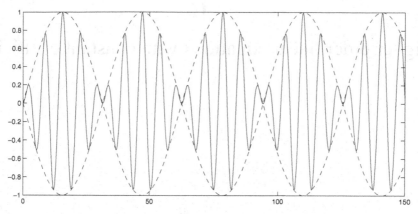

Fig. 15.6. The phenomenon of beats: the graph of $\sin 0.9t \sin 0.1t$ against t (the dashed line shows how the amplitude of the faster oscillation varies like $\sin 0.1t$).

L'Hôpital's rule: if $f(x) \to 0$ as $x \to a$, and $g(x) \to 0$ as $x \to a$ then

$$\lim_{x \to a} \frac{f(x)}{g(x)} = \lim_{x \to a} \frac{f'(x)}{g'(x)} = \frac{f'(a)}{g'(a)}$$

(provided that f and g have continuous derivatives at $x = a$).

15.4 A model for the vibrations of a wine glass is

$$\ddot{x} + \lambda \dot{x} + \omega^2 x = 0,$$

where λ and ω are constants. Suppose that when struck the glass vibrates at 660 Hz (about the second E above middle C on a piano). Show that

$$\sqrt{4\omega^2 - \lambda^2} = 2640\pi.$$

If it takes about 3 seconds for the sound to die away, and this happens when the original vibrations have reduced to $1/100$ of their initial level, show that

$$\lambda = \frac{2 \log 100}{3},$$

and hence that $\lambda = 3.07$ and $\omega = 4.15 \times 10^3$ (both to three significant figures).

The glass can stand deforming only to $x \approx 1$. A pure tone at 660 Hz is produced at D decibels and aimed at the glass, forcing it at its natural frequency, so that the vibrations are now modelled by

$$\ddot{x} + \lambda \dot{x} + \omega^2 x = \frac{10^{(D/10)-8}}{3} \cos(1320\pi t). \tag{E15.3}$$

How loud should the sound be, i.e. how large should D be, in order to shatter the glass? (Decibels are on a logarithmic scale, hence the exponential on the right-hand side of (E15.3). The strange factor in front of the forcing produces roughly the correct volume level.)

16

Higher order linear equations with constant coefficients

The methods that we have developed to treat second order linear equations extend in a straightforward manner to treat higher order linear equations.

A general nth order linear ODE with constant coefficients can be written in the form

$$a_n \frac{\mathrm{d}^n x}{\mathrm{d}t^n} + a_{n-1} \frac{\mathrm{d}^{n-1} x}{\mathrm{d}t^{n-1}} + \cdots + a_1 \frac{\mathrm{d}x}{\mathrm{d}t} + a_0 x = f(t), \qquad (16.1)$$

where we assume that $a_n \neq 0$, cf. (3.5).

The initial value problem in which $x(t)$ and its first $n - 1$ derivatives are specified has a unique solution.

Theorem 16.1 *Given an initial condition*

$$x(t_0) = x_0, \quad \dot{x}(t_0) = x_1, \quad \ddot{x}(t_0) = x_2, \quad \ldots, \quad \frac{\mathrm{d}^{n-1} x}{\mathrm{d}t^{n-1}}(t_0) = x_{n-1} \qquad (16.2)$$

the linear equation (16.1) has a unique solution defined for all $t \in \mathbb{R}$.

Note that in the statement of the theorem it is claimed that the solution exists for all $t \in \mathbb{R}$. This is because for such a linear equation it is possible to guarantee that the solution does not blow up in finite time (cf. Section 6.3 and Exercise 12.3).

16.1 Complementary function and particular integral

In order to find an explicit solution of an nth order linear problem we proceed as we did for second order problems. The only significant difference is that now we need n linearly independent solutions of the homogeneous equation in order to be able to satisfy the n initial conditions in (16.2). See Definition 11.2 for linear independence, and also Section 16.2.

First we find the complementary function by solving the homogeneous equation

$$a_n \frac{d^n y}{dt^n} + a_{n-1} \frac{d^{n-1} y}{dt^{n-1}} + \cdots + a_0 y = 0;$$

we try $y(t) = e^{kt}$. This will give an nth order algebraic equation for k,

$$a_n k^n + a_{n-1} k^{n-1} + \cdots + a_0 = 0.$$

For $n \le 4$ there are methods for finding the roots of such equations,[1] although for higher order equations we would have to use numerical methods.

For each non-repeated real root k of the equation we have a solution e^{kt}; for a real root k that is repeated m times we have m linearly independent solutions,

$$e^{kt}, \qquad te^{kt}, \qquad \ldots, \qquad \text{and} \qquad t^{m-1} e^{kt};$$

for each non-repeated complex conjugate pair $\rho \pm i\omega$ we have the two solutions $e^{\rho t} \sin \omega t$ and $e^{\rho t} \cos \omega t$, while for a complex conjugate pair $\rho \pm i\omega$ that is repeated m times we would have the $2m$ solutions

$$e^{\rho t} \cos \omega t, \ e^{\rho t} \sin \omega t, \quad te^{\rho t} \cos \omega t, \quad te^{\rho t} \sin \omega t, \quad \ldots,$$
$$t^{m-1} e^{\rho t} \cos \omega t, \quad t^{m-1} e^{\rho t} \sin \omega t.$$

Finding a particular integral involves the same style 'guesswork' as in the second order case.

Example 16.2 *Find the general solution of*

$$\frac{d^4 x}{dt^4} - 16x = 64 \sin 2t.$$

---

[1] For a general cubic equation

$$x^3 - bx^2 + cx - d = 0$$

first substitute $x = y + b/3$ which gives

$$x^3 + mx = n \qquad \text{with} \qquad m = c - b^2/3 \quad \text{and} \quad n = d - bc/3 + 2b^3/27.$$

Tartaglia developed a method of solution for this form of cubic (first published by Cardan in 1545); notice that

$$(a - b)^3 + 3ab(a - b) = a^3 - b^3$$

and so if a and b satisfy

$$3ab = m \qquad \text{and} \qquad a^3 - b^3 = n$$

then $a - b$ solves $x^3 + mx = n$. Since $b = m/3a$ the second of these equations is

$$a^3 - \frac{m^3}{27a^3} = n$$

which gives $a^6 - na^3 - m^3/27 = 0$, a quadratic equation for a^3. We can find the roots of this equation using the quadratic formula, from which we get a by taking cube roots. The value of b is given by $b = m/3a$, and then $a - b$ is a solution of the original equation.

To find the complementary function, i.e. the general solution of

$$\frac{d^4 y}{dt^4} - 16y = 0,$$

we try $y(t) = e^{kt}$, and so we need $k^4 = 16$. The solutions of this are $k = \pm 2$ and $k = \pm 2i$, and so the complementary function is

$$y(t) = Ae^{2t} + Be^{-2t} + C\cos 2t + D\sin 2t.$$

To find a particular integral we cannot try a combination of $\sin 2t$ and $\cos 2t$ since they solve the homogeneous equation, so we have to try a combination of $t\sin 2t$ and $t\cos 2t$. If

$$x_p(t) = Et\sin 2t + Ft\cos 2t$$

then after some algebra

$$\frac{d^4 x_p}{dt^4} = E[16t\sin 2t - 32\cos 2t] + F[32\sin 2t + 16t\cos 2t],$$

and so substituting in we want

$$-32E\cos 2t + 32F\sin 2t = 64\sin 2t,$$

i.e. $E = 0$ and $F = 2$, giving a particular integral $x_p(t) = 2t\cos 2t$ and hence the general solution

$$x(t) = Ae^{2t} + Be^{-2t} + (C + 2t)\cos 2t + D\sin 2t.$$

16.2 *The general theory for nth order equations

In Chapter 11 we developed a general theory for second order linear homogeneous equations

$$\frac{d^2 x}{dt^2} + p_1(t)\frac{dx}{dt} + p_2(t)x = 0, \qquad (16.3)$$

showing that given two linearly independent solutions $x_1(t)$ and $x_2(t)$ of this equation we can form the general solution as a linear combination

$$c_1 x_1(t) + c_2 x_2(t).$$

Realising that the collection of all solutions of (16.3) is the same as the kernel of the linear operator $L : C^2(I) \to C^0(I)$ defined by

$$L[x] = \ddot{x} + p_1(t)\dot{x} + p_2(t)x,$$

we could re-express our results as 'dim ker$(L) = 2$'.

It is possible to generalise the results of Chapter 11 to treat the linear nth order equation

$$\frac{d^n x}{dt^n} + p_1(t)\frac{d^{n-1}x}{dt^{n-1}} + \cdots + p_{n-1}(t)\frac{dx}{dt} + p_n(t)x = 0; \qquad (16.4)$$

in particular, similar arguments show that the general solution of (16.4) can be formed from linear combinations of any n linearly independent solutions $x_1(t), \ldots, x_n(t)$ of (16.4),

$$x(t) = c_1 x_1(t) + \cdots + c_n x_n(t).$$

Denoting by $C^n(I)$ the collection of all functions defined on n which have n continuous derivatives, we can define a linear operator $L : C^n(I) \to C^0(I)$ by

$$L[x] = \frac{d^n x}{dt^n} + p_1(t)\frac{d^{n-1}x}{dt^{n-1}} + \cdots + p_{n-1}(t)\frac{dx}{dt} + p_n(t)x$$

(we now need $x \in C^n(I)$ to make sure that the term $d^n x/dt^n$ is sensible). We can then express these results by saying that 'dim ker$(L) = n$'.

Checking whether two functions are linearly independent is straightforward, since we saw that they are linearly independent if and only if they are not proportional. When there are three or more functions involved we have to make more systematic use of the Wronskian, which we discussed only in passing in Chapter 11. Given n functions $f_1(t), \ldots, f_n(t)$, we define their Wronskian to be the matrix determinant

$$W[f_1, \ldots, f_n](t) = \begin{vmatrix} f_1(t) & \cdots & f_n(t) \\ \vdots & \ddots & \vdots \\ d^{n-1}f_1/dt^{n-1}(t) & \cdots & d^{n-1}f_n/dt^{n-1}(t) \end{vmatrix}.$$

Note that for $n = 2$ this reduces to $W[f_1, f_2](t) = f_1 \dot{f_2} - f_2 \dot{f_1}$, as used in Chapter 11. As for the case of two functions, it is possible to show that n solutions of the linear equation (16.4) are linearly independent on an interval I if and only if their Wronskian is never zero on I. Exercises 16.2–16.4 lead you through some of this theory in the case of three functions f_1, f_2 and f_3.

Exercises

16.1 Find the general solution of the following equations:
   (i)

$$\frac{d^3 x}{dt^3} - 6\frac{d^2 x}{dt^2} + 11\frac{dx}{dt} - 6x = e^{-t},$$

(ii)

$$y''' - 3y' + 2y = \sin x,$$

(iii)

$$\frac{d^4x}{dt^4} - 4\frac{d^3x}{dt^3} + 8\frac{d^2x}{dt^2} - 8\frac{dx}{dt} + 4x = \sin t$$

(if $x = e^{kt}$ one solution of the corresponding quartic equation is $k = 1 + i$), and

(iv)

$$\frac{d^4x}{dt^4} - 5\frac{d^2x}{dt^2} + 4x = e^t.$$

16.2 The linear independence of three functions f_1, f_2 and f_3 on an interval I depends on the number of solutions of the equation

$$\alpha_1 f_1(t) + \alpha_2 f_2(t) + \alpha_3 f_3(t) = 0 \qquad \text{for all} \qquad t \in I.$$

By differentiating this equation once, and then once more, show that α_1, α_2 and α_3 satisfy the matrix equation

$$\begin{pmatrix} f_1 & f_2 & f_3 \\ df_1/dt & df_2/dt & df_3/dt \\ d^2 f_1/dt^2 & d^2 f_2/dt^2 & d^2 f_3/dt^2 \end{pmatrix} \begin{pmatrix} \alpha_1 \\ \alpha_2 \\ \alpha_3 \end{pmatrix} = \begin{pmatrix} 0 \\ 0 \\ 0 \end{pmatrix}.$$

Deduce that if $W[f_1, f_2, f_3](t)$, the Wronskian of f_1, f_2 and f_3, defined as

$$W[f_1, f_2, f_3](t) = \begin{vmatrix} f_1 & f_2 & f_3 \\ df_1/dt & df_2/dt & df_3/dt \\ d^2 f_1/dt^2 & d^2 f_2/dt^2 & d^2 f_3/dt^2 \end{vmatrix},$$

is non-zero for any $t \in I$ then f_1, f_2 and f_3 are linearly independent.

16.3 Show that any three solutions of a third order linear differential equation are linearly independent on an interval I if and only if their Wronskian is non-zero on I.

16.4 Suppose that x_1, x_2 and x_3 are three solutions of the third order linear equation

$$\frac{d^3x}{dt^3} + p(t)\frac{d^2x}{dt^2} + q(t)\frac{dx}{dt} + r(t)x = 0, \qquad (E16.1)$$

all defined on some interval I.

We now show that, just as for two solutions of a second order linear equation,

$$\frac{dW}{dt} = -p(t)W \qquad (E16.2)$$

(cf. Exercise 11.2). You will need various properties of determinants, which you can prove by longhand (if you wish) in the next exercise.

(i) By differentiating the determinant form of the Wronskian, show that

$$\dot{W} = \begin{vmatrix} x_1 & x_2 & x_3 \\ \dot{x}_1 & \dot{x}_2 & \dot{x}_3 \\ d^3 x_1/dt^3 & d^3 x_2/dt^3 & d^3 x_3/dt^3 \end{vmatrix}.$$

(You will need parts (i) and (ii) of the next exercise.)

(ii) Substitute in for $d^3 x_j/dt^3$ using the differential equation (E16.1), and hence show that

$$\dot{W} = -p(t) \begin{vmatrix} x_1 & x_2 & x_3 \\ \dot{x}_1 & \dot{x}_2 & \dot{x}_3 \\ \ddot{x}_1 & \ddot{x}_2 & \ddot{x}_3 \end{vmatrix},$$

i.e. that (E16.2) holds. (You will need parts (ii) and (iii) of the next exercise.)

(iii) Solve equation (E16.2) to find an expression for $W(t)$ involving an integral, and deduce that either $W(t) = 0$ for all $t \in I$, or that $W(t) \neq 0$ for all $t \in I$.

16.5 For the previous question you will need the following properties of determinants: you should be able to prove them in the 3×3 case treated here by simple (if laborious) calculation, using the explicit expression for the determinant of a 3×3 matrix

$$\begin{vmatrix} a & b & c \\ r & s & t \\ x & y & z \end{vmatrix} = a(sz - ty) - b(rz - tx) + c(ry - sx).$$

(i)

$$\frac{d}{dt} \begin{vmatrix} a & b & c \\ r & s & t \\ x & y & z \end{vmatrix} = \begin{vmatrix} \dot{a} & \dot{b} & \dot{c} \\ r & s & t \\ x & y & z \end{vmatrix} + \begin{vmatrix} a & b & c \\ \dot{r} & \dot{s} & \dot{t} \\ x & y & z \end{vmatrix} + \begin{vmatrix} a & b & c \\ r & s & t \\ \dot{x} & \dot{y} & \dot{z} \end{vmatrix}$$

(i.e. differentiate one row at a time; this is essentially the product rule),

(ii) If any two rows are proportional then the determinant is zero. Check this for

$$\begin{vmatrix} a & b & c \\ \lambda a & \lambda b & \lambda c \\ x & y & z \end{vmatrix} = 0,$$

(iii) Determinants depend linearly on their rows. Show this for the case

$$\begin{vmatrix} a & b & c \\ r & s & t \\ \alpha x_1 + \beta x_2 & \alpha y_1 + \beta y_2 & \alpha z_1 + \beta z_2 \end{vmatrix}$$

$$= \alpha \begin{vmatrix} a & b & c \\ r & s & t \\ x_1 & y_1 & z_1 \end{vmatrix} + \beta \begin{vmatrix} a & b & c \\ r & s & t \\ x_2 & y_2 & z_2 \end{vmatrix}.$$

Part III

Linear second order equations with variable coefficients

17

Reduction of order

In Chapter 11 we discussed the general theory of second order linear equations. In the intervening chapters we have concentrated on linear equations with constant coefficients, but we now return to the more general case in which the coefficients are allowed to be functions of t,

$$a(t)\frac{d^2x}{dt^2} + b(t)\frac{dx}{dt} + c(t)x = 0. \tag{17.1}$$

We saw in Chapter 11 that in order fully to solve a second order homogeneous linear differential equation we need two linearly independent solutions. In this chapter we show that if we happen to know, or can guess, one solution of an equation like (17.1) then there is a systematic way to find a second, linearly independent, solution.

The method is called 'reduction of order', since it enables us to use our knowledge of one solution to find a *first* order differential equation that we can use to find the second solution.

Suppose we know that $u(t)$ solves the second order linear equation

$$a(t)\frac{d^2x}{dt^2} + b(t)\frac{dx}{dt} + c(t)x = 0. \tag{17.2}$$

The idea is to make the substitution $x(t) = u(t)y(t)$ and then solve the resulting equation for $y(t)$. From $x(t) = u(t)y(t)$ it follows that

$$\dot{x} = \dot{u}y + u\dot{y} \qquad \ddot{x} = \ddot{u}y + 2\dot{u}\dot{y} + u\ddot{y},$$

and substituting these into the original equation gives

$$a(t)(\ddot{u}y + 2\dot{u}\dot{y} + u\ddot{y}) + b(t)(\dot{u}y + u\dot{y}) + c(t)uy = 0.$$

The terms in which the factor of y is not differentiated,

$$a(t)\ddot{u}y + b(t)\dot{u}y + c(t)uy = y[a(t)\ddot{u} + b(t)\dot{u} + c(t)],$$

add to give zero; the expression in square brackets vanishes since u is a solution. So we are left with an equation for $y(t)$,

$$a(t)(2\dot{u}\dot{y} + u\ddot{y}) + b(t)u\dot{y} = 0.$$

Note that although $\ddot{y}$ and $\dot{y}$ occur in this equation, y itself does not. So we can make a second substitution, putting $z = \dot{y}$ to obtain a first order equation for $z(t)$,

$$[a(t)u(t)]\dot{z} + [2a(t)\dot{u}(t) + b(t)u(t)]z = 0. \tag{17.3}$$

Although this equation looks unpleasant, the coefficients of $\dot{z}$ and z are known functions of t, since $a(t)$ and $b(t)$ are from the original problem and $u(t)$ is the solution that we already know. This first order equation can now be solved using the method of integrating factors.[1]

As with many of the techniques developed here, it is not helpful to try to remember the equation (17.3) and produce it from up your sleeve whenever you need to use this idea. Instead you should understand the technique and be ready to apply it to particular examples when it is needed.

Example 17.1 *One solution of*

$$\ddot{x} - 2\lambda\dot{x} + \lambda^2 x = 0$$

can be found by trying $x(t) = e^{kt}$. The resulting quadratic equation has a repeated root $k = \lambda$. Use the reduction of order method to show that there is a second linearly independent solution $te^{\lambda t}$.

To use the reduction of order method we substitute $x(t) = y(t)e^{\lambda t}$. We then have

$$\dot{x} = \lambda e^{\lambda t}y + e^{\lambda t}\dot{y} \qquad \text{and} \qquad \ddot{x} = \lambda^2 e^{\lambda t}y + 2\lambda e^{\lambda t}\dot{y} + e^{\lambda t}\ddot{y},$$

so that

$$\lambda^2 e^{\lambda t}y + 2\lambda e^{\lambda t}\dot{y} + e^{\lambda t}\ddot{y} - 2\lambda^2 e^{\lambda t}y + 2\lambda e^{\lambda t}\dot{y} + \lambda^2 e^{\lambda t}y = 0.$$

After cancelling all the $e^{\lambda t}$s this reads

$$\lambda^2 y + 2\lambda\dot{y} + \ddot{y} - 2\lambda^2 y - 2\lambda\dot{y} + \lambda^2 y = 0,$$

and further cancellations leave just

$$\ddot{y} = 0.$$

---

[1] It is possible to find a general integral expression for the solution of this equation, see Exercise 17.7. However, it is a much better idea to apply the method afresh in each particular case.

Integrating this twice gives $y(t) = A + Bt$, and so

$$x(t) = Ae^{\lambda t} + Bte^{\lambda t}. \tag{17.4}$$

We obtain a constant multiple of our original solution ($e^{\lambda t}$) plus a constant multiple of a new, second solution, $te^{\lambda t}$, as claimed. □

Example 17.2 *For $t > 0$ the function $u(t) = 1/t$ is a solution of*

$$t^2\ddot{x} - 2t\dot{x} - 4x = 0.$$

(This is easy to check: $\dot{x} = -t^{-2}$ and $\ddot{x} = 2t^{-3}$.) Find a second linearly indepen-dent solution.

To apply the reduction of order method, we set $x = y/t$, and so

$$\dot{x} = \frac{\dot{y}}{t} - \frac{y}{t^2} \qquad \text{and} \qquad \ddot{x} = \frac{\ddot{y}}{t} - 2\frac{\dot{y}}{t^2} + 2\frac{y}{t^3}.$$

Substituting these into the equation gives

$$t^2\left[\frac{\ddot{y}}{t} - 2\frac{\dot{y}}{t^2} + 2\frac{y}{t^3}\right] - 2t\left[\frac{\dot{y}}{t} - \frac{y}{t^2}\right] - 4\frac{y}{t} = 0.$$

This simplifies to give

$$t\ddot{y} - 4\dot{y} = 0,$$

which setting $z = \dot{y}$ gives the linear equation

$$\dot{z} - \frac{4}{t}z = 0.$$

We can solve this using an integrating factor,

$$\exp\left(\int -\frac{4}{t}\, dt\right) = \exp(-4\ln t) = t^{-4}.$$

So we have

$$\frac{d}{dt}(t^{-4}z(t)) = 0.$$

Integrating this gives $t^{-4}z(t) = c$, and so

$$z(t) = ct^4.$$

Since $z = \dot{y}$, it follows that $y(t) = at^5 + b$, giving (since $x(t) = y(t)/t$)

$$x(t) = at^4 + \frac{b}{t}.$$

The second term just repeats the solution we already knew, so the new linearly independent solution is t^4.

Exercises

For further examples of the reduction of order method see also Exercises 18.1 (vi), 18.1 (vii) and 20.3.

17.1 One solution of the equation

$$t^2\ddot{y} - (t^2 + 2t)\dot{y} + (t + 2)y = 0$$

is $y(t) = t$. Use the reduction of order method to find a second solution, and hence write down the general solution.

17.2 One solution of

$$(x - 1)y'' - xy' + y = 0$$

that is valid for $x > 1$ is $y(x) = e^x$. Find a second linearly independent solution $z(x)$, and check that the Wronskian of $y(x)$ and $z(x)$ is non-zero for $x > 1$.

17.3 One solution of

$$(t \cos t - \sin t)\ddot{x} + \dot{x}t \sin t - x \sin t = 0$$

is $x(t) = t$. Find a second linearly independent solution.

17.4 One solution of

$$(t - t^2)\ddot{x} + (2 - t^2)\dot{x} + (2 - t)x = 0$$

is $x(t) = e^{-t}$. Find a second linearly independent solution.

17.5 One solution of

$$y'' - xy' + y = 0$$

is $y = x$. Find a second linearly independent solution in the form of an integral. Expanding the integrand in powers of x using the power series form for e^x,

$$e^x = \sum_{n=0}^{\infty} \frac{x^n}{n!},$$

and assuming that the resulting expression can be integrated term-by-term show that this second solution can be written as

$$y(x) = A\left[-1 + \sum_{n=1}^{\infty} \frac{x^{2n}}{2^n(2n - 1)n!}\right]$$

(cf. Exercise 20.2(i)).

17.6 One solution of

$$\tan t \frac{d^2x}{dt^2} - 3\frac{dx}{dt} + (\tan t + 3 \cot t)x = 0$$

is $x(t) = \sin t$. Find a second linearly independent solution.

17.7 (T) If we know one solution $u(t)$ of the equation

$$\frac{d^2x}{dt^2} + p(t)\frac{dx}{dt} + q(t)x = 0 \qquad\qquad (E17.1)$$

then the reduction of order method with $x(t) = u(t)y(t)$ leads to the first order linear equation

$$u(t)\dot{z} + [2\dot{u}(t) + p(t)u(t)]z = 0.$$

for $z = \dot{y}$ (cf. (17.3)). Show that

$$z(t) = \frac{Ae^{-\int p(t)\,dt}}{u(t)^2},$$

and hence find the second linearly independent solution in the form of an integral.

17.8 (T) Suppose that the two solutions of a second order linear differential equation (E17.1) are $u(t)$ and $v(t)$. Use the result of the previous exercise, to show that

$$\frac{d}{dt}\left[\frac{v(t)}{u(t)}\right] = \frac{Ae^{-\int p(t)\,dt}}{u(t)^2},$$

and hence that

$$p(t) = -\frac{u\ddot{v} - v\ddot{u}}{u\dot{v} - v\dot{u}}.$$

Find the function $q(t)$ such that $u(t)$ is a solution of

$$\frac{d^2x}{dt^2} - \frac{u\ddot{v} - v\ddot{u}}{u\dot{v} - v\dot{u}}\frac{dx}{dt} + q(t)x = 0$$

(rearrange the equation for $q(t)$, and substitute $x(t) = u(t)$) and hence show that the second order linear differential equation with solutions $u(t)$ and $v(t)$ can be written as

$$(u\dot{v} - v\dot{u})\frac{d^2x}{dt^2} - (u\ddot{v} - v\ddot{u})\frac{dx}{dt} + (\dot{u}\ddot{v} - \ddot{u}\dot{v})x = 0.$$

This produced Exercises 17.1–17.6 above.

17.9 Using the result of the previous exercise, find a second order linear differential equation whose solutions are e^t and $\cos t$. Check that both of these two functions satisfy the resulting equation.

18

*The variation of constants formula

In the last chapter we saw that knowing one solution $x_1(t)$ of a homogeneous linear second order equation

$$a(t)\frac{d^2x}{dt^2} + b(t)\frac{dx}{dt} + c(t)x = 0 \tag{18.1}$$

enables us to find a second linearly independent solution $x_2(t)$. In this chapter we see that if we know two linearly independent solutions of (18.1) then there is a systematic way to find a particular integral for the inhomogeneous problem

$$a(t)\frac{d^2x}{dt^2} + b(t)\frac{dx}{dt} + c(t)x = f(t). \tag{18.2}$$

For simplicity we assume that $a(t) \neq 0$, divide equation (18.2) by $a(t)$, and rewrite it as

$$\frac{d^2x}{dt^2} + p(t)\frac{dx}{dt} + q(t)x = g(t).$$

Now suppose that we know two linearly independent solutions $x_1(t)$ and $x_2(t)$ of the homogeneous linear problem

$$\ddot{x} + p(t)\dot{x} + q(t)x = 0,$$

which means that its general solution is of the form

$$x(t) = Ax_1(t) + Bx_2(t). \tag{18.3}$$

We will look for a particular integral for the inhomogeneous problem

$$\ddot{x} + p(t)\dot{x} + q(t)x = g(t) \tag{18.4}$$

in the form

$$x(t) = u_1(t)x_1(t) + u_2(t)x_2(t). \tag{18.5}$$

We have replaced the constants in (18.3) by functions of t; this is known as the method of 'variation of constants'. We will obtain a particular solution with u_1 and u_2 given in terms of integrals; in particular examples we may not be able to compute these integrals explicitly.

Soon we will substitute this form for $x(t)$ into equation (18.4), but first notice that this will only provide one equation that has to be satisfied by $u_1(t)$ and $u_2(t)$. Since we have two unknown functions we need two equations to determine them completely; we are therefore free to impose an additional condition of our choice, and we make a choice below that simplifies our calculations.

The first thing to do is to compute the derivative of $x(t)$,

$$\dot{x}(t) = \dot{u}_1(t)x_1(t) + u_1(t)\dot{x}_1(t) + \dot{u}_2(t)x_2(t) + u_2(t)\dot{x}_2(t). \qquad (18.6)$$

When we differentiate again we will get second derivatives of the us and the xs; to get rid of the second derivatives of the functions u_j (which are unknown) we use our 'extra equation', imposing the condition that

$$\dot{u}_1(t)x_1(t) + \dot{u}_2(t)x_2(t) = 0. \qquad (18.7)$$

If this holds then

$$\dot{x}(t) = u_1(t)\dot{x}_1(t) + u_2(t)\dot{x}_2(t),$$

and so

$$\ddot{x} = \dot{u}_1\dot{x}_1 + u_1\ddot{x}_1 + \dot{u}_2\dot{x}_2 + u_2\ddot{x}_2,$$

with no second derivatives of the unknown functions u_1 and u_2.

Substituting these derivatives into equation (18.4) gives

$$\dot{u}_1\dot{x}_1 + u_1\ddot{x}_1 + \dot{u}_2\dot{x}_2 + u_2\ddot{x}_2 + p(t)[u_1\dot{x}_1 + u_2\dot{x}_2] + q(t)[u_1x_1 + u_2x_2] = g(t).$$

This looks unpleasant, but if we group the terms correctly,

$$u_1[\ddot{x}_1 + p(t)\dot{x}_1 + q(t)x_1] + u_2[\ddot{x}_2 + p(t)\dot{x}_2 + q(t)x_2] + \dot{u}_1\dot{x}_1 + \dot{u}_2\dot{x}_2 = g(t),$$

we can use the fact that x_1 and x_2 both solve the homogeneous equation to set both terms in the square brackets to zero, and end up with

$$\dot{u}_1\dot{x}_1 + \dot{u}_2\dot{x}_2 = g(t).$$

We now have two equations for $\dot{u}_1(t)$ and $\dot{u}_2(t)$,

$$\begin{cases} \dot{u}_1x_1 + \dot{u}_2x_2 = 0 \\ \dot{u}_1\dot{x}_1 + \dot{u}_2\dot{x}_2 = g(t). \end{cases}$$

We can solve these equations for $\dot{u}_1(t)$ and $\dot{u}_2(t)$ to give

$$\dot{u}_1(t) = -\frac{x_2(t)g(t)}{x_1(t)\dot{x}_2(t) - x_2(t)\dot{x}_1(t)}$$

and

$$\dot{u}_2(t) = \frac{x_1(t)g(t)}{x_1(t)\dot{x}_2(t) - x_2(t)\dot{x}_1(t)}.$$

As a shorthand we will write

$$W(t) = x_1(t)\dot{x}_2(t) - x_2(t)\dot{x}_1(t)$$

for the denominator in these equations.[1] Now to find $u_1(t)$ and $u_2(t)$ we integrate, and therefore obtain

$$x(t) = -x_1(t) \int \frac{x_2(t)g(t)}{W(t)} \, dt + x_2(t) \int \frac{x_1(t)g(t)}{W(t)} \, dt. \qquad (18.8)$$

We will now do two examples, one for which we already know the particular integral, and one for which we do not.

Example 18.1 *Use the formula (18.8) to find a particular integral for the equation*

$$\ddot{x} + \dot{x} - 6x = 5e^{-3t}$$

(this was Example 14.8).

Note that generally it is not a good idea to try to remember the formula (18.8). It is much better to apply the method itself, which will lead naturally to the same expression, and we will do this in the next example.

Two linearly independent solutions of the homogeneous equation are $x_1(t) = e^{-3t}$ and $x_2(t) = e^{2t}$ (we found these in Chapter 12), for which

$$W(t) = x_1(t)\dot{x}_2(t) - x_2(t)\dot{x}_1(t) = 2e^{-3t}e^{2t} - -3e^{2t}e^{-3t} = 5e^{-t}.$$

(Note that $W(t)$ is never zero.) The formula (18.8) gives

$$x(t) = -e^{-3t} \int \frac{5e^{2t}e^{-3t}}{5e^{-t}} dt + e^{2t} \int \frac{5e^{-3t}e^{-3t}}{5e^{-t}} dt,$$

$$= -e^{-3t} \int 1 \, dt + 5e^{2t} \int e^{-5t} \, dt$$

$$= -te^{-3t} - e^{-3t}.$$

---

[1] In fact this is the Wronskian of $x_1(t)$ and $x_2(t)$, as defined in Section 11.4; we saw there that if x_1 and x_2 are linearly independent then their Wronskian is never zero, and so our expressions for $\dot{u}_1$ and $\dot{u}_2$ make sense.

We have once again found the particular integral $-te^{-3t}$. The second term here is just a multiple of one of the solutions of the homogeneous equation and so can be absorbed into the complementary function; for the general solution

$$x(t) = Ae^{-3t} + Be^{2t} - te^{-3t} - e^{-3t} = (A - 1)e^{-3t} + Be^{2t} - te^{-3t},$$

and $A - 1$ is just another arbitrary constant. $\qquad\square$

Example 18.2 *Find a particular integral for the equation*

$$\ddot{x} + x = \tan t. \qquad (18.9)$$

For this example we will follow the method outlined in general above, rather than just plugging functions into the resulting formula.

Two linearly independent solutions of the homogeneous equation

$$\ddot{x} + x = 0$$

are $x_1(t) = \sin t$ and $x_2(t) = \cos t$, so for a particular integral we try

$$x(t) = u(t)\sin t + v(t)\cos t.$$

The first derivative of $x(t)$ is given by

$$\dot{x} = \dot{u}\sin t + u\cos t + \dot{v}\cos t - v\sin t,$$

and here we impose an additional condition to make sure that there are no second derivatives of u or v in $\ddot{x}$,

$$\dot{u}\sin t + \dot{v}\cos t = 0. \qquad (18.10)$$

This means that $\dot{x}$ is given by

$$\dot{x} = u\cos t - v\sin t,$$

and we can differentiate to find

$$\ddot{x} = \dot{u}\cos t - u\sin t - \dot{v}\sin t - v\cos t.$$

Substituting for x and $\ddot{x}$ in (18.9) gives (after some cancellation)

$$\dot{u}\cos t - \dot{v}\sin t = \tan t. \qquad (18.11)$$

Equations (18.10) and (18.11) are a pair of simultaneous equations for $\dot{u}$ and $\dot{v}$,

$$\begin{cases} \dot{u}\sin t + \dot{v}\cos t = 0 \\ \dot{u}\cos t - \dot{v}\sin t = \tan t, \end{cases}$$

with solution

$$\dot{u} = \sin t \qquad \text{and} \qquad \dot{v} = \frac{\sin^2 t}{\cos t} = \cos t - \frac{1}{\cos t}.$$

Integrating these two gives[2]

$$u = -\cos t \qquad v = \sin t - \ln|\sec t + \tan t|,$$

and so a particular integral is

$$x(t) = -\cos t \sin t + \sin t \cos t - \ln|\sec t + \tan t| \cos t$$
$$= -\ln|\sec t + \tan t| \cos t.$$

You can check this by substitution (this is less work than it looks if you remember that $\ln|\sec t + \tan t|$ is the integral of $\sec t$).

Exercises

18.1 Use the method of variation of constants to find a particular integral for the following equations:

   (i) $y'' - y' - 6y = e^x$ (you could use the method of undetermined coefficients for this example, which would be much more sensible);

   (ii) $\ddot{x} - x = t^{-1}$ (you can leave the answer as an integral);

   (iii) $y'' + 4y = \cot 2x$. Hint: $\int \csc x \, dx = \ln|\csc x - \cot x|$;

   (iv) $t^2\ddot{x} - 2x = t^3$ (to find the solutions of the homogeneous equation try $x = t^k$, see next chapter);

   (v) $\ddot{x} - 4\dot{x} = \tan t$ (leave your answer as an integral);

   (vi)

$$(\tan^2 x - 1)\frac{d^2y}{dx^2} - 4\tan^3 x \frac{dy}{dx} + 2y\sec^4 x = (\tan^2 x - 1)(1 - 2\sin^2 x),$$

one solution of the homogeneous equation is $y(x) = \sec^2 x$, and the reduction of order method, which is somewhat painful, can be used (if you wish) to show

---

[2] The integral of $\sec t = 1/\cos t$ is the ungainly $\ln|\sec t + \tan t|$. You can check this by differentiating,

$$\frac{d}{dt}\ln(\sec t + \tan t) = \frac{1}{\sec t + \tan t} \times \left(\frac{\sin t}{\cos^2 t} + \frac{\cos^2 t + \sin^2 t}{\cos^2 t}\right)$$

$$= \frac{1}{(1 + \sin t)/\cos t} \times \frac{1 + \sin t}{\cos^2 t}$$

$$= \frac{1}{\cos t}.$$

that a second linearly independent solution is $\tan x$. You should be able to find a particular integral explicitly for this example;

(vii)

$$(1 + \sin^2 t)\ddot{x} - (2\tan t + \sin t \cos t)\dot{x} + (1 - 2\tan^2 t)x = f(t),$$

one solution of the homogeneous equation is $\tan t$, and again the reduction of order method will provide a second solution, $\cos t$, after some effort. You should leave your final answer as an integral.

19

*Cauchy–Euler equations

In this chapter we look at another general class of linear second order equations that we can solve in a systematic way. These are the Cauchy–Euler equations,

$$ax^2\frac{d^2y}{dx^2} + bx\frac{dy}{dx} + cy = 0. \tag{19.1}$$

We will see in the next chapter that an understanding of these equations provides insight that is useful when we try to find the solutions of more complicated equations in the form of power series.

There are two possible approaches that yield the solution of this problem. One method uses the substitution $x = e^z$ to reduce the equation to the more familiar

$$a\frac{d^2y}{dz^2} + (b-a)\frac{dy}{dz} + cy = 0,$$

which can then be solved by trying $y(z) = e^{kz}$, see Exercise 19.2.

However, the method we will use here is similar to the one that we would use to solve the constant coefficient equation

$$a\frac{d^2y}{dx^2} + b\frac{dy}{dx} + cy = 0.$$

For this equation we try $y(x) = e^{kx}$, because for this guess every term in the equation is a multiple of e^{kx}. Now, note that (19.1) has a very special form; whenever there is a derivative the corresponding term is also multiplied by x, i.e. we have x^2y'' and xy'. The kind of function that when differentiated and multiplied by x is a multiple of itself[1] is simply a power of x, $y(x) = x^k$ for some k.

If we try $y(x) = x^k$ in (19.1) then, since

$$y' = kx^{k-1} \qquad \text{and} \qquad y'' = k(k-1)x^{k-2},$$

---

[1] If this does not seem obvious then the correct form for $y(x)$ can be found by solving the separable equation $xy' = ky$, see Exercise 8.3.

we have

$$ak(k-1)x^k + bkx^k + cx^k = 0.$$

Cancelling the factor of x^k that occurs in each term we obtain the *indicial equation*, a quadratic equation for the index k,

$$ak(k-1) + bk + c = 0. \tag{19.2}$$

As with the second order constant coefficient case, the types of solution that we obtain using this approach depend on whether the roots of the indicial equation (19.2) are real and distinct, repeated or complex.

19.1 Two real roots

If there are two distinct real roots k_1 and k_2 of (19.2) then this implies that $y(x) = x^{k_1}$ and $y(x) = x^{k_2}$ are both solutions of (19.1), and so the general solution is a linear combination of these,

$$y(x) = Ax^{k_1} + Bx^{k_2}.$$

Example 19.1 *Find the general solution of the equation*

$$2x^2 y'' + 3xy' - y = 0$$

We try $y(x) = x^k$, and so

$$2k(k-1)x^k + 3kx^k - x^k = 0,$$

which gives the indicial equation for k,

$$2k^2 + k - 1 = 0.$$

This can be factorised as $(2k-1)(k+1) = 0$, and so $k = \frac{1}{2}$ or $k = -1$, and the general solution is

$$y(x) = Ax^{1/2} + Bx^{-1}. \qquad \square$$

19.2 A repeated root

If the indicial equation (19.2) has a repeated real root k then this provides only one solution $y(x) = x^k$. However, we can use the reduction of order method to find a second, linearly independent solution. This will turn out to be $y(x) = x^k \ln x$, and when actually solving an equation like this you should just write down the general solution

$$y(x) = Ax^k + Bx^k \ln x.$$

Any equation that results in a repeated root can be rewritten as

$$x^2 y'' + (1 - 2\lambda)xy' + \lambda^2 y = 0. \tag{19.3}$$

First we check that this equation really does produce a repeated root for the indicial equation; trying $y(x) = x^k$ yields (after cancelling the factors of x^k)

$$k(k-1) + (1 - 2\lambda)k + \lambda^2 = 0,$$

which is $k^2 - 2\lambda k + \lambda^2 = 0$, or $(k - \lambda)^2 = 0$. So we only obtain the one solution $y(x) = x^\lambda$.

Now we use the reduction of order method (see Chapter 17), trying $y(x) = x^\lambda u(x)$. This gives

$$y' = x^\lambda u' + \lambda x^{\lambda-1} u \qquad \text{and} \qquad y'' = x^\lambda u'' + 2\lambda x^{\lambda-1} u' + \lambda(\lambda - 1)x^{\lambda-2} u,$$

and substituting into (19.3), remembering that all the terms in which $u(x)$ has not been differentiated will cancel, we obtain

$$x^2[x^\lambda u'' + 2\lambda x^{\lambda-1} u'] + (1 - 2\lambda)x[x^\lambda u'] = 0.$$

After cancelling a factor of $x^{\lambda+1}$ this gives

$$xu'' + u' = 0.$$

If $v = u'$ then $v' = -v/x$; the solution of this equation can be found by separating variables,

$$\frac{dv}{v} = -\frac{dx}{x}.$$

Integrating both sides gives

$$\ln v = -\ln x + C,$$

and so $v(x) = A/x$. Since $v = u'$, this implies that

$$\frac{du}{dx} = \frac{A}{x},$$

and so $u(x) = A \ln x + B$, which finally yields

$$y(x) = Ax^\lambda \ln x + Bx^\lambda,$$

and we can identify the new solution as $x^\lambda \ln x$.

Example 19.2 *Find the general solution of the equation*

$$x^2 y'' + 3xy' + y = 0.$$

If we try $y(x) = x^k$ then the indicial equation for k is

$$k(k-1) + 3k + 1 = 0$$

which is $k^2 + 2k + 1 = 0$, or $(k+1)^2 = 0$. So $k = -1$ is a repeated root. It follows that the general solution is given by

$$y(x) = Ax^{-1} + Bx^{-1}\ln x.$$

Note that since $(\ln x)/x \to 0$ as $x \to 0$, all solutions tend to zero as $x \to \infty$. $\square$

19.3 Complex roots

If the indicial equation has complex roots, $k = \rho \pm i\omega$ then the solution is

$$y(x) = x^\rho[A\cos(\omega \ln x) + B\sin(\omega \ln x)].$$

Although you should just write down the solution in this case, we now see how it can be derived from

$$y(x) = Cx^{\rho+i\omega} + Dx^{\rho-i\omega}.$$

We can understand x^k when k is complex if we use the identity

$$x^k = e^{k\ln x}.$$

While this is clearly true if x is real (since $x = e^{\ln x}$), it can also be used as a definition of x^k if k is complex. If $k = \rho + i\omega$ then we have

$$x^{\rho+i\omega} = x^\rho x^{i\omega} = x^\rho e^{i\omega \ln x}$$
$$= x^\rho[\cos(\omega \ln x) + i\sin(\omega \ln x)],$$

since $e^{i\theta} = \cos\theta + i\sin\theta$. From this formula it follows that $x^{\rho-i\omega}$ is the complex conjugate of $x^{\rho+i\omega}$, so in order to make our solution real we want

$$y(x) = Cx^{\rho+i\omega} + C^*x^{\rho-i\omega},$$

where now C is complex, $C = \alpha + i\beta$. We therefore have

$$y(x) = 2\,\mathrm{Re}[Cx^{\rho+i\omega}]$$
$$= 2\,\mathrm{Re}[(\alpha + i\omega)x^\rho[\cos(\omega \ln x) + i\sin(\omega \ln x)]]$$
$$= 2x^\rho[\alpha\cos(\omega \ln x) - \beta\sin(\omega \ln x)],$$

which, choosing $A = 2\alpha$ and $B = -2\beta$ (they are both arbitrary constants) shows that

$$y(x) = x^\rho[A\cos(\omega \ln x) + B\sin(\omega \ln x)].$$

Example 19.3 *Find the general solution of the equation*

$$x^2y'' - xy' + 5y = 0.$$

We try $y(x) = x^k$, which yields

$$k(k - 1) - k + 5 = 0,$$

i.e. $k^2 - 2k + 5 = 0$. The roots of this equation are

$$k = \frac{2 \pm \sqrt{4 - 20}}{2} = 1 \pm 2i.$$

So the solution of the equation is

$$y(x) = x[A \cos(2 \ln x) + B \sin(2 \ln x)]. \qquad \Box$$

Exercises

19.1 Find the general solution of the following equations, and also the particular solution satisfying the two specified conditions.
   (i) $x^2y'' - 4xy' + 6y = 0$, $y(1) = 0$ and $y'(1) = 1$;
   (ii) $4x^2y'' + y = 0$, $y(1) = 1$ and $y'(1) = 0$;
   (iii) $t^2\ddot{x} - 5t\dot{x} + 10x = 0$; $x(1) = 2$ and $\dot{x}(1) = 1$;
   (iv) $t^2\ddot{x} + t\dot{x} - x = 0$, $x(1) = \dot{x}(1) = 1$;
   (v) $x^2z'' + 3xz' + 4z = 0$, $z(1) = 0$ and $z'(1) = 5$;
   (vi) $x^2y'' - xy' - 3y = 0$, $y(1) = 1$ and $y'(1) = -1$;
   (vii) $4t^2\ddot{x} + 8t\dot{x} + 5x = 0$, $x(1) = 2$ and $\dot{x}(1) = 0$;
   (viii) $x^2y'' - 5xy' + 5y = 0$, $y(1) = -2$ and $y'(1) = 1$;
   (ix) $3x^2z'' + 5xz' - z = 0$, $z(1) = 3$ and $z'(1) = -1$; and
   (x) $t^2\ddot{x} + 3t\dot{x} + 13x = 0$, $x(1) = -1$ and $\dot{x}(1) = 2$.

19.2 If $x = e^z$ then

$$\frac{d}{dx} = e^{-z}\frac{d}{dz}.$$

Show that

$$\frac{d^2y}{dx^2} = e^{-2z}\left(\frac{d^2y}{dz^2} - \frac{dy}{dz}\right),$$

and hence that substituting $x = e^z$ in

$$ax^2\frac{d^2y}{dx^2} + bx\frac{dy}{dx} + cy = 0 \qquad\qquad (E19.1)$$

yields the linear equation

$$a\frac{d^2y}{dz^2} + (b - a)\frac{dy}{dz} + cy = 0. \qquad\qquad (E19.2)$$

By solving (E19.2) find the solution of (E19.1) when the auxiliary equation

$$ak^2 + (b - a)k + c = 0$$

has

(i) two distinct real roots k_1 and k_2;

(ii) a repeated real root k; and

(iii) a complex conjugate pair of roots $\rho \pm i\omega$.

20

*Series solutions of second order linear equations

We now consider how we might go about finding a solution of the second order linear equation

$$\frac{d^2 y}{dx^2} + p(x)\frac{dy}{dx} + q(x)y = 0$$

in the form of a power series[1]

$$y(x) = \sum_{n=0}^{\infty} a_n x^n. \tag{20.1}$$

20.1 Power series

Before we see how power series can be used to find solutions of differential equations, we briefly recall, without proof, some of their basic properties.

Whenever we consider infinite series, the issue of convergence becomes important. A power series is said to *converge* at a point x if the finite sums

$$\sum_{n=0}^{N} a_n x^n$$

---

[1] Throughout this chapter we only consider power series solutions in this form. Although it is possible to consider more general series solutions like

$$y(x) = \sum_{n=0}^{\infty} a_n (x - x_0)^n$$

(an expansion 'about $x = x_0$') it is always possible to convert such series to something in the form of (20.1) by making the substitution $\tilde{x} = x - x_0$ in the differential equation.

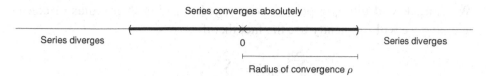

Series converges absolutely

Series diverges 0 Series diverges

Radius of convergence ρ

Fig. 20.1. Within the radius of convergence the power series converges absolutely; outside the series diverges, and on the boundary it may converge or diverge.

tend to a limit as $N \to \infty$, and it is said to converge *absolutely* if

$$\sum_{n=0}^{N} |a_n x^n|$$

tends to a limit as $N \to \infty$. If a series converges absolutely then it must also converge, but the converse is not true.

For every power series there is a number $\rho \geq 0$, known as the *radius of convergence*, such that the series in (20.1) converges absolutely for any x with $|x| < \rho$, and does not converge for $|x| > \rho$. When $|x| = \rho$ the series may converge, or may diverge, see Figure 20.1.

Most important for us is the fact that while x lies within the radius of convergence it is possible to differentiate $y(x)$ by differentiating the power series term-by-term so that

$$y'(x) = \sum_{n=0}^{\infty} n a_n x^{n-1}.$$

The resulting power series for $y'(x)$ has the same radius of convergence as the original power series for $y(x)$.

Although there is no surefire way to find this radius of convergence, one extremely useful method is based on the ratio test. The ratio test guarantees that a series converges absolutely provided that (the modulus of) the ratio of successive terms is eventually less than one,

$$\lim_{n \to \infty} \left| \frac{a_{n+1} x^{n+1}}{a_n x^n} \right| = |x| \lim_{n \to \infty} \left| \frac{a_{n+1}}{a_n} \right| < 1.$$

It follows that the radius of convergence is given by

$$\rho = \lim_{n \to \infty} \left| \frac{a_n}{a_{n+1}} \right|,$$

provided that the limit on the right-hand side exists.

We can add and multiply power series together within their radius of convergence, and we will frequently use the fact that if

$$\sum_{n=0}^{\infty} a_n x^n = \sum_{n=0}^{\infty} b_n x^n$$

then $a_n = b_n$ for every n. In particular, if the right-hand side is zero then $a_n = 0$ for all n.

20.2 Ordinary points

We first suppose that $p(x)$ and $q(x)$ are *analytic*. Essentially this means that they can both be expanded as convergent power series,

$$p(x) = \sum_{n=0}^{\infty} p_n x^n \qquad \text{and} \qquad q(x) = \sum_{n=0}^{\infty} q_n x^n.$$

In such a situation, $x = 0$ is referred to as an *ordinary point* for the equation. In this case we can try a power series solution for $y(x)$ of the form

$$y(x) = \sum_{n=0}^{\infty} a_n x^n. \tag{20.2}$$

Assuming that we are within the radius of convergence of the power series we can differentiate term-by-term to find y' and y'', then substitute into the equation and compare terms involving the same powers of x.

We will start by finding the solution of a familiar equation in this new way.

Example 20.1 *Find the general solution of the equation*

$$y'' = -y$$

as a power series

$$y(x) = \sum_{n=0}^{\infty} a_n x^n,$$

and hence identify two linearly independent solutions.

Assuming that x lies within the radius of convergence of the power series, we have, differentiating term-by-term,

$$y'(x) = \sum_{n=0}^{\infty} n a_n x^{n-1} = \sum_{n=1}^{\infty} n a_n x^{n-1}$$

(since the first term in the first sum is zero), and then

$$y''(x) = \sum_{n=1}^{\infty} n(n-1)a_n x^{n-2} = \sum_{n=2}^{\infty} n(n-1)a_n x^{n-2}.$$

Substituting into the equation $y'' = -y$ we get

$$\sum_{n=2}^{\infty} n(n-1)a_n x^{n-2} = -\sum_{n=0}^{\infty} a_n x^n.$$

It is useful to rewrite both sums so that the index of the power of x is the same,

$$\sum_{n=0}^{\infty} (n+2)(n+1)a_{n+2} x^n = -\sum_{n=0}^{\infty} a_n x^n.$$

(The 'summation variable' n is a dummy variable, just like the variable in an integration, so we can change it without affecting the value of the sum.)

All terms contain every power of x from zero upwards, and equating the coefficients of x^n gives

$$(n+1)(n+2)a_{n+2} = -a_n,$$

or more usefully

$$a_{n+2} = -\frac{a_n}{(n+1)(n+2)}.$$

This provides a *recurrence relation* that tells us a_{n+2} if we know a_n.

The way that this recurrence relation works means that if we know a_0 then we know a_n for all even n, and if we know a_1 then we know a_n for all odd n. So, as we would expect for a second order linear equation, there will be two arbitrary constants (a_0 and a_1) in the general solution.

First we consider the even coefficients:

$$a_2 = -\frac{a_0}{2} \qquad a_4 = -\frac{1}{3 \times 4} \times -\frac{a_0}{2} = \frac{a_0}{4!} \qquad a_6 = -\frac{1}{5 \times 6} \frac{a_0}{4!} = -\frac{a_0}{6!}$$

so in general it looks like we have

$$a_{2n} = (-1)^n \frac{a_0}{(2n)!}. \tag{20.3}$$

We should really check these general coefficients using induction, although we will only do so twice in this chapter, once here, and once in our final (and significantly more unpleasant) example. If we assume that (20.3) is correct for $n = k$, then the recurrence relation implies that

$$a_{2(k+1)} = -\frac{a_{2k}}{(2k+1)(2k+2)} = -\frac{1}{(2k+1)(2k+2)} \times \frac{(-1)^k a_0}{(2k)!}$$

$$= (-1)^{k+1}\frac{a_0}{(2(k+1))!},$$

as required. Since (20.3) is correct when $n = 0$ it is therefore correct for all n.
For odd values of n

$$a_3 = -\frac{a_1}{2 \times 3} \qquad a_5 = -\frac{1}{4 \times 5} \times -\frac{a_1}{3!} = \frac{a_1}{5!} \qquad a_7 = -\frac{1}{6 \times 7}\frac{a_1}{5!} = -\frac{a_1}{7!},$$

and in general we have

$$a_{2n+1} = (-1)^n \frac{a_1}{(2n+1)!}.$$

It follows that our full series solution is

$$y(x) = a_0\left[1 - \frac{x^2}{2} + \frac{x^4}{24} + \cdots + (-1)^n\frac{x^{2n}}{(2n)!} + \cdots\right]$$

$$+ a_1\left[x - \frac{x^3}{6} + \frac{x^5}{120} - \cdots + (-1)^n\frac{x^{2n+1}}{(2n+1)!} + \cdots\right].$$

We can recognise the power series in the square brackets as those for $\cos x$ and
$\sin x$ (see Appendix C), and so

$$y(x) = a_0 \cos x + a_1 \sin x,$$

as we might have expected. These power series are known to converge for every
x; we can easily check this using the ratio test. The ratio of two successive terms
in the series for $\cos x$ is

$$(-1)^{n+1}\frac{x^{2n+2}}{(2n+2)!} \bigg/ (-1)^n\frac{x^{2n}}{(2n)!} = \frac{-x^2}{(2n+2)(2n+1)}$$

which, for each fixed x, tends to zero as $n \to \infty$. Similarly, for the ratio of two
successive terms of the series for $\sin x$ we have

$$(-1)^{n+1}\frac{x^{2n+3}}{(2n+3)!} \bigg/ (-1)^n\frac{x^{2n+1}}{(2n+1)!} = \frac{-x^2}{(2n+3)(2n+1)}.$$

The graph of $\cos x$, along with the result of taking a finite number of terms in
their power series expansions, is shown in Figure 20.2.   □

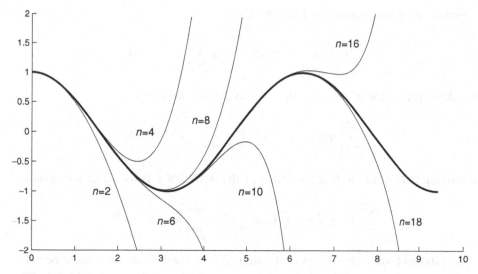

Fig. 20.2. A graph of $\cos x$ against x (the bold line) along with the sum of a finite number of terms of the series solution (n gives the highest power of x in the expansion).

We now use the series solution method on a less familiar example.[2]

Example 20.2 *Find the general solution of the Airy equation*

$$y'' - xy = 0 \qquad (20.4)$$

in the form

$$y(x) = \sum_{n-0}^{\infty} a_n x^n.$$

Assuming that x lies within the radius of convergence of the power series, we have, differentiating term-by-term,

$$y'(x) = \sum_{n=0}^{\infty} n a_n x^{n-1} = \sum_{n=1}^{\infty} n a_n x^{n-1}$$

(since the first term in the first sum is zero), and then

$$y''(x) = \sum_{n=1}^{\infty} n(n-1) a_n x^{n-2} = \sum_{n=2}^{\infty} n(n-1) a_n x^{n-2}.$$

---

[2] Although both the Airy equation and Bessel's equation (the subject of Section 20.4) may be unfamiliar, their solutions are standard functions, just as are sine and cosine. Indeed, MATLAB has built in definitions of both the Airy functions (`airy`) and of the Bessel functions (`besselj` and `bessely`).

Substituting into equation (20.4) we have

$$\sum_{n=2}^{\infty} n(n-1)a_n x^{n-2} - x \sum_{n=0}^{\infty} a_n x^n = 0$$

or, taking the factor of x inside the sum in the second term,

$$\sum_{n=2}^{\infty} n(n-1)a_n x^{n-2} - \sum_{n=0}^{\infty} a_n x^{n+1} = 0.$$

Rewriting both sums so that the index of the power of x is the same we obtain

$$\sum_{n=0}^{\infty} (n+2)(n+1)a_{n+2} x^n - \sum_{n=1}^{\infty} a_{n-1} x^n = 0.$$

A constant term only occurs in the first sum on the left, and this must be zero to match the right-hand side, from which we obtain $2a_2 = 0$, so $a_2 = 0$. Otherwise, setting the coefficient of x^n on the left-hand side to zero (in order to match the right-hand side) we have

$$(n+2)(n+1)a_{n+2} - a_{n-1} = 0,$$

which we can rewrite as

$$a_{n+3} = \frac{a_n}{(n+3)(n+2)}.$$

Now the recurrence relation tells us a_{n+3} if we know a_n, so will give a_3, a_6, a_9, etc. in terms of a_0, and a_4, a_7, a_{10}, etc. in terms of a_1. Since $a_2 = 0$, it follows that $a_5 = a_8 = a_{11} = \cdots = 0$.

For the coefficients based on a_0 we have

$$a_3 = \frac{a_0}{2 \cdot 3} \qquad a_6 = \frac{a_3}{5 \cdot 6} = \frac{a_0}{2 \cdot 3 \cdot 5 \cdot 6} \qquad a_9 = \frac{a_6}{8 \cdot 9} = \frac{a_0}{2 \cdot 3 \cdot 5 \cdot 6 \cdot 8 \cdot 9},$$

and so in general

$$a_{3n} = \frac{a_0}{2 \cdot 3 \cdot 5 \cdot 6 \cdots (3n-1) \cdot 3n}.$$

Similarly for the coefficients based on a_1 we have

$$a_4 = \frac{a_1}{3 \cdot 4} \qquad a_7 = \frac{a_4}{6 \cdot 7} = \frac{a_1}{3 \cdot 4 \cdot 6 \cdot 7} \qquad a_{10} = \frac{a_7}{9 \cdot 10} = \frac{a_1}{3 \cdot 4 \cdot 6 \cdot 7 \cdot 9 \cdot 10},$$

and in general

$$a_{3n+1} = \frac{a_1}{3 \cdot 4 \cdot 6 \cdot 7 \cdots 3n \cdot (3n+1)}.$$

So the solution of the equation is

$$y(x) = a_0 \left[1 + \frac{x^3}{2 \cdot 3} + \frac{x^6}{2 \cdot 3 \cdot 5 \cdot 6} + \cdots + \frac{x^{3n}}{2 \cdot 3 \cdots (3n-1) \cdot 3n} + \cdots \right]$$

$$+ a_1 \left[x + \frac{x^4}{3 \cdot 4} + \frac{x^7}{3 \cdot 4 \cdot 6 \cdot 7} + \cdots + \frac{x^{3n+1}}{3 \cdot 4 \cdots 3n \cdot (3n+1)} + \cdots \right].$$

Note that this gives the solution in the form

$$y(x) = a_0 A_1(x) + a_1 A_2(x),$$

i.e. as a superposition of two (we presume) linearly independent solutions $A_1(x)$ and $A_2(x)$, each of which is given as a power series:

$$A_1(x) = 1 + \sum_{n=1}^{\infty} \frac{x^{3n}}{2 \cdot 3 \cdots (3n-1) \cdot 3n}$$

and

$$A_2(x) = x + \sum_{n=1}^{\infty} \frac{x^{3n+1}}{3 \cdot 4 \cdots 3n \cdot (3n+1)}.$$

The ratio of successive terms in $A_1(x)$ is

$$\left| \frac{x^{3(n+1)}}{2 \cdot 3 \cdots (3n-1) \cdot 3n \cdot (3n+2) \cdot (3n+3)} \middle/ \frac{x^{3n}}{2 \cdot 3 \cdots (3n-1) \cdot 3n} \right|$$

$$= \frac{|x|^3}{(3n+2)(3n+3)},$$

so that whatever the value of x this expression tends to zero as $n \to \infty$. It follows that the series converges for every value of x. A similar analysis shows that $A_2(x)$ also converges for every x.

The graph of $A_1(x)$ is shown in Figure 20.3, along with the approximations given by taking a finite number of terms in the corresponding power series. □

The standard forms of the Airy functions, known as Ai(x) and Bi(x), are linear combinations of $A_1(x)$ and $A_2(x)$; Ai(x) is chosen such that Ai$(x) \to 0$ as $x \to \infty$ and Bi(x) such that the Wronskian $W[\text{Ai}, \text{Bi}](x) = 1/\pi$.

20.3 Regular singular points

There are many important equations that can be written in the form

$$y'' + p(x)y' + q(x)y = 0 \tag{20.5}$$

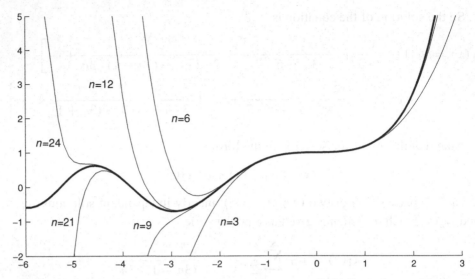

Fig. 20.3. The graph of $A_1(x)$ against x shown as a bold line, along with the values of some of the finite sums.

in which $p(x)$ and $q(x)$ are *not* analytic, i.e. cannot be expressed as a power series. The simplest example is the Cauchy–Euler equation

$$x^2 y'' + p_0 x y' + q_0 y = 0 \tag{20.6}$$

from the previous chapter. Dividing through by x^2 gives

$$y'' + \frac{p_0}{x} y' + \frac{q_0}{x^2} y = 0, \tag{20.7}$$

and so $p(x) = p_0/x$ and $q(x) = q_0/x^2$ cannot be expanded as power series.

However, we saw in the previous chapter that (20.6) has solutions of the form $y(x) = A x^\sigma$, for an arbitrary constant A, where σ has to satisfy the indicial equation

$$\sigma(\sigma - 1) + p_0 \sigma + q_0 = k. \tag{20.8}$$

We can extend the power series method to equations in which $p(x)$ is 'no worse than p_0/x' and $q(x)$ is 'no worse than q_0/x^2', which are the 'bad' factors occurring in (20.7).

A point x is known as a *regular singular point* for equation (20.6) if $xp(x)$ and $x^2q(x)$ are analytic (i.e. have convergent power series expansions). In this case $p(x)$ and $q(x)$ can be written as

$$p(x) = \frac{p_0}{x} + p_1 + p_2 x + p_3 x^2 + \cdots$$

$$q(x) = \frac{q_0}{x^2} + \frac{q_1}{x} + q_2 + q_3 x + q_4 x^2 + \cdots. \tag{20.9}$$

To find a power series solution in this case we replace the arbitrary constant in our solution $y(x) = Ax^\sigma$ of the Cauchy–Euler equation by a power series, and look for a solution in the form

$$y(x) = x^\sigma \sum_{n=0}^{\infty} a_n x^n = \sum_{n=0}^{\infty} a_n x^{\sigma+n}. \tag{20.10}$$

The appropriate values of σ will once again be determined by an indicial equation, as we now see.

Provided that we are within the radius of convergence of the series we have

$$y'(x) = \sum_{n=0}^{\infty} a_n (n + \sigma) x^{\sigma+n-1}$$

and

$$y''(x) = \sum_{n=0}^{\infty} a_n (n + \sigma)(n + \sigma - 1) x^{\sigma+n-2}.$$

Substituting these into $y'' + p(x)y' + q(x)y = 0$ and using the expansion of $p(x)$ and $q(x)$ in (20.9) gives

$$\sum_{n=0}^{\infty} a_n (n + \sigma)(n + \sigma - 1) x^{\sigma+n-2}$$

$$+ \left[\frac{p_0}{x} + p_1 + p_2 x + p_3 x^2 + \cdots \right] \sum_{n=0}^{\infty} a_n (n + \sigma) x^{\sigma+n-1} \tag{20.11}$$

$$+ \left[\frac{q_0}{x^2} + \frac{q_1}{x} + q_2 + q_3 x + q_4 x^2 + \cdots \right] \sum_{n=0}^{\infty} a_n x^{\sigma+n} = 0.$$

Looking at the coefficient of the lowest power of x, $x^{\sigma-2}$, and setting this to zero, we can show that we need σ to satisfy the familiar indicial equation

$$\sigma(\sigma - 1) + p_0 \sigma + q_0 = 0, \tag{20.12}$$

cf. (20.8). We might expect that if we have two distinct roots $\sigma_1 > \sigma_2$ of (20.12) then this would give us two linearly independent power series solutions of our differential equation, and this will be the case provided that σ_1 and σ_2 do not differ by an integer.

To see what the problem is if the roots differ by an integer, we look at the coefficient of $x^{\sigma+n-2}$ from equation (20.11), which is

$$[(n + \sigma)(n + \sigma - 1) + p_0(n + \sigma) + q_0]a_n + \text{terms involving } a_0, \ldots, a_{n-1} = 0.$$

If we rearrange this to find a recurrence relation for a_n we get

$$a_n = \frac{\text{terms involving } a_0, \ldots, a_{n-1}}{(n+\sigma)(n+\sigma-1) + p_0(n+\sigma) + q_0}.$$

The denominator of this equation will be zero if $n + \sigma$ solves the indicial equation (20.12). For the larger root, σ_1, the denominator will never be zero, so we will obtain a solution. However, if the roots differ by an integer N then $\sigma_2 + N = \sigma_1$, and so the recurrence relation for the series involving $\sigma = \sigma_2$ will generally run into a problem when $n = N$ since the denominator will be zero.

Thus if there is a repeated real root, or (in general) if the roots differ by an integer, we will only be able to find one solution in the form of the power series (20.10),

$$y_0(x) = \sum_{n=0}^{\infty} a_n x^{n+\sigma}.$$

However, by making our 'guess' a little more complicated it is possible to find the second solution. Our analysis of the Cauchy–Euler equation in the previous chapter provides a clue as to the form of this guess. There we found that if we had a repeated root σ of the indicial equation then the general solution was

$$y(x) = Ax^\sigma \ln x + Bx^\sigma. \tag{20.13}$$

Once again we replace the arbitrary constants by power series; we replace Ax^σ by our expansion for $y_0(x)$, and B by a new power series whose coefficients we have to find,

$$y_1(x) = y_0(x) \ln x + \sum_{n=0 \text{ or } 1}^{\infty} b_n x^{\sigma+n}. \tag{20.14}$$

If there is a repeated root σ then the sum on the right-hand side is taken from 1, while if the roots differ by an integer $\sigma = \sigma_2$ is the smaller root, and the sum is taken from zero.[3] The algebra involved in substituting (20.14) into the equation is usually fairly daunting, and we will only consider one relatively simple example in what follows.

---

[3] Taking the sum from zero in the case of a repeated root will add a multiple of $y_0(x)$ to the solution $y_1(x)$ and complicate the algebra.

20.4 Bessel's equation

We consider how this method applies to 'Bessel's equation of order v',

$$x^2\frac{d^2y}{dx^2} + x\frac{dy}{dx} + (x^2 - v^2)y = 0. \tag{20.15}$$

By choosing various different values of v we will be able to produce examples of all the different possibilities outlined above.

If we divide by x^2 then we obtain an equation in the standard form,

$$\frac{d^2y}{dx^2} + \underbrace{\frac{1}{x}}_{p(x)}\frac{dy}{dx} + \underbrace{\left(1 - \frac{v^2}{x^2}\right)}_{q(x)}y = 0,$$

from which we can see that $x = 0$ is a regular singular point, since although neither $p(x)$ nor $q(x)$ is analytic, $xp(x)$ and $x^2q(x)$ are. However, it is more convenient to work with the equation in the form (20.15).

We try a solution

$$y(x) = \sum_{n=0}^{\infty} a_n x^{\sigma+n},$$

and so within the radius of convergence

$$y'(x) = \sum_{n=0}^{\infty} a_n(\sigma + n)x^{\sigma+n-1}$$

and

$$y''(x) = \sum_{n=0}^{\infty} a_n(\sigma + n)(\sigma + n - 1)x^{\sigma+n-2}.$$

Substituting this into (20.15) gives

$$\sum_{n=0}^{\infty} a_n(\sigma + n)(\sigma + n - 1)x^{\sigma+n} + \sum_{n=0}^{\infty} a_n(\sigma + n)x^{\sigma+n}$$

$$+ \sum_{n=2}^{\infty} a_{n-2}x^{\sigma+n} - v^2\sum_{n=0}^{\infty} a_n x^{\sigma+n} = 0.$$

The coefficient of x^σ gives

$$a_0\sigma(\sigma - 1) + a_0\sigma - v^2 a_0 = 0,$$

which yields the indicial equation $\sigma^2 = v^2$. The coefficient of $x^{\sigma+1}$ gives

$$a_1(\sigma + 1)\sigma + a_1(\sigma + 1) - v^2 a_1 = 0 \tag{20.16}$$

and so $a_1 = 0$ whenever $\nu \neq 1/2$. For $n \geq 2$ the coefficient of $x^{\sigma+n}$ is

$$a_n(\sigma + n)(\sigma + n - 1) + a_n(\sigma + n) + a_{n-2} - \nu^2 a_n = 0,$$

yielding the recurrence relation

$$a_n = -\frac{a_{n-2}}{(n+\sigma)^2 - \nu^2}.$$

Since $\sigma = \pm\nu$ this gives

$$a_n = -\frac{a_{n-2}}{n(n+2\sigma)}. \tag{20.17}$$

Example 20.3 *Find two series solutions of Bessel's equation when* $\nu = \frac{1}{3}$.

The indicial equation has roots $\sigma = \pm\frac{1}{3}$, and these differ by $2/3$, which is not an integer. So we should be able to find two power series solutions,

$$\sum_{n=0}^{\infty} a_n x^{n+\frac{1}{3}} \quad \text{and} \quad \sum_{n=0}^{\infty} b_n x^{n-\frac{1}{3}}.$$

For the first series $\sigma = \frac{1}{3}$ and the recurrence relation in (20.17) becomes

$$a_n = -\frac{a_{n-2}}{n(n+2/3)} = -\frac{9a_{n-2}}{3n(3n+2)}.$$

Since $a_1 = 0$ the recurrence relation shows that $a_n = 0$ for all odd values of n, and we have

$$a_2 = -\frac{9a_0}{6 \cdot 8} \qquad a_4 = -\frac{9a_2}{12 \cdot 14} = \frac{9^2 a_0}{6 \cdot 8 \cdot 12 \cdot 14} \qquad a_6 = -\frac{9^3 a_0}{6 \cdot 8 \cdot 12 \cdot 14 \cdot 18 \cdot 20},$$

and in general

$$a_{2n} = (-1)^n \frac{9^n a_0}{6 \cdot 8 \cdots 6n \cdot (6n+2)} = (-1)^n \left(\frac{3}{2}\right)^{2n} \frac{a_0}{3 \cdot 4 \cdots 3n \cdot (3n+1)}.$$

So we have one solution

$$j_{1/3}(x) = x^{1/3} \sum_{n=0}^{\infty} (-1)^n \frac{(3x/2)^{2n}}{3 \cdot 4 \cdots 3n \cdot (3n+1)}.$$

For the second solution with $\sigma = -\frac{1}{3}$ the recurrence relation is

$$b_n = -\frac{b_{n-2}}{n(n-2/3)} = -\frac{9b_{n-2}}{3n(3n-2)},$$

and so

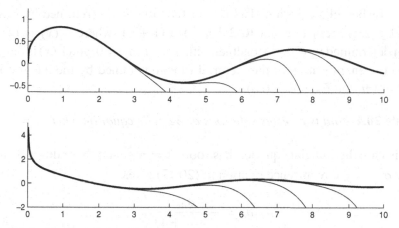

Fig. 20.4. Graphs of $j_{1/3}(x)$ (top) and $j_{-1/3}(x)$ (bottom) against x along with their approximations from series with 4, 6, 8 and 10 terms.

$$b_2 = -\frac{9b_0}{6 \cdot 4} \qquad b_4 = -\frac{9b_2}{12 \cdot 10} = \frac{9^2 b_0}{4 \cdot 6 \cdot 10 \cdot 12} \qquad b_6 = -\frac{9^3 b_0}{4 \cdot 6 \cdot 10 \cdot 12 \cdot 16 \cdot 18},$$

and in general

$$b_{2n} = (-1)^n \frac{9^n b_0}{4 \cdot 6 \cdot \cdots \cdot (6n-2) \cdot 6n} = (-1)^n \left(\frac{3}{2}\right)^n \frac{b_0}{2 \cdot 3 \cdot \cdots \cdot (3n-1) \cdot 3n},$$

giving a second solution

$$j_{-1/3}(x) = x^{-1/3} \sum_{n=0}^{\infty} (-1)^n \frac{(3x/2)^{2n}}{2 \cdot 3 \cdot \cdots \cdot (3n-1) \cdot 3n}.$$

Graphs of $j_{1/3}(x)$ and $j_{-1/3}(x)$ are shown in Figure 20.4.　　□

It is interesting to note that these functions are related to the solutions of the Airy equation. For example, we had

$$A_1(x) = 1 + \frac{x^3}{2 \cdot 3} + \frac{x^6}{2 \cdot 3 \cdot 5 \cdot 6} + \cdots + \frac{x^{3n}}{2 \cdot 3 \cdot \cdots \cdot (3n-1) \cdot 3n} + \cdots$$

$$= 1 - \frac{(ix^{3/2})^2}{2 \cdot 3} + \frac{(ix^{3/2})^4}{2 \cdot 3 \cdot 5 \cdot 6} + \cdots + (-1)^n \frac{(ix^{3/2})^{2n}}{2 \cdot 3 \cdot \cdots \cdot (3n-1) \cdot 3n} + \cdots$$

$$= j_{-1/3}\left(\frac{2}{3} ix^{3/2}\right).$$

Just as there are standard choices for the two linearly independent solutions of the Airy equation, there are standard normalisations for the series solutions of Bessel's equation. The Bessel function $J_\nu(x)$ (returned by MATLAB's besselj(nu, x)) is equal to $2^{-\nu} j_\nu(x)/\Gamma(1+\nu)$, where $j_\nu(x)$ is the series solution determined by (20.17) together with $a_0 = 1$, $a_1 = 0$, and $\Gamma(x)$ is the gamma function (a generalisation of the factorial function defined by the integral $\Gamma(z) = \int_0^\infty t^{z-1} e^{-t} dt$, see Exercise 20.8).

Example 20.4 *Find two series solutions of Bessel's equation when* $\nu = \frac{1}{2}$.

In this case the indicial equation has roots $\sigma = \pm \frac{1}{2}$, and these differ by an integer. For $\sigma = \frac{1}{2}$ the recurrence relation in (20.17) gives

$$a_n = -\frac{a_{n-2}}{n(n+1)}.$$

As before all odd coefficients are zero,

$$a_2 = -\frac{a_0}{2 \cdot 3} \qquad a_4 = \frac{a_0}{2 \cdot 3 \cdot 4 \cdot 5} \qquad a_6 = \frac{a_0}{7!},$$

and the general coefficient is

$$a_{2n} = (-1)^n \frac{a_0}{(2n+1)!}.$$

It follows that

$$y(x) = a_0 \left[\sum_{n=0}^\infty (-1)^n \frac{x^{2n} + \frac{1}{2}}{(2n+1)!} \right]$$

$$= \frac{a_0}{\sqrt{x}} \left[\sum_{n=0}^\infty (-1)^n \frac{x^{2n+1}}{(2n+1)!} \right]$$

$$= a_0 \frac{\sin x}{\sqrt{x}},$$

and one solution of the equation is $j_{1/2}(x) = \sin x / \sqrt{x}$.

We might suspect that the other solution is $\cos x / \sqrt{x}$ and this in fact turns out to be the case. Indeed, in this case we can find a second solution using the series method even though σ_1 and σ_2 differ by an integer. For $\sigma = -\frac{1}{2}$ the recurrence relation in (20.17) becomes

$$a_n = \frac{a_{n-2}}{n(n-1)}, \qquad (20.18)$$

while the potential 'problem with zero' occurs only in (20.16), which requires $0 \times a_1 = 0$. So in this case rather than being forced to be infinite, the occurrence of zero coefficient means that a_1 is not determined. Choosing a_1 non-zero would simply add a multiple of $j_{1/2}(x)$ to our second solution, so instead we choose $a_1 = 0$.

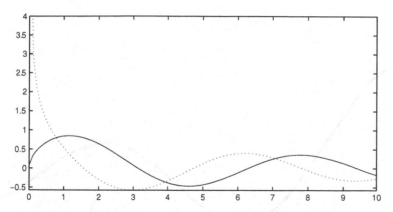

Fig. 20.5. The two solutions $j_{1/2}(x) = \sin x/\sqrt{x}$ (solid), and $j_{-1/2}(x) = \cos x/\sqrt{x}$ (dashed), plotted against x.

Then for our second linearly independent solution all the odd coefficients are zero and for the even coefficients we get

$$a_2 = -\frac{a_0}{2} \qquad a_4 = \frac{a_0}{4!} \qquad a_6 = -\frac{a_0}{6!},$$

and in general $a_{2n} = (-1)^n/(2n)!$; we have

$$y(x) = a_0 \left[\sum_{n=0}^{\infty} \frac{x^{2n-\frac{1}{2}}}{(2n)!} \right]$$

$$= \frac{a_0}{\sqrt{x}} \left[\sum_{n=0}^{\infty} \frac{x^{2n}}{(2n)!} \right]$$

$$= a_0 \frac{\cos x}{\sqrt{x}}.$$

As we suspected, there is a second solution $j_{-1/2}(x) = \cos x/\sqrt{x}$.

These solutions are illustrated in Figure 20.5.

Example 20.5 *Find a series solution of Bessel's equation when $v = 0$,*

$$x^2 y'' + xy' + x^2 y = 0. \tag{20.19}$$

In this case the roots of the indicial equation are both $\sigma = 0$, so we can only find one solution in the form of a simple power series. The recurrence relation in (20.17) becomes

$$a_n = -\frac{a_{n-2}}{n^2}.$$

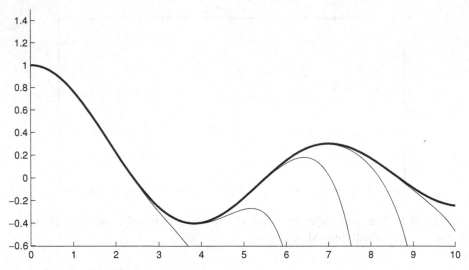

Fig. 20.6. The graph of $J_0(x)$, plotted against x, along with the sum of the series with 4, 6, 8, 10 and 12 terms. The series with 14 terms cannot be distinguished from $J_0(x)$ over this range of x values.

Since $a_1 = 0$ all odd coefficients are zero, and

$$a_2 = -\frac{a_0}{2^2} \qquad a_4 = \frac{a_0}{2^2 \cdot 4^2} \qquad a_6 = -\frac{a_0}{2^2 \cdot 4^2 \cdot 6^2}.$$

It follows that

$$a_{2n} = \frac{(-1)^n a_0}{2^2 4^2 \cdots (2n)^2} = \frac{(-1)^n a_0}{2^{2n}(n!)^2}.$$

Thus we have

$$y_0(x) = a_0 \left[\sum_{n=0}^{\infty} \frac{(-1)^n}{2^{2n}(n!)^2} x^{2n} \right],$$

with the convention that $0! = 1$. The quantity in the square brackets is $J_0(x)$, the Bessel function of the first kind of order zero. A graph of $J_0(x)$, along with its approximations by taking a finite number of terms from the power series, is shown in Figure 20.6. □

We end this chapter with a somewhat more painful calculation of the second solution for Bessel's equation of order zero.

Example 20.6 *Find a solution for Bessel's equation of order zero in the form*

$$y(x) = J_0(x) \ln x + \sum_{n=1}^{\infty} b_n x^n.$$

First we calculate

$$y'(x) = J_0'(x) \ln x + \frac{J_0(x)}{x} + \sum_{n=1}^{\infty} n b_n x^{n-1}$$

and

$$y''(x) = J_0''(x) \ln x + 2\frac{J_0'(x)}{x} - \frac{J_0(x)}{x^2} + \sum_{n=2}^{\infty} n(n-1) b_n x^{n-2}.$$

When we substitute these expressions into Bessel's equation of order zero, $x^2 y'' + x y' + x^2 y = 0$, all the terms involving $J_0(x)$ cancel except one, and we end up with

$$2x J_0'(x) + \sum_{n=2}^{\infty} n(n-1) b_n x^n + \sum_{n=1}^{\infty} n b_n x^n + \sum_{n=1}^{\infty} b_n x^{n+2} = 0.$$

Since

$$J_0(x) = \sum_{n=0}^{\infty} \frac{(-1)^n}{2^{2n}(n!)^2} x^{2n}$$

it follows that

$$x J_0'(x) = \sum_{n=1}^{\infty} \frac{(-1)^n 2n}{2^{2n}(n!)^2} x^{2n}.$$

Therefore we have

$$b_1 x + 4 b_2 x^2 + \sum_{n=3}^{\infty} [n^2 b_n + b_{n-2}] x^n = -2 \sum_{n=1}^{\infty} \frac{(-1)^n 2n}{2^{2n}(n!)^2} x^{2n}.$$

Since only even powers occur on the right-hand side it follows that $b_1 = 0$, and that for n odd

$$n^2 b_n + b_{n-2} = 0,$$

which implies that $b_n = 0$ for all odd indices n. When n is even we have $b_2 = 1/4$ and for $n \geq 2$ the coefficients of x^{2n} give

$$(2n)^2 b_{2n} + b_{2(n-1)} = -\frac{(-1)^n n}{2^{2(n-1)}(n!)^2}$$

which is

$$b_{2n} = \frac{1}{2^2 n^2} \left(-b_{2(n-1)} - \frac{(-1)^n n}{2^{2(n-1)}(n!)^2} \right).$$

Thus

$$b_4 = \frac{1}{2^2 \cdot 2^2}\left(-\frac{1}{4} - \frac{2}{2^2 2^2}\right) = -\frac{1}{2^2 4^2}\left(1 + \frac{1}{2}\right),$$

then

$$b_6 = \frac{1}{2^2 \cdot 3^2}\left[\frac{1}{2^2 4^2}\left(1 + \frac{1}{2}\right) + \frac{3}{2^4 (3!)^2}\right] = \frac{1}{2^2 4^2 6^2}\left(1 + \frac{1}{2} + \frac{1}{3}\right).$$

It looks as though we have

$$b_{2n} = \frac{(-1)^{n+1}}{2^{2n}(n!)^2}\left(1 + \frac{1}{2} + \cdots + \frac{1}{n}\right),$$

although this is one solution that we should check by induction. Assuming that this is correct for b_{2k}, we have

$$b_{2(k+1)} = \frac{1}{2^2(k+1)^2}\left[-\frac{(-1)^{k+1}}{2^{2k}(k!)^2}\left(1 + \frac{1}{2} + \cdots + \frac{1}{k}\right) - \frac{(-1)^{k+1}(k+1)}{2^{2k}((k+1)!)^2}\right]$$

$$= \frac{(-1)^{k+2}}{2^2(k+1)^2}\left[\frac{1}{2^{2k}(k!)^2}\left(1 + \frac{1}{2} + \cdots + \frac{1}{k}\right) + \frac{1}{2^{2k}(k!)^2(k+1)}\right]$$

$$= \frac{(-1)^{k+2}}{2^2 2^{2k}(k+1)^2(k!)^2}\left[\left(1 + \frac{1}{2} + \cdots + \frac{1}{k}\right) + \frac{1}{k+1}\right]$$

$$= \frac{(-1)^{k+2}}{2^{2(k+1)}((k+1)!)^2}\left(1 + \frac{1}{2} + \cdots + \frac{1}{k+1}\right),$$

as required. Writing H_n for the sum of the first n terms of the harmonic series

$$H_n = 1 + \frac{1}{2} + \cdots + \frac{1}{n}$$

we therefore have the second solution

$$y_0(x) = J_0(x)\ln x + \sum_{n=1}^{\infty}(-1)^{n+1}\frac{H_n}{2^{2n}(n!)^2}x^{2n}. \tag{20.20}$$

This second solution, which blows up as $x \to 0$ like $\ln x$, is shown in Figure 20.7.

The standard normalisation of this second solution means that the 'Bessel function of the second kind of order zero', $Y_0(x)$, returned by the MATLAB function bessely(0,x), is somewhat surprisingly given by

$$Y_0(x) = \frac{2}{\pi}[y_0(x) + (\gamma - \ln 2)J_0(x)],$$

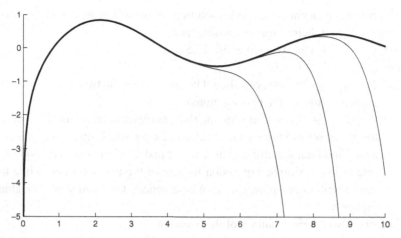

Fig. 20.7. A second linearly independent solution for Bessel's equation of order zero plotted against x, and its approximations by $J_0(x) \ln x + S_n$, where S_n is the series part of (20.20) with 2, 4 and 6 terms.

where γ is the Euler–Mascheroni constant defined by

$$\gamma = \lim_{n \to \infty} (H_n - \ln n) \approx 0.5772.$$

Exercises

You may find the following two identities useful for these exercises:

$$2 \cdot 4 \cdot 6 \cdots \cdot 2n = 2^n n!$$

and

$$1 \cdot 3 \cdot 5 \cdots \cdot (2n - 1) = \frac{(2n)!}{2^n n!}.$$

20.1 Legendre's equation is

$$(1 - x^2)y'' - 2xy' + l(l + 1)y = 0. \qquad \text{(E20.1)}$$

If $y(x)$ is given by a power series,

$$y(x) = \sum_{n=0}^{\infty} a_n x^n,$$

find the recurrence relation satisfied by the coefficients a_n. Show that if l is a positive integer then there is a solution given by a power series that has only a finite number of terms, i.e. a polynomial. For each value $l = 1, 2, 3$, and 4 find the polynomial solution that has $y(1) = 1$ (these are the 'Legendre polynomials' $P_l(x)$).

20.2 Find two independent power series solutions of the following equations, and use the ratio test to find their radius of convergence.

(i) $y'' - xy' + y = 0$ (cf. Exercise 17.5),

(ii) $(1 + x^2)y'' + y = 0$,

(iii) $2xy'' + y' - 2y = 0$ (you should be able to sum the two power series to obtain explicit forms for the two solutions),

(iv) $y'' - 2xy' + 2ky = 0$. By finding the recurrence relation for the coefficients in the power series identify those values of k for which one solution is a polynomial. Find both solutions when $k = -2$ and $k = 2$; in each case you should be able to find a simple expression for one of the two solutions, while the other can be written as a power series whose general term you should be able to find explicitly.

20.3 Find one power series solution of the equation

$$x(1 - x)y'' - 3xy' - y = 0.$$

You should be able to sum this power series to write down the solution explicitly. Now use the reduction of order method to find a second solution.

20.4 Find one series solution of the 'modified Bessel equation'

$$x^2 y'' + xy' - x^2 y = 0.$$

20.5 Find a series solution for Bessel's equation of order one,

$$x^2 y'' + xy' + (x^2 - 1)y = 0. \tag{E20.2}$$

You should obtain

$$y(x) = cx \sum_{n=0}^{\infty} \frac{(-1)^n x^{2n}}{2^{2n}(n+1)! \, n!};$$

with the choice $c = 1/2$ this gives the standard form of the Bessel function $J_1(x)$,

$$J_1(x) = \sum_{n=0}^{\infty} \frac{(-1)^n}{(n+1)! \, n!} \left(\frac{x}{2}\right)^{2n+1}.$$

20.6 In order to find a second solution of (E20.2), substitute

$$y(x) = J_1(x) \ln x + \frac{1}{x} \left[\sum_{n=0}^{\infty} b_n x^n \right],$$

where $J_1(x)$ is the series solution from the previous question, to show that

$$b_1 + b_0 x + \sum_{n=2}^{\infty} [(n^2 - 1)b_{n+1} + b_{n-1}]x^n = -2 \sum_{k=0}^{\infty} \frac{(-1)^k (2k+1)}{(k+1)! \, k!} \left(\frac{x}{2}\right)^{2k+1}.$$

Hence show that $b_0 = -1$, $b_1 = 0$, and that b_n obeys the recurrence relation

$$(n^2 - 1)b_{n+1} + b_{n-1} = 0$$

if n is even and, for $k = 1, 2, 3, \ldots,$

$$[(2k+1)^2 - 1]b_{2(k+1)} + b_{2k} = -\frac{(-1)^k (2k+1)}{2^{2k}(k+1)!\,k!}. \qquad (E20.3)$$

Deduce that $b_j = 0$ for all odd values of j.

Denoting by H_n the sum

$$H_n = \sum_{j=1}^{n} \frac{1}{j},$$

verify that

$$b_{2k} = \frac{(-1)^k (H_k + H_{k-1})}{2^{2k} k!\,(k-1)!}$$

solves (E20.3) and hence write down a second solution of (E20.2).

20.7 Show that when n is a positive integer one solution of Bessel's equation

$$x^2 \frac{d^2 y}{dx^2} + x \frac{dy}{dx} + (x^2 - n^2)y = 0$$

can be written as the power series

$$J_n(x) = \sum_{j=0}^{\infty} (-1)^j \frac{1}{j!(n+j)!} \left(\frac{x}{2}\right)^{n+2j}. \qquad (E20.4)$$

20.8 (T) The gamma function generalises the factorial function to values that are not integers. For any real number z we define

$$\Gamma(z) = \int_0^{\infty} t^{z-1} e^{-t} \, dt.$$

Integrate by parts in order to show that for a positive integer n

$$\Gamma(n+1) = n\Gamma(n).$$

Since $\Gamma(1) = 1$, deduce that $\Gamma(n+1) = n!$. (Using the gamma function in place of one of the factorials in the power series (E20.4) gives

$$J_\nu(x) = \sum_{j=0}^{\infty} (-1)^j \frac{1}{j!\,\Gamma(\nu+j+1)} \left(\frac{x}{2}\right)^{\nu+2j},$$

and this formula now applies for any real number ν. This is where the strange normalisation of J_ν for non-integer ν comes from, see comments after Example 20.3.)

20.9 (C) Write a short program to generate the coefficients in the power series expansion of $J_\nu(x)$ for any value of ν using the recurrence relation (20.17). Investigate how many terms of the expansion you need to take in order to approximate the solution well on a fixed interval ($0 \le x \le 10$, say). (You might like to look at the M-file besselseries.m, which produced the Bessel function figures in this chapter.)

20.10 (T) The Bessel functions might seem exotic, but they arise very naturally in problems that have radial symmetry. For example, the vibrations of a circular drum satisfy

$$\frac{\partial^2 u}{\partial t^2} = \frac{1}{r}\frac{\partial}{\partial r}\left(r\frac{\partial u}{\partial r}\right) + \frac{1}{r^2}\frac{\partial^2 u}{\partial \theta^2}, \tag{E20.5}$$

where $u(r, \theta, t)$ is the displacement of the circular skin of the drum at a point expressed in polar coordinates. In the method of separation of variables we look for a solution of the form

$$u(r, \theta, t) = R(r)\Theta(\theta)T(t),$$

and try this guess in the equation. Substitute this in to (E20.5) and show that

$$\frac{1}{T}\frac{d^2 T}{dt^2} = \frac{1}{rR}\frac{d}{dr}\left(r\frac{dR}{dr}\right) + \frac{1}{r^2\Theta}\frac{d^2\Theta}{d\theta^2}. \tag{E20.6}$$

The left-hand side of this equation is a function of t alone, and the right-hand side a function of r and θ, so in order to be always equal they must both be constants. Choosing

$$\frac{1}{T}\frac{d^2 T}{dt^2} = -k^2$$

(there are good physical reasons for choosing this constant to be negative) show that we can rearrange (E20.6) to give

$$-\frac{1}{\Theta}\frac{d^2\Theta}{d\theta^2} = \frac{r}{R}\frac{d}{dr}\left(r\frac{dR}{dr}\right) + r^2 k^2. \tag{E20.7}$$

Now the left-hand side is a function of θ alone, while the right-hand side is a function of r alone; so both sides must be equal to a constant. Now we choose

$$-\frac{1}{\Theta}\frac{d^2\Theta}{d\theta^2} = \nu^2$$

(again there are good physical reasons why this constant should be positive); show that in this case (E20.7) can be rearranged to give

$$r^2\frac{d^2 R}{dr^2} + r\frac{dR}{dr} + (r^2 k^2 - \nu^2)R = 0.$$

Finally substitute $x = rk$ to show that R satisfies Bessel's equation of order ν,

$$x^2\frac{d^2 R}{dx^2} + x\frac{dR}{dx} + (x^2 - \nu^2)R = 0.$$

Part IV

Numerical methods and difference equations

21

Euler's method

If we have a differential equation that we cannot solve analytically then we can always try to treat the problem numerically. In this chapter we consider one numerical way to approximate the equation

$$\frac{dx}{dt} = f(x, t) \qquad x(0) = x_0. \tag{21.1}$$

The first step is to give up the idea of finding a solution for all values of the independent variable ($x(t)$ for any $t \in \mathbb{R}$) and instead try to find an approximation to the solution at a discrete set of values of t. This is illustrated in Figure 21.1.

In the simplest case we try to approximate the solution at equally spaced values of the independent variable. This means, for example, that we want to approximate $x(h)$, $x(2h)$, $x(3h)$, $x(4h)$, etc. The difference between two successive times, here h, is called the *timestep* (or more generally, when the independent variable is not necessarily time, the *step size*).

If we have a method for approximating $x(t + h)$ given $x(t)$, then we can apply the method repeatedly to find approximations for $x(nh)$ for any n, if $x(0)$ is specified initially; first we approximate $x(h) = x(0 + h)$, then using our approximation for $x(h)$ we approximate $x(2h) = x(h + h)$, then $x(3h) = x(2h + h)$, etc.

In this chapter we will introduce the simplest numerical method for producing approximations to the solutions of (21.1).

21.1 Euler's method

The differential equation

$$\frac{dx}{dt} = f(x, t) \tag{21.2}$$

tells us that at time $t = s$, the rate of change of $x(t)$ is $f(x(s), s)$. If the timestep h is small enough for us to assume that this derivative changes little between $t = s$

201

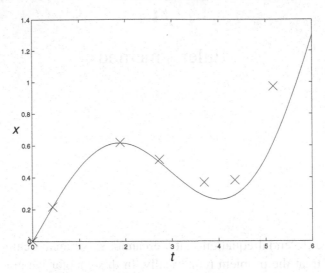

Fig. 21.1. The curve shows a notional 'true solution', and the crosses show the result of numerically approximating the solution at a discrete set of values of t.

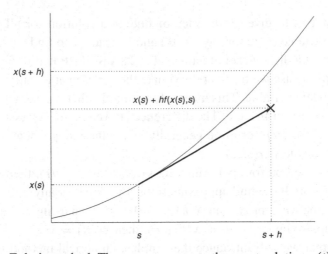

Fig. 21.2. Euler's method. The curve represents the exact solution $x(t)$, the bold line has slope $f(x(s), s)$, and the cross gives the Euler approximation to $x(s + h)$ given $x(s)$.

and $t = s + h$, then we can pretend that $\dot{x}(t)$ is actually constant over this interval, and so we can make the approximation

$$x(s + h) \approx x(s) + h\, f(x(s), s). \qquad (21.3)$$

The effect of doing this is shown in Figure 21.2.

The expression in (21.3) is also what you would get from keeping only the first two terms in the Taylor expansion of x near time $t = s$,

$$x(s + h) \approx x(s) + h\dot{x}(s) = x(s) + hf(x(s), s),$$

since $\dot{x}(s) = f(x(s), s)$ (see Appendix C).

In order to have a more compact notation, we can write $t_n = nh$ and $x_n = x(t_n)$. The approximation in (21.3) then gives rise to *Euler's method* of numerical solution,

$$x_{n+1} = x_n + h\, f(x_n, t_n) \qquad \text{with} \qquad x_0 = x(0). \tag{21.4}$$

This is a simple example of a difference equation, where the continuous variable (t) has been replaced by a discrete index (n). We will treat difference equations more systematically in the next chapter, but for now we will consider the application of Euler's method to some simple examples.

21.2 An example

Because Euler's method is so simple, it is possible to apply it 'by hand'. Suppose that we want to approximate the solution of

$$\frac{dx}{dt} = t - x^2 \qquad x(0) = 0, \tag{21.5}$$

at time $t = 2$. This is an equation whose solution cannot be found explicitly, hence the need for a numerical method. An accurate numerical value of $x(2)$ is $x(2) = 1.1936$ (correct to 4 decimal places), and we will try to reproduce this with Euler's method.

Using the method with a timestep $h = 1$ we have $t_n = n$, $x_n = x(n)$, and

$$\begin{aligned}
x_{n+1} &= x_n + hf(x_n, t_n) \\
&= x_n + h\left(t_n - x_n^2\right) \\
&= x_n + \left(n - x_n^2\right).
\end{aligned}$$

So we have

$$\begin{aligned}
x_1 &= x_0 + \left(0 - x_0^2\right) = 0 \\
x_2 &= x_1 + \left(1 - x_1^2\right) = 1,
\end{aligned}$$

which gives the value $x(2) \simeq x_2 = 1$. That this approximation is not very accurate is unsurprising, since we have assumed that the derivative of $x(t)$ is constant between $t = 0$ and $t = 1$, and between $t = 1$ and $t = 2$.

However, we can do much better if we use the timestep $h = 1/2$. In this case we have $t_n = n/2$, $x_n = x(n/2)$, and

$$
\begin{aligned}
x_{n+1} &= x_n + hf(x_n, t_n) \\
&= x_n + h(t_n - x_n^2) \\
&= x_n + \tfrac{1}{2}(\tfrac{n}{2} - x_n^2).
\end{aligned}
$$

So we have

$$
\begin{aligned}
x_1 &= x_0 + \tfrac{1}{2}(0 - x_0^2) = 0 \\
x_2 &= x_1 + \tfrac{1}{2}(\tfrac{1}{2} - x_1^2) = 0 + \tfrac{1}{2}(\tfrac{1}{2} - 0^2) = \tfrac{1}{4} \\
x_3 &= x_2 + \tfrac{1}{2}(1 - x_2^2) = \tfrac{1}{4} + \tfrac{1}{2}(1 - \tfrac{1}{16}) = \tfrac{23}{32} \\
x_4 &= x_3 + \tfrac{1}{2}(\tfrac{3}{2} - x_3^2) = \tfrac{23}{32} + \tfrac{1}{2}(\tfrac{3}{2} - (\tfrac{23}{32})^2) \\
&= \tfrac{23}{32} + \tfrac{1}{2}(\tfrac{3}{2} - \tfrac{529}{1024}) = \tfrac{23}{32} + \tfrac{1007}{2048} \\
&= \tfrac{834}{689}.
\end{aligned}
$$

The final stage here, $x_4 = 834/689 \simeq 1.2104$, is our approximation to $x(2)$, and significantly more accurate than the result for $h = 1$.

Although this is already slightly beyond the limit of what is comfortable to calculate by hand, a computer can happily apply the method with much smaller timesteps. As the timestep is made smaller, the distance between successive times decreases. As this happens the assumption that $\dot{x}$ is constant between times t_n and t_{n+1} becomes more and more accurate, and so the approximation to $x(t)$ becomes better and better. Figure 21.3 shows the approximations of the true solution of (21.5) for various values of h over the time interval $0 \le t \le 2$.

It appears that the Euler method has served us extremely well, even with the relatively large timestep $h = 1/2$. However, if we try to use this method to approximate the solution on a longer time interval, say $0 \le t \le 11$, then Figure 21.4 shows that we can start to run into problems. The Euler method continues to approximate the solution extremely well until around $t = 8$, but then the errors begin to accumulate and the numerical solutions starts to oscillate wildly, while the true solution continues to increase relatively gently.

21.3 *MATLAB implementation of Euler's method

Euler's method is easy to implement computationally. The MATLAB code required to apply the method to equation (21.5) is given below (and is available as euler.m on the web). Also included are two commands to plot the numerical results graphically, as in Figures 21.3 and 21.4.

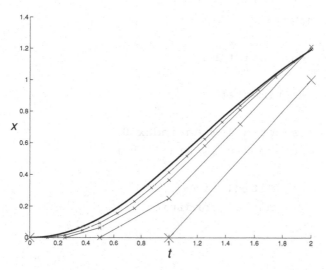

Fig. 21.3. The solution of equation (21.5) as calculated by Euler's method, for $h = 1$, $h = 1/2$, $h = 1/4$, $h = 1/8$; the numerical values are marked by crosses. Also shown is the 'exact solution' in bold. (This 'exact solution' is in fact the result of Euler's method with $h = 2^{-10}$.)

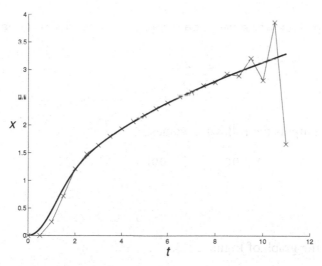

Fig. 21.4. The solution of equation (21.5) as calculated by Euler's method with $h = 1/2$. As before the numerical values are marked by crosses, and the exact solution is shown in bold.

```
%% Euler's method

T=2;           %% final time

h=0.5;         %% timestep

%% MATLAB does not allow an index 0
%% on a vector, so x_n is x(n+1) here

t(1)=0;        %% initial time
x(1)=0;        %% initial condition

for n=1:T/h;

    t(n+1)=n*h;
    x(n+1)=x(n) + h * (t(n)-x(n)^2);

end

[t x]          %% display values

%% Plot crosses at numerical values, and join these

plot(t,x,'x','MarkerSize',20)
hold on
plot(t,x)
```

This program outputs the following values:

```
t =          0    0.5000    1.0000    1.5000    2.0000


x =          0    0         0.2500    0.7188    1.2104
```

and produces the graph of Figure 21.5.

21.4 Convergence of Euler's method

We can investigate how Euler's method behaves when we make the timestep smaller if we apply the method to an equation whose solution we already know. In this case we can more easily compare our approximate numerical solution with the exact solution.

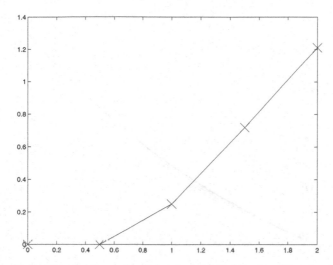

Fig. 21.5. The graph produced by the MATLAB code of Section 21.3.

We will apply the method to the simple linear equation

$$\frac{dx}{dt} = x \qquad \text{with} \qquad x(0) = 1,$$

for which we know the exact solution $x(t) = e^t$.

If we use a timestep h then $t_n = nh$, $x_n = x(nh)$, and Euler's method gives

$$x_{n+1} = x_n + h x_n = (1+h)x_n \qquad \text{with} \qquad x_0 = 1.$$

It is easy to find the solution of this equation, since

$$x_1 = (1+h)x_0 = (1+h)$$
$$x_2 = (1+h)x_1 = (1+h)(1+h) = (1+h)^2$$
$$x_3 = (1+h)x_2 = (1+h)(1+h)^2 = (1+h)^3,$$

and so in general[1] $x_n = (1+h)^n$.

Since x_n is an approximation to $x(nh)$, the approximation we obtain for $x(t)$ is $x_{t/h}$ (where we assume that t/h is an integer),

$$x(t) \simeq (1+h)^{t/h}.$$

Since $t = nh$ we can replace h by t/n, and so

$$x(t) \simeq \left(1 + \frac{t}{n}\right)^n.$$

---

[1] This can be checked using induction. The induction hypothesis is that $x_n = (1+h)^n$. Assuming that this is true for $n = m$, it follows that $x_{m+1} = (1+h)x_m = (1+h)(1+h)^m = (1+h)^{m+1}$, and so the hypothesis holds for $n = m + 1$. Since $x_0 = (1+h)^0 = 1$, induction shows that this is the correct solution.

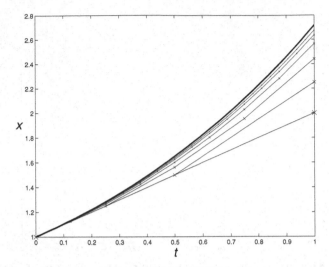

Fig. 21.6. The bold line is the exact solution $x(t) = e^t$, while the other lines show the results of the Euler method with successively smaller values of h, from $h = 1$ to $h = 1/32$, with crosses marking the numerical values.

If we keep x fixed and let $h \to 0$ then since $n = t/h$ we need to let $n \to \infty$. A standard result from analysis (see Exercise 21.4) guarantees that

$$\left(1 + \frac{t}{n}\right)^n \to e^t \tag{21.6}$$

as $n \to \infty$, and so as the timestep is refined our numerical approximation does tend to the exact solution. This is illustrated in Figure 21.6, which shows the exact solution and the solution obtained using Euler's method with various values of h.

Although we have shown that the Euler method works for this particular example, if it is to be a truly reliable method then we should have a proof that whatever the equation, if the timestep is small enough then the numerical solution will be a good approximation. The mathematical discipline of numerical analysis deals with such problems. For example, suppose that $f(x)$ is a function that satisfies[2]

$$|f(x) - f(y)| \le L|x - y|$$

for some constant L, and $x(t)$ is the exact solution of

$$\frac{dx}{dt} = f(x) \qquad x(0) = y_0.$$

---

[2] This condition, that f be a *Lipschitz* function, is in fact what is required to ensure that the equation $\dot{x} = f(x)$ has unique solutions, see equation (6.5).

Now, if x_n is the solution of the Euler method

$$x_{n+1} = x_n + hf(x_n) \qquad \text{with} \qquad x_0 = y_0$$

then x_n should approximate $x(t_n)$ (where $t_n = nh$). Writing $\tilde{x}(t_n)$ for x_n, given any $T > 0$ it is possible to prove the *error estimate*

$$\max_{0 \le t_n \le T} |\tilde{x}(t_n) - x(t_n)| \le Kh, \tag{21.7}$$

for some constant K (depending on T). This says that as h is made smaller, the error between the approximation and the true solution can be guaranteed to decrease over the whole time interval.

Because the Euler method is relatively simple, as h is made smaller the error decreases fairly slowly; halving h will only halve the error. Numerical methods that are used in practice generally have much better error properties. One popular method is the Runge–Kutta scheme, where

$$x_{n+1} = x_n + \frac{h}{6}(f_1 + 2f_2 + 2f_3 + f_4)$$

with $f_1, \ldots, f_4$ given by

$$f_1 = f(x_n, t_n)$$
$$f_2 = f\left(x_n + \tfrac{1}{2}hf_1, t_n + \tfrac{1}{2}h\right)$$
$$f_3 = f\left(x_n + \tfrac{1}{2}hf_2, t_n + \tfrac{1}{2}h\right)$$
$$f_4 = f(x_n + hf_3, t_n + h).$$

Although this method appears much more complicated, implementing such a scheme computationally is fairly straightforward, and now the error satisfies

$$\max_{0 \le t_n \le T} |\tilde{x}(t_n) - x(t_n)| \le Kh^4.$$

This means that halving the timestep will increase the accuracy of the method by a factor of 16. MATLAB's `ode45` routine uses a refined version of this method.

Exercises

21.1 Apply Euler's method to the general linear equation $\dot{x} = \lambda x$. Find the approximation x_n, and using (21.6) show that as $h \to 0$ the numerical solution converges to the true solution.

21.2 There are variants of the Euler method that have the advantage of better stability properties, but have the disadvantage of no longer being explicit schemes. For example, the *backwards Euler* method is

$$x_{n+1} = x_n + hf(x_{n+1}, t_{n+1}),$$

which has to be solved at each stage to find x_{n+1} in terms of x_n. Apply this method to the linear equation $\dot{x} = x$, and show that once again the method converges to the true solution $x(t) = e^t$ as $t \to \infty$.

21.3 Another variant of the standard Euler method is the trapezoidal Euler method. If $x(t)$ is the solution of $\dot{x} = f(x, t)$ then we have

$$x(t + h) = x(t) + \int_t^{t+h} f(x(s), s)\, ds.$$

Use the trapezium rule to approximate the integral to derive this scheme,

$$x_{n+1} = x_n + h\left[\tfrac{1}{2} f(x_n, t_n) + \tfrac{1}{2} f(x_{n+1}, t_{n+1})\right].$$

21.4 (T) Since $(d/dx)e^x = e^x$, if we calculate the derivative of e^x at $x = 0$ as a limit it follows that

$$\lim_{h \to 0} \frac{e^h - 1}{h} = 1.$$

By rearranging this (note that $1 = \lim_{h \to 0} 1$) show that

$$e = \lim_{h \to 0} (1 + h)^{1/h},$$

and hence that

$$e^x = \lim_{n \to \infty} \left(1 + \frac{x}{n}\right)^{1/n}.$$

(Hint: if

$$\lim_{h \to 0} f(h) = \lim_{h \to 0} g(h) = y,$$

and $\kappa(x)$ is continuous at $x = y$, then

$$\lim_{h \to 0} \kappa[f(h)] = \lim_{h \to 0} \kappa[g(h)].$$

You will need to use this once for each step.)

21.5 (T) In this question we suppose that f satisfies the Lipschitz condition

$$|f(x) - f(y)| \le L|x - y|$$

and consider the Euler θ-method for approximating solutions of $\dot{x} = f(x)$,

$$x_{n+1} = x_n + h[(1 - \theta) f(x_n) + \theta f(x_{n+1})].$$

For $\theta = 0$ this is the standard Euler method; for $\theta = \tfrac{1}{2}$ this is the trapezoidal method; and for $\theta = 1$ this is the 'backwards Euler' method. Since x_{n+1} is not given explicitly as a function of x_{n+1}, we need a reliable way of calculating it numerically.

(i) The first thing we must check is that there is a unique solution for x_{n+1}. Suppose that

$$y = x_n + h[(1 - \theta) f(x_n) + \theta f(y)] \qquad \text{and}$$

$$z = x_n + h[(1 - \theta) f(x_n) + \theta f(z)],$$

i.e. that both y and z satisfy the equation. By subtracting these two equations show that

$$y - z = \theta h[f(y) - f(z)],$$

and hence deduce that

$$|y - z| \leq hL\theta|y - z|,$$

and therefore that $y = z$ provided that $h < 1/L\theta$.

(ii) Suppose therefore that $h < 1/L\theta$. Given an initial guess y_0 for x_{n+1}, we can refine this guess successively by setting

$$y_{j+1} = x_n + h[(1 - \theta)f(x_n) + \theta f(y_j)]; \qquad \text{(E21.1)}$$

if $y_{j+1} = y_j = y$ then

$$y = x_n + h[(1 - \theta)f(x_n) + \theta f(y)],$$

and so y would be the required value for x_{n+1}. Show that

$$|y_{j+1} - y_j| \leq hL\theta|y_j - y_{j-1}|, \qquad \text{(E21.2)}$$

and hence that successive values of y_j are closer together. Thus, for large j, we would expect that $y_{j+1} \approx y_j$ and that y_j is a good approximation to x_{n+1}.

(iii) Still assuming that $h < 1/L\theta$, use (E21.2) to show that

$$|y_{j+1} - y_j| \leq (hL\theta)^j|y_1 - y_0|,$$

and hence that

$$|y_j - y_k| \leq \frac{(hL\theta)^J}{1 - hL\theta}|y_1 - y_0|$$

for any $j, k \geq J$.

It follows that $\{y_j\}$ is a Cauchy sequence, and so converges to a limit y. Taking limits as $j \to \infty$ on both sides of (E21.1) we get

$$y = x + \tfrac{1}{2}h[f(x_n) + f(y)],$$

and thus $x_{n+1} = y$.

21.6 (C) For a number of values of t and h compare the exact solution of $\dot{x} = x$ with the solution from Euler's method, and verify the error estimate in (21.7).

21.7 (C) Implement the backwards Euler scheme of Exercise 21.2 numerically, and apply it to the equation $\dot{x} = x(1 - x)$ to find the solution when $x(0) = \frac{1}{2}$ for $0 \leq t \leq 8$. In order to find x_{n+1} given x_n you can use the approach of Exercise 21.5, and iterate

$$g_{k+1} = x_n + hg_k(1 - g_k)$$

to give a succession of 'guesses' g_k for x_{n+1} until g_k appears to stabilise (e.g. until $|g_{k+1} - g_k| < h^3$). You will need to choose h carefully to ensure that your sequence of guesses converges. (Can you work out, using the theoretical results of

Exercise 21.5, what value of h should suffice?) The MATLAB M-file `backeuler.m`, implementing this scheme, can be downloaded from the web.

21.8 (C) Write a MATLAB program to implement the Runge–Kutta method introduced at the end of the chapter. Apply this method to $\dot{x} = t - x^2$ when $h = 0.5$, and compare this to the solution obtained using Euler's method with the same timestep. (You can download the MATLAB M-file `rungekutta.m` from the web if you wish.)

22

Difference equations

It is not only for numerical approximations that it is more appropriate to have an independent variable that only takes discrete values. For example, an experiment may take measurements at equally spaced time intervals, we may be interested in the size of a population in successive generations, or perhaps we want to compare the value of the Financial Times index of the top one hundred UK shares (the FTSE) at the end of trading on a sequence of successive days (see Figure 22.1). In all these cases, it is much more natural to have a dependent variable indexed by n, where n is an integer (x_n), than a continuous function of t.

An equation that relates the values of x_n for different values of n is called a *difference equation*, and the order of a difference equation is the largest difference between any two of the indices (attached to x) occurring in the equation, i.e. Euler's method

$$x_{n+1} = x_n + hf(x_n, nh)$$

is a first order difference equation, as is

$$x_{n+1} = 2^{n+7} + \cos x_n$$

(the $n + 7$ is not an index of x), while

$$x_{n+2} = x_{n+1}^2 - \exp(x_{n-3})$$

is fifth order. We have already seen some examples of difference equations in the recurrence relations for the coefficients in our power series expansions in Chapter 20.

22.1 First order difference equations

A first order difference equation relates the 'next' value of x and its current value,

$$F(x_n, x_{n+1}, n) = 0.$$

213

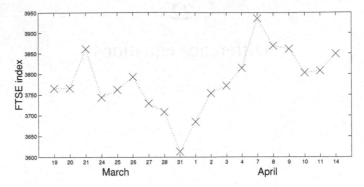

Fig. 22.1. The FTSE index during the period of the second Gulf War, from the start of hostilities on the evening of 19 March 2003 until the fall of Tikrit on 14 April 2003. The data consists of the values marked by crosses; these are joined by a dotted line to make the trends easier to follow.

We will only consider here[1] equations in which x_{n+1} can be given explicitly in terms of x_n,

$$x_{n+1} = f(x_n, n).$$

For such equations we do not need a theorem to show us that there is a unique solution, since we have a fixed rule by which we can construct the solution. In this chapter we consider only linear difference equations, and turn to nonlinear difference equations in the next chapter.

If we consider a simple linear difference equation

$$x_{n+1} = kx_n$$

and suppose that we know x_0 then it is simple to find x_n for any n. We have

$$x_1 = kx_0,$$
$$x_2 = kx_1 = k(kx_0) = k^2 x_0,$$
$$x_3 = kx_2 = k(k^2 x_0) = k^3 x_0,$$

and it is easy to see that in general $x_n = k^n x_0$ (we did something very similar in Section 21.4 of the previous chapter).

The analysis of second order linear difference equations that we will now pursue is entirely analogous to that of second order linear differential equations, except

---

[1] There are very natural difference equations for which this in not the case. For example, the 'backwards Euler method', which has nicer properties than the Euler method introduced in the previous chapter, is

$$x_{n+1} = x_n + hf(x_{n+1}).$$

This 'implicit' scheme has to be solved for x_{n+1} at every timestep as in Exercise 21.5.

that the exponential function $x(t) = e^{kt}$ (the solution of $\dot{x} = kx$) that we used for differential equations is now replaced by $x_n = k^n$ (the solution of $x_{n+1} = kx_n$).

22.2 Second order difference equations: complementary function and particular solution

We will concentrate now on linear second order difference equations,

$$x_{n+2} + ax_{n+1} + bx_n = f_n, \tag{22.1}$$

although the techniques that we will introduce extend easily to higher orders.

Just as we could split the problem of solving a linear second order *differential* equation into finding the complementary function and then a particular integral, we can split the problem of solving (22.1) into finding a solution of the corresponding homogeneous problem

$$y_{n+2} + ay_{n+1} + by_n = 0, \tag{22.2}$$

and then of finding one particular solution x_n of (22.1).

22.3 The homogeneous equation

First we deal with the homogeneous equation,

$$ax_{n+2} + bx_{n+1} + cx_n = 0. \tag{22.3}$$

We guess that the solution is of the form $x_n = k^n$, just as we guessed a solution of the form $x(t) = e^{kt}$ for the second order differential equation $a\ddot{x} + b\dot{x} + cx = 0$.
Trying $x_n = k^n$ in (22.3) we get

$$ak^{n+2} + bk^{n+1} + ck^n = 0,$$

and cancelling k^n then gives the auxiliary equation

$$ak^2 + bk + c = 0.$$

We obtain a quadratic equation for k, and the form of solution for (22.2) will depend on the nature of its roots.

22.3.1 Distinct real roots

If the auxiliary equation has two distinct real roots k_1 and k_2 then $x_n = k_1^n$ and $x_n = k_2^n$ are both solutions of (22.3), and so the general solution is

$$x_n = Ak_1^n + Bk_2^n.$$

Particular values of x_0 and x_1 will enable us to determine the constants A and B.

As an example, we find an expression for the *nth* Fibonacci number. These are the numbers that satisfy

$$x_n = x_{n-1} + x_{n-2}, \qquad\qquad (22.4)$$

and start with $x_0 = 1$ and $x_1 = 1$. The first few are

$$1 \quad 1 \quad 2 \quad 3 \quad 5 \quad 8 \quad 13 \quad 21 \quad 34 \quad 55 \quad \cdots$$

These numbers crop up frequently in nature, and have fascinated artists for hundreds of years. A modern example is shown in Figure 22.2.

Fig. 22.2. Mario Merz' 'Fibonacci Sequence 1–55' on the chimney of the power station in Turku, Finland. Photograph by Dr Ching-Kuang Shene of Michigan Technological University, and reproduced with his kind permission.

To solve (22.4) we try $x_n = k^n$, and find

$$k^2 = k + 1.$$

This equation has roots

$$k = \frac{1 \pm \sqrt{5}}{2},$$

so that the general solution of (22.4) is

$$x_n = \alpha \left(\frac{1 + \sqrt{5}}{2} \right)^n + \beta \left(\frac{1 - \sqrt{5}}{2} \right)^n.$$

The initial conditions require

$$\alpha + \beta = 1 \qquad (1 + \sqrt{5})\alpha + (1 - \sqrt{5})\beta = 2,$$

and thus, solving for α and β,

$$\alpha = \frac{1 + \sqrt{5}}{2\sqrt{5}} \qquad \beta = \frac{\sqrt{5} - 1}{2\sqrt{5}}.$$

The nth Fibonacci number is therefore given by

$$x_n = \frac{1}{\sqrt{5}} \left(\frac{1 + \sqrt{5}}{2} \right)^{n+1} - \frac{1}{\sqrt{5}} \left(\frac{1 - \sqrt{5}}{2} \right)^{n+1}. \qquad (22.5)$$

It is somewhat surprising that this formula always gives an integer. Since the second term is always smaller than $1/\sqrt{5} \approx 0.4472$, we can also write

$$x_n = \left[\left[\frac{1}{\sqrt{5}} \left(\frac{1 + \sqrt{5}}{2} \right)^{n+1} \right] \right],$$

where $[[x]]$ denotes the nearest integer to x.

22.3.2 Repeated roots

If the auxiliary equation has a repeated real root k, then we have a similar problem to the one we had when considering second order differential equations, since we will only obtain a single solution $x_n = k^n$. Thankfully the resolution of this difficulty is also similar: we introduce the extra factor of n where before we had an extra factor of t. The general solution is therefore

$$x_n = Ak^n + Bnk^n.$$

To check that this second solution is correct we will substitute it into

$$x_{n+2} - 2kx_{n+1} + k^2 x_n = 0,$$

since any second order linear difference equation whose auxiliary equation has k as a repeated root can be rewritten in this form (see Exercise 22.2). Trying $x_n = nk^n$ in the left-hand side gives

$$(n+2)k^{n+2} - 2k(n+1)k^{n+1} + k^2 nk^n = (n+2 - 2(n+1) + n)k^{n+2} = 0$$

as required.

Example 22.1 *Find the general solution of*

$$x_n - 4x_{n-1} + 4x_{n-2} = 0.$$

Trying $x_n = k^n$ yields the auxiliary equation

$$k^2 - 4k + 4 = 0,$$

so that $k = 2$ 'twice'. The general solution is therefore

$$x_n = \alpha 2^n + \beta n 2^n.$$ □

22.3.3 Complex roots

We can also have complex roots, for which things are a little more involved. If we have $k = a \pm ib$, then we need to write k in modulus and argument form,

$$k = re^{\pm i\theta},$$

where

$$r^2 = a^2 + b^2 \qquad \text{and} \qquad \theta = \tan^{-1}(b/a)$$

(see Appendix A). Then the solution is

$$x_n = r^n[A \cos n\theta + B \sin n\theta].$$

To see that this solution is consistent with our guess $x_n = k^n$, first write

$$x_n = C[re^{i\theta}]^n + C^*[re^{-i\theta}]^n,$$

where we take the coefficients to be $C = \alpha + i\beta$ and C^* (its complex conjugate) to ensure that x_n is real. This gives

$$x_n = r^n[Ce^{in\theta} + C^*e^{-in\theta}].$$

Since $z^* + z = 2\,\mathrm{Re}(z)$, and

$$Ce^{in\theta} = (\alpha + i\beta)(\cos n\theta + i \sin n\theta)$$
$$= (\alpha \cos n\theta - \beta \sin n\theta) + i(\beta \cos n\theta + \alpha \sin n\theta),$$

we get

$$x_n = r^n[A \cos n\theta + B \sin n\theta]$$

if we take $A = 2\alpha$ and $B = -2\beta$.

Example 22.2 *Find the general solution of the difference equation*

$$x_{n+2} - 2x_{n+1} + 2x_n = 0.$$

Trying $x_n = k^n$ gives the quadratic equation

$$k^2 - 2k + 2 = 0,$$

and so

$$k = \frac{2 \pm \sqrt{4-8}}{2} = 1 \pm i.$$

Since

$$1 \pm i = \sqrt{2}e^{\pm i\pi/4}$$

the solution is

$$x_n = 2^{n/2}[A \cos(n\pi/4) + B \sin(n\pi/4)]. \qquad \square$$

22.4 Particular solutions

When we have an equation with a non-zero right-hand side,

$$ax_{n+2} + bx_{n+1} + cx_n = f_n,$$

we use the same method that we had for differential equations, i.e. we guess the form of the particular solution and then substitute in to determine the constants in our guess.

22.4.1 Right-hand side f_n is a polynomial in n

When the right-hand side is a polynomial depending on n the appropriate 'guess' for a particular solution is a general polynomial of the same order as the right-hand side. If our guess solves the homogeneous problem then we have to multiply by an additional factor of n.

We start with a simple first order example.

Example 22.3 *Find the general solution of*

$$x_{n+1} = kx_n + a.$$

The solution of the homogeneous equation

$$y_{n+1} = ky_n,$$

is $y_n = Ak^n$. To find a particular solution we can try $x_n = c$ and then we require

$$c = kc + a,$$

so we take $c = a/(1 - k)$ and obtain the general solution

$$x_n = Ak^n + a/(1 - k). \qquad \square$$

Example 22.4 *Find the general solution of*

$$x_n - x_{n-1} - 6x_{n-2} = -36n.$$

First we solve the homogeneous equation $y_n - y_{n-1} - 6y_{n-2} = 0$ by trying $y_n = k^n$; we need k to solve the equation

$$k^2 - k - 6 = 0 \qquad \Rightarrow \qquad (k - 3)(k + 2) = 0,$$

and so the complementary function is $y_n = A3^n + B(-2)^n$. For the particular solution we try a general first order polynomial in n, $x_n = \alpha n + \beta$. Substituting in we get

$$\alpha n + \beta - (\alpha(n - 1) + \beta) - 6(\alpha(n - 2) + \beta) = -6\alpha n + 13\alpha - 6\beta.$$

So we need $\alpha = 6$ and $\beta = 13$ which gives the particular solution $x_n = 6n + 13$; the general solution is therefore

$$x_n = A3^n + B(-2)^n + 6n + 13. \qquad \square$$

Example 22.5 *Find the general solution of*

$$x_{n+1} - 2x_n + x_{n-1} = 8.$$

To solve the homogeneous equation $y_{n+1} - 2y_n + y_{n-1} = 0$ we try $y_n = k^n$ and obtain the auxiliary equation

$$k^2 - 2k + 1 = 0.$$

This equation has the repeated root $k = 1$, and so the complementary function is

$$y_n = A + Bn.$$

We cannot try $x_n = c$ for our particular solution (A is part of the complementary function), nor can we try $x_n = cn$ (since Bn is also part of the complementary function), so we have to try $x_n = cn^2$. Then we need

$$c(n+1)^2 - 2cn^2 + c(n-1)^2 = c[n^2 + 2n + 1 - 2n^2 + n^2 - 2n + 1] = 2c = 8,$$

i.e. $c = 4$. So a particular solution is $x_n = 4n^2$, and the general solution is

$$x_n = 4n^2 + A + Bn. \qquad \square$$

22.4.2 Right-hand side $f_n = \lambda^n$

This case is similar to having an exponential on the right-hand side of a differential equation. If λ is not a solution of the auxiliary equation we try $x_n = \alpha\lambda^n$; if λ is a non-repeated root of the auxiliary equation we try $x_n = \alpha n\lambda^n$, while if λ is a repeated root we have to try $x_n = \alpha n^2\lambda^n$.

Example 22.6 *Find the general solution of the equation*

$$x_{n+2} + x_{n+1} - 6x_n = 12(-2)^n.$$

To find the solution of the homogeneous equation $y_{n+2} + y_{n+1} - 6y_n = 0$ we try $y_n = k^n$ and obtain the auxiliary equation

$$k^2 + k - 6 = 0 \quad \Rightarrow \quad (k+3)(k-2) = 0,$$

so that $k = 2$ or $k = -3$ and the complementary function is

$$y_n = A2^n + B(-3)^n.$$

Since $(-2)^n$ is not a solution of the homogeneous equation we can try $x_n = \alpha(-2)^n$ for a particular solution; we need

$$\alpha(-2)^{n+2} + \alpha(-2)^{n+1} - 6\alpha(-2)^n = 12(-2)^n.$$

Cancelling a factor of $(-2)^n$ we require

$$(-2)^2\alpha + (-2)\alpha - 6\alpha = 12,$$

or $\alpha = -3$. So a particular solution is $x_n = -3(-2)^n$, and the general solution is

$$x_n = A2^n + B(-3)^n - 3(-2)^n. \qquad \square$$

Example 22.7 *Find the general solution of the equation*

$$x_{n+2} + x_{n+1} - 6x_n = 30 \times 2^n.$$

We found the complementary function above, $y_n = A2^n + B(-3)^n$. Since the right-hand side occurs in the complementary function we have to try $x_n = \alpha n2^n$

for a particular integral. Substituting in we need

$$\alpha(n+2)2^{n+2} + \alpha(n+1)2^n - 6\alpha n2^n = 30 \times 2^n,$$

or, cancelling a factor of 2^n,

$$\alpha[4(n+2) + 2(n+1) - 6n] = 10\alpha \quad = \quad 30.$$

So $\alpha = 3$; a particular solution is $x_n = 8n2^n$, and the general solution is

$$x_n = A2^n + B(-3)^n + 3n2^n. \qquad \square$$

Exercises

22.1 Find the solutions of the following difference equations satisfying the given initial conditions.

(i) $x_{n+2} - 4x_{n+1} + 3x_n = 0$ with $x_0 = 0$ and $x_1 = 1$;

(ii) $2x_{n+1} - 3x_n - 2x_{n-1} = 0$ with $x_1 = x_2 = 1$;

(iii) $x_{n+2} = 2x_{n+1} - 2x_n$ with $x_0 = 1$ and $x_1 = 2$;

(iv) $x_{n+2} + 6x_{n+1} + 9x_n = 0$ with $x_0 = 1$ and $x_1 = 6$;

(v) $2x_n = 3x_{n-1} - x_{n-2}$ with $x_0 = 3$ and $x_1 = 2$; and

(vi) $x_{n+2} - 2x_{n+1} + 5x_n = 0$ with $x_0 = \sqrt{5}$ and $x_1 = 5 \cos \tan^{-1} 2$.

22.2 Show that if the auxiliary equation

$$ak^2 + bk + c = 0$$

has a repeated root $k = \lambda$ then the difference equation

$$ax_{n+2} + bx_{n+1} + cx_n = 0$$

can be rewritten in the form

$$x_{n+2} - 2\lambda x_{n+1} + \lambda^2 x_n = 0.$$

22.3 The 'golden ratio' is the ratio (greater than one) of the sides of a rectangle with the following property: remove a square whose sides are the length of the shorter side of the rectangle, and the remaining rectangle is similar to the original one (its sides are in the same ratio), see Figure 22.3. This ratio was used by the Greeks in constructing the Parthenon (among many other monuments), and has been a favourite tool of artists ever since.

Suppose that $\{x_n\}$ is a sequence of numbers satisfying the recurrence relation

$$x_{n+2} = x_{n+1} + x_n.$$

Show that if all the elements of the sequence are integers then the ratio of consecutive terms, x_{n+1}/x_n, converges to the golden ratio. Show that the same result is true if all the terms in the sequence have the same sign. (In particular this is true for the Fibonacci numbers, which have $x_0 = 0$ and $x_1 = 1$.)

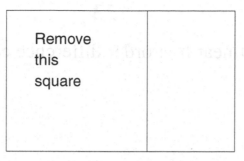

Fig. 22.3. The golden rectangle.

22.4 Find the general solution of the following difference equations, and then find the solution that satisfies the specified initial conditions.

(i) $x_{n+2} - 4x_n = 27n^2$, with $x_0 = 1$ and $x_1 = 3$;

(ii) $x_{n+1} - 4x_n + 3x_{n-1} = 36n^2$, with $x_0 = 12$ and $x_1 = 0$;

(iii) $x_{n+1} - 4x_n + 3x_{n-1} = 2^n$, with $x_0 = -4$ and $x_1 = -6$;

(iv) $x_{n+1} - 4x_n + 3x_{n-1} = 3^n$, with $x_0 = 2$ and $x_1 = 13/2$;

(v) $x_{n+2} - 2x_{n+1} + x_n = 1$, with $x_0 = 3$ and $x_1 = 6$;

(vi) $x_{n+2} + x_n = 2^n$, with $x_0 = x_1 = 0$;

(vii) $x_{n+2} + x_{n+1} + x_n = c$ (the general solution is enough here).

22.5 Find the solution of the difference equation

$$x_{n+1} = x_n(1 + n)$$

with $x_1 = 1$. Now show that if $x_1 = c$ then

$$x_n = c \frac{\Gamma(c + n)}{\Gamma(c + 1)}$$

where the Γ function, which was defined in Exercise 20.8, satisfies $\Gamma(x + 1) = x\Gamma(x)$.

23

Nonlinear first order difference equations

In general the solutions of a nonlinear difference equation

$$x_{n+1} = f(x_n) \tag{23.1}$$

can have very complicated behaviour. To find the solution of such a difference equation we have to iterate (apply repeatedly) the map f ('map' is just another word for function, and is frequently used in this context). The solution is given by the sequence of iterates

$$x_0, \quad x_1 = f(x_0), \quad x_2 = f(f(x_0)), \quad x_3 = f(f(f(x_0))), \quad \ldots,$$

called the 'orbit' of x_0. Since these nested fs rapidly become unmanageable we adopt the notation $f^n(x)$ to mean f applied n times to x,

$$f^n(x) = \underbrace{f(f(f(\cdots f(x) \cdots)))}_{n \text{ times}}.$$

We can write the 'solution' of (23.1) that has $x_0 = y_0$ as $x_n = f^n(y_0)$, but this is clearly no more descriptive of the solution than (23.1) itself.

23.1 Fixed points and stability

In order to describe the dynamics of solutions we make use of similar concepts as we used in Chapter 7 for the one-dimensional dynamical systems that arise from autonomous differential equations. In particular it is often useful to concentrate on what happens to solutions 'eventually'.

For an iterated map such as (23.1) a *fixed point* is a point x^* such that

$$f(x^*) = x^*,$$

so that if $x_n = x^*$ then $x_{n+1} = x^*$. (The fixed points of $x_{n+1} = f(x_n)$ are analogous to the 'stationary points' of $\dot{x} = f(x)$.)

We describe a fixed point as *stable* if you stay close to it provided that you start sufficiently near: x^* is stable if for any $\epsilon > 0$ there exists a $\delta > 0$ such that

$$\underbrace{|x_0 - x^*| < \delta}_{\text{start near}} \quad \Rightarrow \quad \underbrace{|f^n(x_0) - x^*| < \epsilon \quad \text{for all} \quad n = 0, 1, 2, \ldots}_{\text{stay near}}.$$

Again, we can also introduce the related but distinct concept of being attracting ('start near tend to'): x^* is *attracting* if there is a $\delta > 0$ such that

$$\underbrace{|x_0 - x^*| < \delta}_{\text{start close enough}} \quad \Rightarrow \quad \underbrace{f^n(x_0) \to x^* \quad \text{as} \quad n \to \infty}_{\text{tend to}}.$$

Negating 'stable' we get unstable: x^* is *unstable* if there exists an ϵ such that no matter how small we make δ, we can find an x_0 with

$$|x_0 - x^*| < \delta \quad \text{but} \quad |f^n(x_0) - x^*| > \epsilon \quad \text{for some } n > 0.$$

In order to discover analytically whether or not a fixed point is stable, suppose that $x_n = x^* + \delta_n$ where δ_n is small; then, using a Taylor series expansion about $x = x^*$,

$$x_{n+1} = f(x^* + \delta_n)$$
$$\approx f(x^*) + f'(x^*)\delta_n$$
$$= x^* + f'(x^*)\delta_n.$$

So if we write $x_{n+1} = x^* + \delta_{n+1}$ we have

$$\delta_{n+1} \approx f'(x^*)\delta_n. \tag{23.2}$$

The solution of (23.2) is $\delta_n = [f'(x^*)]^n \delta_0$, so it is clear that successive values of δ_j will decrease if $|f'(x^*)| < 1$, and increase if $|f'(x^*)| > 1$. So x^* is stable and attracting if $|f'(x^*)| < 1$, and unstable if $|f'(x^*)| > 1$. (This can be made rigorous, see Exercise 23.4.)

23.2 Cobweb diagrams

In order to work out what happens when we iterate f we can use a graphical method which produces a picture known as a 'cobweb diagram'. First we draw the graph of $f(x)$ against x, and then the graph of $y = x$. Now, given x_n we can find x_{n+1} by drawing a line up to $f(x_n)$, then over to $y = x$, and back down to the x-axis to give x_{n+1}. To find x_{n+2} we do the same, but starting at x_{n+1} – we start now from the point on the diagonal, rather than the point on the x-axis, since this simplifies the picture – and so on. This is illustrated in Figure 23.1, for the example $f(x) = 3x(1 - x)$.

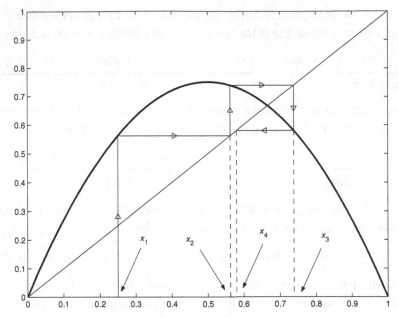

Fig. 23.1. The first few steps of the 'cobweb' method for $x_{n+1} = 3x_n(1 - x_n)$.

23.3 Periodic orbits

For any choice of x_0, the linear difference equation

$$x_{n+1} = -x_n$$

generates an orbit that flips between the two values x_0 and $-x_0$. This is a simple example of a *periodic orbit of period 2*, or more concisely a *period 2 orbit*. In general if $x_{n+1} = f(x_n)$ then a periodic orbit of period 2 is a pair of values x_1 and x_2 such that

$$f(x_1) = x_2 \qquad \text{and} \qquad f(x_2) = x_1,$$

so that $f^2(x_1) = x_1$, see Figure 23.2.

A periodic orbit of period k (or a period k orbit) is a sequence of k values $\{x_1, \ldots, x_k\}$ such that

$$f(x_j) = x_{j+1} \qquad \text{for} \qquad j = 1, \ldots, n-1 \qquad \text{and} \qquad f(x_n) = x_1,$$

so that iterates of x_1 cycle around these k values for ever. (Strictly we also need to make sure that $f(x_j) \neq x_1$ for $j = 1, \ldots, n-1$, so that k is the 'minimal period' of the orbit.) We will see several examples of such periodic orbits in the next chapter.

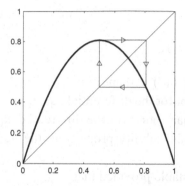 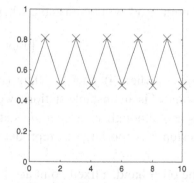

Fig. 23.2. A period 2 orbit for the map $x_{n+1} = (1 + \sqrt{5})x_n(1 - x_n)$. On the left is the cobweb diagram, while the right-hand picture shows successive values of x_n against n.

23.4 Euler's method for autonomous equations

In the remainder of this chapter we apply some of the above ideas in order to understand how well the qualitative behaviour of the differential equation

$$\mathrm{d}x/\mathrm{d}t = f(x) \tag{23.3}$$

(which we studied in detail in Chapter 7) can be captured using the numerical Euler method introduced in Chapter 21. Recall that the stationary points x^* of (23.3) occur when $f(x^*) = 0$, and that they are stable if $f'(x^*) < 0$ and unstable if $f'(x^*) > 0$.

If we apply the Euler method to (23.3) with a timestep h then we have

$$x_{n+1} = x_n + hf(x_n), \tag{23.4}$$

where x_n is an approximation to $x(t_n)$ with $t_n = nh$. We will write $g(x) = x + hf(x)$, so that (23.4) can be written more concisely as

$$x_{n+1} = g(x_n).$$

The fixed points of the map g are those x values x^* for which $x^* = g(x^*)$, i.e. for which

$$x^* = x^* + hf(x^*).$$

This means that the fixed points of g occur when $f(x^*) = 0$, i.e. they are the same as the stationary points of the differential equation (23.3).

To determine the stability of the fixed points we have to consider the modulus of

$$g'(x) = 1 + hf'(x).$$

A fixed point at x^* will be unstable whenever

$$|g'(x^*)| > 1.$$

This happens when (i) $f'(x^*) > 0$, or (ii) $f'(x^*) < 0$ and $h > 2/|f'(x^*)|$. In case (i) the point x^* is an unstable stationary point for the differential equation, but case (ii) says that although x^* is a *stable* stationary point for the differential equation, the timestep h is too large to reproduce this stability property in the numerical method.

On the other hand, a fixed point at x^* is stable provided that

$$|g'(x^*)| < 1$$

which happens when $f'(x^*) < 0$ and $h < 2/|f'(x^*)|$. So x^* has to be a stable stationary point of the differential equation *and* the timestep h has to be sufficiently small.

It follows that the stationary points of a differential equation such as (23.3), along with their stability properties, will be reproduced correctly by the Euler method provided that h is small enough so that $h < 2/|f(x^*)|$ for every stationary point x^*.

We now look at these phenomena as they occur in a particular example.

23.4.1 An example

We will apply Euler's method to the equation

$$dx/dt = x(k - x^2) \qquad \text{with} \qquad k > 0, \tag{23.5}$$

which gives

$$x_{n+1} = x_n + hx_n(k - x_n^2). \tag{23.6}$$

We will write $f(x) = x(k - x^2)$ and $g(x) = x + hx(k - x^2)$.

For $k > 0$ equation (23.5) has an unstable stationary point at the origin and stable stationary points at $x = \pm\sqrt{k}$, as shown in the phase diagram in Figure 23.3 (cf. Figure 7.13).

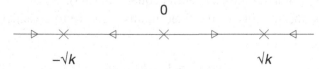

Fig. 23.3. Phase diagram for equation (23.5).

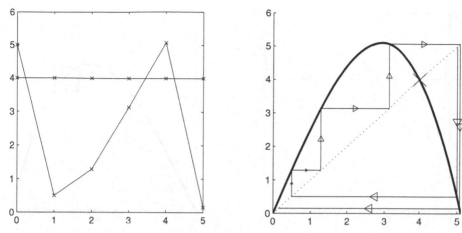

Fig. 23.4. Euler's method applied to $\dot{x} = x(16 - x^2)$ with $h = 0.1$ and initial conditions $x_0 = 4$ and $x_0 = 5$. The timestep is too large to preserve the correct stability properties of the point $x = 4$. [In this and the following three figures, the left-hand side shows successive values of x_n against n, while the right-hand side shows the cobweb diagram of the same orbit.]

The Euler scheme has fixed points at 0, $-\sqrt{k}$, and $\sqrt{k}$, and their stability is determined by the size of

$$|g'(x)| = |1 + h(k - 3x^2)|.$$

Since we have

$$|g'(0)| = 1 + kh \qquad \text{and} \qquad |g'(\pm\sqrt{k})| = |1 - 2hk|,$$

the fixed point at the origin is always unstable, while the fixed points at $x = \pm\sqrt{k}$ are stable if $h < 1/k$ and unstable if $h > 1/k$.

We now fix $k = 16$ and choose three values for h, concentrating on the behaviour of positive solutions. First, if $h = 0.1$ then $h = 0.1 > 1/16 = 1/k$, and the fixed point at $x = 4$ is unstable. The successive values of x_n, along with the cobweb diagram, are shown in Figure 23.4.

If we decrease h a little to $h = 0.065$, then it is still greater than $1/16$. Figure 23.5 shows that the orbit has settled down to a period 2 orbit rather than to the fixed point.

If we now reduce h so that it is less than $1/16$ then we would expect $x = 4$ to become stable. Figure 23.6 shows the results of the method with $h = 0.06$. Although the fixed point at $x = 4$ is now stable, the sequence x_n oscillates about $x = 4$ as it converges. This is in contrast to the behaviour of the differential equation, where $x(t)$ would approach $x = 4$ monotonically.

Finally, Figure 23.7 shows the result of the method when $h = 0.01$. The timestep is now small enough so that the Euler method yields a very good approximation to the solution of the ordinary differential equation.

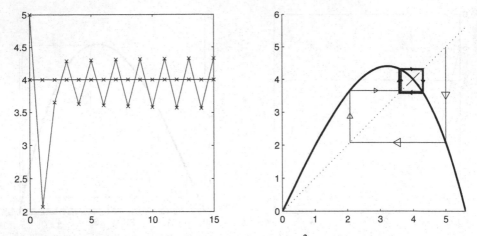

Fig. 23.5. Euler's method applied to $\dot{x} = x(16 - x^2)$ with $h = 0.065$. The solution has ended up switching between two values of x on a period 2 orbit.

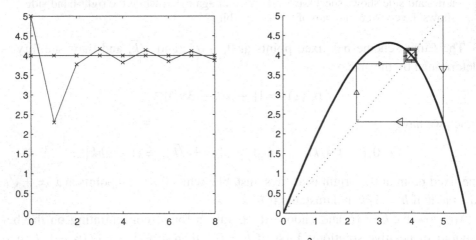

Fig. 23.6. Euler's method applied to $\dot{x} = x(16 - x^2)$ with $h = 0.06$ and initial conditions $x_0 = 4$ and $x_0 = 5$. The timestep is small enough so that $x = 4$ is stable, but solutions oscillate as they approach this fixed point.

Exercises

23.1 Show that there is an orbit of period 3 containing the point $x = 1$ for the difference equation

$$x_{n+1} = \frac{14}{3}x_n^2 - \frac{13}{2}x_n + \frac{7}{3}.$$

23.2 Suppose that the differential equation $\dot{x} = f(x)$ has a stationary point x^* where $f'(x^*) < 0$. We saw that the point x^* is a stable fixed point for

$$x_{n+1} = x_n + hf(x_n),$$

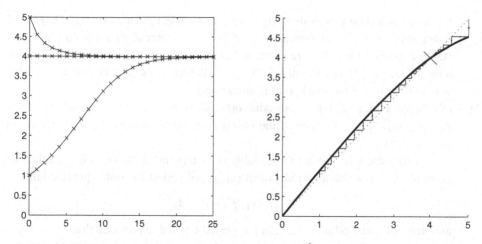

Fig. 23.7. Euler's method applied to $\dot{x} = x(16 - x^2)$ with $h = 0.01$ and initial conditions $x_0 = 1$, $x_0 = 4$ and $x_0 = 5$. The timestep is small enough so that the method approximates the solution of the differential equation very well. (Two orbits in addition to the fixed point are now shown in both diagrams.)

provided that $h < 2/|f'(x^*)|$. Assuming that x_0 is sufficiently close to x^*, show that if $h > 1/|f'(x^*)|$ then x_n is alternately greater than and less than x^*, while if $h < 1/|f'(x^*)|$ the orbit x_n approaches x^* monotonically.

23.3 In this question we consider the trapezoidal Euler method

$$x_{n+1} = x_n + \tfrac{1}{2}h[f(x_n) + f(x_{n+1})].$$

Show that $x_{n+1} = x_n = x^*$ if and only if $f(x^*) = 0$, i.e. that the fixed points of the numerical scheme coincide with the stationary points of the differential equation $\dot{x} = f(x)$.

Using the chain rule show that

$$\frac{dx_{n+1}}{dx_n} = \frac{1 + \tfrac{1}{2}hf'(x_n)}{1 - \tfrac{1}{2}hf'(x_{n+1})},$$

and hence that a fixed point x^* is stable if $f'(x^*) < 0$ and unstable if $f'(x^*) > 0$, i.e. that whatever the timestep the stability coincides with that of the corresponding stationary point in the differential equation.

23.4 (T) It follows from the definition of the derivative that

$$f(x^* + h) = f(x^*) + f'(x^*)h + o(h),$$

where $o(h)$ indicates that the remainder terms satisfy

$$\frac{o(h)}{h} \to 0 \qquad \text{as} \qquad h \to 0.$$

In particular, given $\epsilon > 0$ there exists a $\delta > 0$ such that

$$|o(h)| \le \epsilon h$$

for all $|h| \leq \delta$. Use this to show rigorously that a fixed point x^* of $x_{n+1} = f(x_n)$ is stable if $|f'(x^*)| < 1$ and unstable if $|f'(x^*)| > 1$. (Recall that a fixed point x^* is stable if given an $\epsilon > 0$ there exists a $\delta > 0$ such that whenever $|x_0 - x^*| < \delta$ we have $|f^n(x_0) - x^*| < \epsilon$ for all $n = 0, 1, \ldots$. In fact you should be able to show that when $|f'(x^*)| < 1$ the fixed point is attracting.)

23.5 (T) Suppose that f has a periodic orbit of period k consisting of the points $\{x_1, \ldots, x_k\}$. Show that each of the points on the orbit is a fixed point for the map $g(x) = f^k(x)$.

A periodic orbit is said to be stable if each point on the orbit is a stable fixed point of f^k. Show that a periodic orbit $\{x_1, x_2\}$ of period 2 is stable provided that

$$|f'(x_1)f'(x_2)| < 1,$$

and that a periodic orbit $\{x_1, \ldots, x_k\}$ of period k is stable provided that

$$|f'(x_1)f'(x_2)\cdots f'(x_{k-1})f'(x_k)| < 1.$$

Note in particular that if one point on the orbit is a stable fixed point of f^k then so are all the others.

23.6 (T) Consider the iterated map

$$y_{n+1} = f(y_n) = ry_n + y_n^2$$

for $r \leq 0$. Find the two fixed points, and show that the fixed point at $y = 0$ is stable for $-1 < r \leq 0$ and unstable for $r < -1$.

Show that if y lies on an orbit of period 2 then

$$y^2 + (r+1)y + (r+1) = 0,$$

and deduce that there is a period 2 orbit if $r < -1$. Hint: we must have $f^2(y) = y$, and you can factorise the resulting equation since $f(0) = 0$ and $f(1-r) = 1 - r$.

If y_1 and y_2 are the points on this orbit, show that

$$f'(y_1)f'(y_2) = 4 + 2r - r^2,$$

and hence that this orbit is stable for $1 - \sqrt{6} < r < -1$.

23.7 (C) Apply Euler's method with timestep h to the equation $\dot{x} = x(k - x)$ (cf. Exercise 7.6). Investigate how the stability of the fixed points depends on k and h. Now implement this Euler scheme numerically and verify your results (e.g. compare the cases $k = 1$ and $k = 3$ with timestep $h = 1$). (You could adapt the MATLAB M-file euler.m, which is available on the web.)

24

The logistic map

In this chapter we consider a particular example of a nonlinear difference equation, the logistic map

$$x_{n+1} = rx_n(1 - x_n). \tag{24.1}$$

Despite its simplicity the orbits of this equation can be extremely complicated, and this has made it one of the standard models in the theory of dynamical systems and chaos.

We will vary the parameter r between 0 and 4, since then if x_n lies between zero and one, so does x_{n+1} (the maximum value of $rx(1 - x)$ occurs when $x = \frac{1}{2}$ and is $r/4$). The graphs of $f(x) = rx(1 - x)$ are shown in Figure 24.1 for various values of r.

The equation can be thought of as a discrete model of a population (we saw a very similar differential equation earlier) with limited resources. If the population is small (measured, presumably, in thousands or millions to give an approximately continuous variable) then (24.1) predicts that its size will increase, since when $x_n \approx 0$,

$$x_{n+1} \approx rx_n.$$

But when x_n approaches the maximum size sustainable by the resources available (which here is 1), the population dies out rapidly; if $x_n = 1 - y_n$ with $y_n \approx 0$ then

$$x_{n+1} \approx ry_n,$$

so that most of the population dies out.

What one would naïvely expect from such a model is that the population size would settle down to some steady state (as in the solution of the differential equation $\dot{x} = rx(1 - x)$), or perhaps oscillate between two nearby states, and indeed this does happen when r is small. However, as r increases towards 4 this intuition can be very wrong.

233

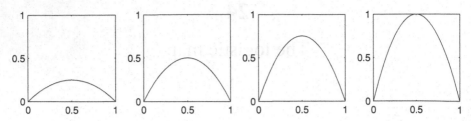

Fig. 24.1. Graphs of $rx(1-x)$ for (left to right) $r = 1, 2, 3$ and 4.

We will base our discussion around the existence of fixed points and periodic orbits. However, we will also make use of the idea of the 'attractor'; rather than defining it precisely here, the best way to think of it is as the points around which the orbits of $x_{n+1} = f(x_n)$ will move 'eventually'.

24.1 Fixed points and their stability

First note that there are at most two fixed points lying between zero and one. Fixed points are the solutions of

$$x = rx(1-x);$$

these are $x = 0$ and $x = 1 - (1/r)$. Since we want to consider positive populations, the non-zero fixed point will only be interesting when $r > 1$.

If we calculate $f'(x) = r(1 - 2x)$ then $f'(0) = r$; the fixed point at $x = 0$ will be stable while $r < 1$, and unstable once $r > 1$.

If $0 < r < 1$ then $1 - (1/r) < 0$ and so there is no positive fixed point, and over time the population decreases to zero (the interpretation of this being that the reproductive rate is not high enough to sustain the population). In this case the dynamics are very simple; $x = 0$ is stable and attracting, see Figure 24.2.

When $r > 1$ the origin is no longer stable, and there is another positive fixed point. Since

$$f'(1 - (1/r)) = r(1 - 2 + (2/r)) = 2 - r,$$

this fixed point is stable while $r < 3$. So for $1 < r < 3$ all orbits are attracted to $1 - (1/r)$, as shown in Figure 24.3.

24.2 Periodic orbits

When r increases beyond 3, things become more complicated. The fixed point at $1 - (1/r)$ is now unstable, since the derivative of f in that case has modulus larger

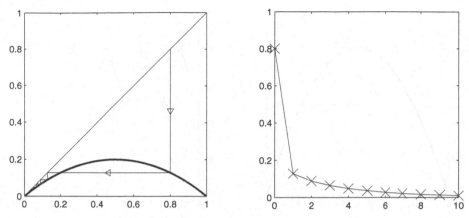

Fig. 24.2. For $0 < r < 1$ (here $r = 0.8$) the origin is a stable fixed point, and the population dies out. This, and all similar figures in this chapter, show the cobweb diagram of a representative orbit on the left, and the successive values of x_n against n for the same orbit on the right.

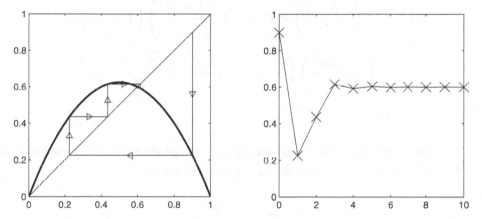

Fig. 24.3. For $1 < r < 3$ the origin is unstable, and there is an attracting non-zero fixed point. The pictures here are for $r = 2.5$.

than 1. If r is just a little larger than 3, almost every choice of initial condition (apart from either fixed point) ends up cycling between two different values of x on a period 2 orbit, as shown in Figure 24.4.

In the figure, $r = 1 + \sqrt{5}$, and one point on the period 2 orbit is $x = \frac{1}{2}$. To see that this really does give a period 2 orbit, we first calculate

$$f\left(\tfrac{1}{2}\right) = (1 + \sqrt{5})\tfrac{1}{2}\left(1 - \tfrac{1}{2}\right) = \tfrac{1+\sqrt{5}}{4},$$

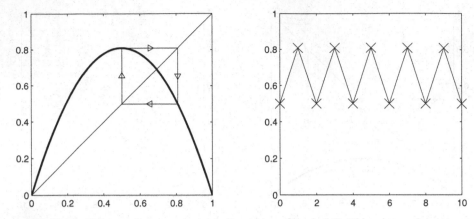

Fig. 24.4. For $r > 3$ there are periodic orbits of period 2. This picture has $r = 1 + \sqrt{5} \approx 3.2361$.

and then

$$f^2(\tfrac{1}{2}) = f\left(\frac{1+\sqrt{5}}{4}\right) = (1+\sqrt{5})\left(\frac{1+\sqrt{5}}{4}\right)\left(1-\frac{1+\sqrt{5}}{4}\right)$$

$$= \left(\frac{6+2\sqrt{5}}{4}\right)\left(\frac{3-\sqrt{5}}{4}\right) = \frac{18-6\sqrt{5}+6\sqrt{5}-10}{16}$$

$$= \tfrac{1}{2}.$$

If we try to find a period 2 orbit analytically for general r then we want to find a value of y such that if $x_n = y$ then $x_{n+2} = y$. So we want

$$y = f^2(y)$$

which is $y = f(ry(1-y))$ or

$$y = r[ry(1-y)][1 - ry(1-y)]. \tag{24.2}$$

This is a quartic (fourth order) equation for y; but since we know that $y = 0$ and $y = 1 - (1/r)$ must be solutions (they are fixed points with $f(y) = y$, so certainly $f(f(y)) = f(y) = y$) we can remove a factor $y(y - [1 - (1/r)])$. If y is a period 2 point it must therefore solve the equation

$$ry^2 - (1+r)y + \left(1 + \frac{1}{r}\right) = 0. \tag{24.3}$$

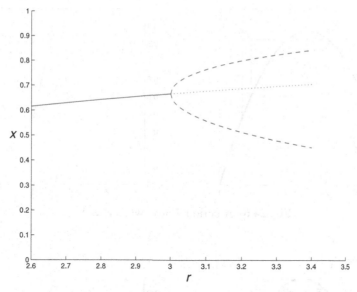

Fig. 24.5. As r passes through 3 the interior fixed point becomes unstable, and a stable orbit of period 2 appears. The stable fixed point is shown as a solid line, the unstable fixed point as a dotted line, and the two points on the stable period 2 orbit as a dashed curve.

This equation only has real roots if the discriminant is positive ($b^2 - 4ac > 0$), i.e. if

$$(1 + r)^2 - 4r \left(1 + \frac{1}{r}\right) = (1 + r)(r - 3) > 0.$$

Since we have restricted to the parameter range $0 \le r \le 4$ the factor $(1 + r)$ is positive; for a period 2 orbit we must therefore have $r > 3$.

Note that the two points on the periodic orbit (the solutions of (24.3)) are given by

$$\frac{(1 + r) \pm \sqrt{(1 + r)(r - 3)}}{2r}.$$

When $r = 3$ this would give $2/3$, which is the position of the positive fixed point. When $r > 3$ the fixed point at $1 - (1/r)$ is unstable, and the two points on the orbit 'split off' on either side of the fixed point, see Figure 24.5.

24.3 The period-doubling cascade

As r is increased a little further, this attracting periodic orbit of period 2 becomes unstable – two points break off from each point on the orbit, and we end up with an attracting orbit of period 4, as shown in Figure 24.6.

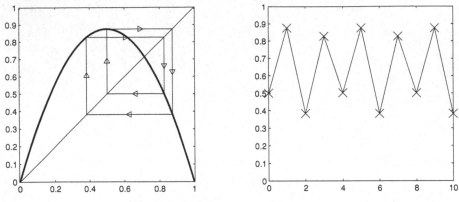

Fig. 24.6. A period 4 orbit when $r = 3.5$.

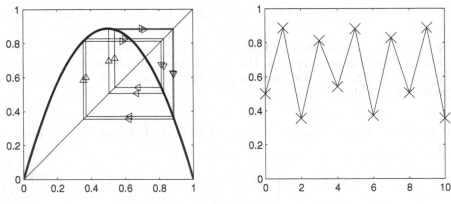

Fig. 24.7. A period 8 orbit when $r = 3.55$.

Increasing r to $r = 3.55$ we obtain an attracting period 8 orbit, as in Figure 24.7.

When r is a little larger the period of the orbit doubles again to 16, then again to 32, then again to 64 ... This is known as the 'period doubling cascade'. The parameter values at which these successive period doublings occur get closer and closer together, and converge towards a critical parameter value $r \approx 3.5701$.

24.4 The bifurcation diagram and more periodic orbits

When r is increased beyond this critical parameter value the behaviour of solutions becomes extremely complicated. One way to try to keep track of it is to draw the bifurcation diagram, as shown in Figure 24.8. The horizontal axis represents the parameter r, and for each value of r the attracting set is plotted vertically; this is done by choosing an initial condition, applying f a large number of times (here

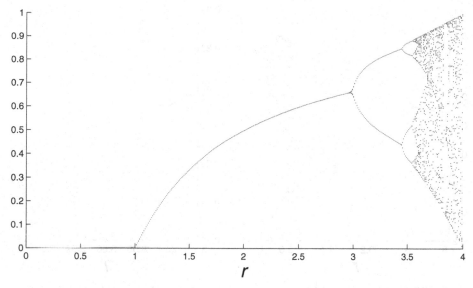

Fig. 24.8. The bifurcation diagram for $0 \leq r \leq 4$. After one hundred iterations, the next thirty are plotted vertically for a number of r values.

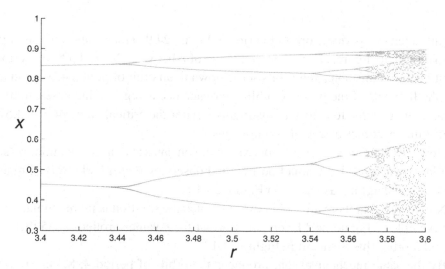

Fig. 24.9. The bifurcation diagram for $3.4 \leq r \leq 3.6$.

100) until the orbit has 'settled' down, and then plotting more points of the orbit on the vertical axis.

The diagram shows the stable fixed point at $x = 0$ while $r \leq 1$. For $1 < r < 3$ the fixed point at the origin is no longer stable, and the fixed point at $x = 1 - (1/r)$ is attracting. For $r > 3$ this fixed point becomes unstable, and instead all orbits

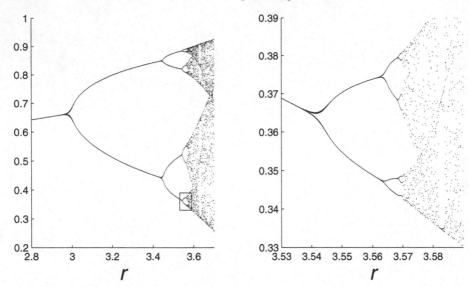

Fig. 24.10. The portion of the bifurcation diagram in the small box of the left-hand figure is magnified in the right-hand figure, and proves strikingly similar to the original picture.

are attracted to a periodic orbit of period 2. Figure 24.9, a magnified version of part of the previous figure, shows that at a value of r between 3.4 and 3.5 the period 2 orbit becomes unstable, and orbits settle down to an orbit of period 4. We can also see the first part of the period doubling cascade occurring as r increases, until the parameters become too close to distinguish. After the critical value of $r \approx 3.5701$ everything becomes extremely complicated.

However, there is still order. For example, you can see that this diagram is 'self-similar' by magnifying a small portion and observing that it looks very similar to the original diagram, as shown in Figure 24.10.

Notice also that there are 'windows' in which the solution is more regular again, for example in Figure 24.11 you can see that for r values around 3.835 there is a period 3 orbit; this is shown in Figure 24.12.

By the same mechanism that produces the orbits of period 4, 8, etc. from the initial period 2 orbit, this period 3 orbit will period double to 6, 12, 24, 48, etc. and lead to another chaotic region.

24.5 Chaos

The phenomenon of chaos consists, essentially, of *deterministic* motion in which the motion *appears* to be random, see Figure 24.13.

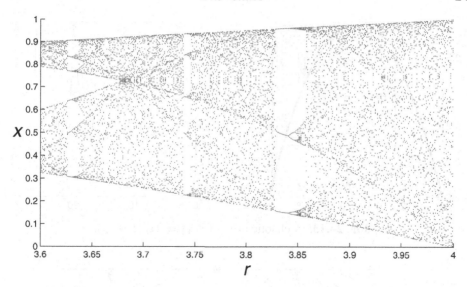

Fig. 24.11. The bifurcation diagram for $3.6 \le r \le 4$.

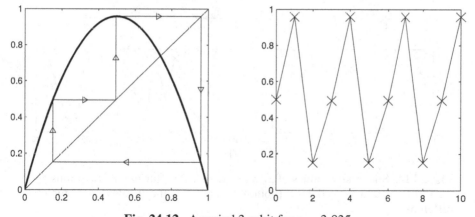

Fig. 24.12. A period 3 orbit for $r = 3.835$.

A more rigorous defining feature of chaotic systems is that small changes in the initial conditions will produce wildly different behaviour if we wait long enough. This phenomenon is known as *sensitive dependence on initial conditions*, and is illustrated in Figure 24.14, which shows successive values of x_n when $r = 4$ for initial conditions that agree to the fourth decimal place.

When a model exhibits such sensitive dependence on initial conditions it will be of little use for predicting the future, since tiny errors in the initial conditions will lead to very different outcomes. However, this pessimistic observation has a more optimistic converse; even if the behaviour of a system appears very complicated,

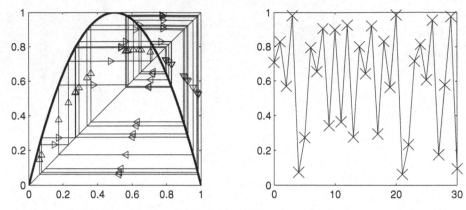

Fig. 24.13. A chaotic orbit of $x_{n+1} = 4x_n(1 - x_n)$.

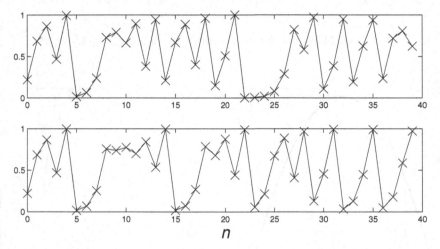

n

Fig. 24.14. Successive values of x_n vs n when $r = 4$, for initial conditions $x_0 = 0.2189$ (top) and $x_0 = 0.2188$ (bottom). For $n \geq 9$ the iterates are completely different.

or even random, it may still be subject to a very simple underlying rule. Since the advent of chaos theory, much experimental data that was once discarded as spurious and useless has been re-analysed and found to contain a high degree of order.

24.6 *Analysis of $x_{n+1} = 4x_n(1 - x_n)$

When $r = 4$ the map is

$$x_{n+1} = 4x_n(1 - x_n),$$

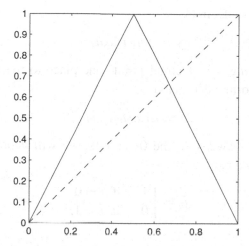

Fig. 24.15. The 'tent map' obtained from our original map by a substitution.

and we can simplify this problem by means of a careful substitution. Since $x_n \in [0, 1]$, we can set $x_n = \sin^2 \theta_n$, with $\theta_n \in [0, \pi/2]$. The equation for θ_n is then

$$\sin^2 \theta_{n+1} = 4 \sin^2 \theta_n \cos^2 \theta_n$$
$$= \sin^2(2\theta_n).$$

Since we want $\theta \in [0, \pi/2]$, we can take

$$\theta_{n+1} = \begin{cases} 2\theta_n & 0 \le \theta_n \le \pi/4 \\ \pi - 2\theta_n & \pi/4 < \theta_n \le \pi/2. \end{cases}$$

If we rescale θ, setting $y_n = 2\theta_n/\pi$, to give $y \in [0, 1]$, we obtain

$$y_{n+1} = \begin{cases} 2y_n & 0 \le \frac{1}{2} \\ 2(1 - y_n) & \frac{1}{2} < y_n \le 1. \end{cases}$$

This new map (the 'tent map') is shown in Figure 24.15.

The easiest way to consider the dynamics of this new map is by writing down the 'binary decimal' expansion of y_n,

$$y_n = a_0.a_1a_2a_3a_4\ldots,$$

where

$$y_n = \sum_{j=0}^{\infty} a_j 2^{-j}.$$

Although we appear to have made the problem significantly more complicated this way, it will enable us to understand the dynamics much more easily.

Doubling the binary decimal

$$y = 0.a_1 a_2 a_3 a_4 \ldots$$

corresponds to shifting the 'decimal point' one place to the right (analogous to multiply by 10 for normal decimals),

$$2y = a_1.a_2 a_3 a_4 a_5 \ldots.$$

Subtracting y from 2 swaps all the 0s and 1s: we will denote the operation of swapping 0 and 1 by a bar,

$$\bar{a} = \begin{cases} 1 & \text{if } a = 0 \\ 0 & \text{if } a = 1, \end{cases}$$

and so

$$2 - a_0.a_1 a_2 a_3 a_4 \ldots = \bar{a}_0.\bar{a}_1 \bar{a}_2 \bar{a}_3 \bar{a}_4 \ldots.$$

In the range of y that we are considering ($y \in [0, 1]$) we always have $a_0 = 0$ provided that we represent 1 by 0.1^∞ (we use the notation $(r_1 \ldots r_n)^\infty$ to mean $r_1 \ldots r_n$ repeated ad infinitum). If $y_n < 1/2$ then we also have $a_1 = 0$, and so

$$y_{n+1} = 0.a_2 a_3 a_4 a_5 \ldots.$$

If $y_n \geq 1/2$ then $a_1 = 1$, and we have

$$y_{n+1} = 2 - 1.a_2 a_3 a_4 a_5 \ldots = 0.\bar{a}_2 \bar{a}_3 \bar{a}_4 \bar{a}_5 \ldots.$$

Therefore we can rewrite our map as

$$\text{if } y_n = 0.a_1 a_2 a_3 a_4 \ldots \quad \text{then} \quad y_{n+1} = \begin{cases} 0.a_2 a_3 a_4 a_5 \ldots & \text{for } a_1 = 0 \\ 0.\bar{a}_2 \bar{a}_3 \bar{a}_4 \bar{a}_5 \ldots & \text{for } a_1 = 1. \end{cases}$$

With the map written in this way it is possible to understand its dynamics fairly easily.

Suppose that y_0 is rational. Then just as in base 10, its decimal expansion will either be finite, or eventually repeat. If its decimal expansion is finite,

$$y_0 = 0.a_1 \ldots a_n,$$

then $y_j = 0$ for all $j \geq n$, and the orbit ends up at zero after a finite number of iterations. If the decimal expansion eventually repeats then

$$y_0 = 0.b_1 b_2 b_3 \ldots b_n (a_1 a_2 \ldots a_m)^\infty.$$

In this case we will have

$$y_1 = 0.b_2 b_3 b_4 b_5 \ldots b_n (a_1 a_2 \ldots a_m)^\infty \quad \text{or} \quad 0.\overline{b_2 b_3 b_4 b_5 \ldots b_n (a_1 a_2 \ldots a_m)^\infty},$$

$$y_2 = 0.b_3 b_4 b_5 \ldots b_n (a_1 a_2 \ldots a_m)^\infty \quad \text{or} \quad 0.\overline{b_3 b_4 b_5 \ldots b_n (a_1 a_2 \ldots a_m)^\infty},$$

until after n iterations we have

$$y_n = 0.(a_1 a_2 \dots a_m)^\infty \qquad \text{or} \qquad 0.(\overline{a_1 a_2 \dots a_m})^\infty.$$

Without loss of generality we can consider what happens when

$$y_n = 0.(a_1 a_2 \dots a_m)^\infty.$$

There are now two possibilities. Either $f^m(y_n) = y_n$, in which case the orbit repeats every m iterations; or all the zeros and ones of y_n are interchanged and $f^m(y_n) = \bar{y}_n$. In the second case, since $f(\bar{y}) = f(y)$ for any y,

$$f^m(y_{n+1}) = f^m(f(y_n)) = f(f^m(y_n)) = f(\bar{y}_n) = f(y_n) = y_{n+1},$$

and so the orbit repeats every m iterations after just one more application of f. Thus every rational number lies on an orbit that is eventually periodic.

However, none of these periodic orbits can be stable. Indeed, all initial conditions that are not equal will eventually separate, falling on different sides of $x = \frac{1}{2}$. Suppose that

$$y_0 = 0.a_1 a_2 \dots a_n a_{n+1} \dots \qquad \text{and} \qquad z_0 = 0.a_1 a_2 \dots a_n b_{n+1} \dots$$

with $a_{n+1} \neq b_{n+1}$; then the first decimal places of $f^n(y_0)$ and $f^n(z_0)$ will not be equal.

If y_0 is irrational then the decimal expansion will go on for ever and never repeat. It follows that the orbit starting at y_0 will also never repeat, and so cannot be periodic.

We have a strange situation, then. There are no stable orbits, but all rational numbers are eventually periodic or reach zero after a finite number of iterations. Any number that starts irrational will have a binary expansion, and hence an orbit, that never repeats itself; one can also show that the distribution of points along such an orbit is effectively random, even though the evolution is deterministic.

The idea of following an orbit by seeing which side of $x = \frac{1}{2}$ it falls (which is what the binary expansion here does very simply) can be generalised into a useful technique for analysing the original logistic map for other parameter values. This model is now very well understood, and is one of the standard examples used in the theory of dynamical systems.

Exercises

24.1 Consider the iterated map

$$x_{n+1} = r x_n (1 - x_n^2).$$

Show that for $0 < r < 3\sqrt{3}/2$ if $0 \leq x_n \leq 1$ then $0 \leq x_{n+1} \leq 1$. Show that if $r < 1$ then the only fixed point in $[0, 1]$ is zero, and that this is stable.

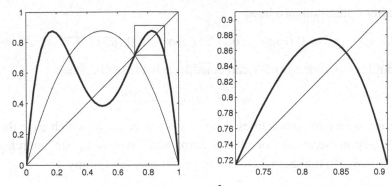

Fig. 24.16. On the left is the graph of f and f^2 (in bold) for $r = 3.5$. On the right is a magnified version of the box in the left-hand figure, showing that f^2 has a fixed point.

When $r > 1$ there is another fixed point in $[0, 1]$. Find the value of this fixed point (as a function of r). For which values of r is it stable, and for which values is it unstable?

What would you expect to happen when $r > 2$?

24.2 (C) Use the M-files `logistic.m` (which draws cobweb diagrams), `xnvsn.m` (which plots successive values of x_n vs n) and `bifurcation.m` (which draws the bifurcation diagram for a given range of r) to investigate the dynamics of the logistic map. Modify the programs to investigate the dynamics of the map in the previous exercise.

24.3 (C and T) The M-file `f2.m` plots the graph of $f(x)$ and of $f^2(x)$ in the left-hand figure, and the graph of f^2 restricted to the little box in a blown-up version on the right (see Figure 24.16). By looking at a succession of pictures as r increases from 0 to 4, observe that the rescaled version of f^2 behaves in the same way that f does as r increases. This can be made precise, and explains the period doubling cascade. Since the fixed point of f becomes unstable and gives rise to a period 2 orbit, the same thing happens to f^2; its fixed point (a period 2 orbit for f) becomes unstable and gives rise to a period 2 orbit (a period 4 orbit for f). Since whatever happens to f happens to f^2, whatever happens to f^2 happens to $(f^2)^2 = f^4$; its fixed point (a period 4 orbit for f) will become unstable and give rise to a period 2 orbit (a period 8 orbit for f). Similar reasoning holds for each orbit of period 2^k, showing that it becomes unstable and produces an orbit of period 2^{k+1}. The map formed by restricting f^2 to the little box, and then rescaling to the interval $[0, 1]$, is known as the *renormalisation of f*. You can investigate the dynamics of the renormalised map as r changes using the M-file `renormalised.m`.

Part V

Coupled linear equations

25

*Vector first order equations and higher order equations

All the equations that we have considered so far have been first order equations in which there was only one dependent variable (e.g. $x(t)$ or $y(x)$, where x and y are scalars). If we were restricted to equations in which there is only one dependent variable then this would exclude the vast majority of applications: for example, specifying the position of something in the three-dimensional space in which we live requires three coordinates.

Although it is much harder to find solution methods for equations involving a number of dependent variables, the theoretical ideas are straightforward generalisations of what we did for scalar equations in Chapter 6. Here we make precise what we mean by a solution, and state the theorem that guarantees the existence and uniqueness of solutions under easily checked conditions.

Suppose that we have n dependent variables $x_1, \ldots, x_n$, and each of these obeys a differential equation with the right-hand side (perhaps) depending on some of the other variables,

$$
\begin{aligned}
\dot{x}_1 &= f_1(x_1, x_2, \ldots, x_n, t) \\
\dot{x}_2 &= f_2(x_1, x_2, \ldots, x_n, t) \\
\vdots &= \vdots \\
\dot{x}_n &= f_n(x_1, x_2, \ldots, x_n, t).
\end{aligned}
\tag{25.1}
$$

This is a set of n coupled first-order equations which we can write in a much more convenient way if we make use of vector notation.

We write the n dependent variables $x_1, \ldots, x_n$ as a vector $\mathbf{x}$,

$$
\mathbf{x} = \begin{pmatrix} x_1 \\ \vdots \\ x_n \end{pmatrix},
$$

249

and define a vector function $\mathbf{f}(\mathbf{x}, t)$ by

$$\mathbf{f}(\mathbf{x}, t) = \mathbf{f}(x_1, \ldots, x_n, t) = \begin{pmatrix} f_1(x_1, \ldots, x_n, t) \\ f_2(x_1, \ldots, x_n, t) \\ \vdots \\ f_n(x_1, \ldots, x_n, t) \end{pmatrix}.$$

With this notation we can rewrite the coupled equations in (25.1) as

$$\frac{d\mathbf{x}}{dt} = \mathbf{f}(\mathbf{x}, t),$$

which encourages us to find a theory that unifies the treatment of scalar equations and coupled equations. On the theoretical level this is possible, and here we give a general existence and uniqueness result which reduces to the scalar existence and uniqueness theorem (Theorem 6.2) when $n = 1$.

First we define a solution of the appropriate initial value problem, cf. Definition 6.1.

Definition 25.1 A solution *of the initial value problem*

$$\frac{d\mathbf{x}}{dt}(t) = \mathbf{f}(\mathbf{x}, t) \qquad \text{with} \qquad \mathbf{x}(t_0) = \mathbf{x}_0 \qquad \mathbf{x} \in \mathbb{R}^n, \tag{25.2}$$

on an open interval I that contains t_0 is a differentiable function $\mathbf{x} : I \to \mathbb{R}^n$, *with* $\mathbf{x}(t_0) = \mathbf{x}_0$ *and* $\dot{\mathbf{x}}(t) = \mathbf{f}(\mathbf{x}, t)$ *for all $t \in I$.*

We now have essentially the same existence and uniqueness theorem as before (Theorem 6.2). We use the notation $D\mathbf{f}$ to denote the matrix of partial derivatives of $\mathbf{f}$,

$$D\mathbf{f} = \begin{pmatrix} \partial f_1/\partial x_1 & \cdots & \partial f_1/\partial x_n \\ \vdots & \ddots & \vdots \\ \partial f_n/\partial x_1 & \cdots & \partial f_n/\partial x_n \end{pmatrix}. \tag{25.3}$$

Theorem 25.2 *If $\mathbf{f}(\mathbf{x}, t)$ and $D\mathbf{f}(\mathbf{x}, t)$ are continuous functions of $\mathbf{x}$ (i.e. of x_1, $x_2, \ldots, x_n$) and t for*

$$\mathbf{x} \in U = (a_1, b_1) \times (a_2, b_2) \times \cdots \times (a_n, b_n),$$

and for $c < t < d$ then for any $\mathbf{x}_0 \in U$ and $t_0 \in (c, d)$ the equation (25.2) has a unique solution on some open interval containing t_0.

Note that the only real change is that we have to check that all the partial derivatives $\partial f_i/\partial x_j$ are continuous, rather than just the one derivative we get in the scalar case. You should also check that you are happy that the above definition and theorem reduce to the scalar case when $n = 1$.

We will spend some time with pairs of coupled first order equations ($n = 2$) in Chapters 26–36, and consider a particular example of three coupled equations ($n = 3$) in Chapter 37.

25.1 Existence and uniqueness for second order equations

Theorem 25.2 can also be used to show that second (or higher) order scalar equations have unique solutions. If we consider, for example, the second order equation

$$\ddot{x} = f(\dot{x}, x, t),$$

then we recast this as a vector equation by defining another variable representing $\dot{x}$. We let $x_1 = x$ and $x_2 = \dot{x}$, so that[1] $\mathbf{x} = (x_1, x_2) \equiv (x, \dot{x})$, and then

$$\dot{x}_1 = \dot{x} = x_2$$
$$\dot{x}_2 = \frac{\mathrm{d}\dot{x}}{\mathrm{d}t} = \ddot{x} = f(\dot{x}, x, t) = f(x_2, x_1, t),$$

which we can rewrite as

$$\frac{\mathrm{d}\mathbf{x}}{\mathrm{d}t} = \mathbf{g}(\mathbf{x}, t), \qquad (25.4)$$

with

$$\mathbf{g}(\mathbf{x}, t) = \begin{pmatrix} x_2 \\ f(x_2, x_1, t) \end{pmatrix}.$$

Existence and uniqueness are guaranteed by Theorem 25.2 for equation (25.4) as long as we provide an initial condition $\mathbf{x}(t_0) = \mathbf{x}_0$; because $\mathbf{x}(t_0) = (x(t_0); \dot{x}(t_0))$ this is equivalent to specifying both $x(t_0)$ and $\dot{x}(t_0)$.

Theorem 25.2 requires that $\mathbf{g}$ and $D\mathbf{g}$ are continuous. Since x_2 is continuous, we need $f(x_2, x_1, t)$ to be a continuous function of x_1 and x_2; and since

$$D\mathbf{g} = \begin{pmatrix} 0 & 1 \\ \partial f/\partial x_1 & \partial f/\partial x_2 \end{pmatrix}$$

we need the partial derivatives of f with respect to x_1 and x_2 to be continuous. So we can rewrite Theorem 25.2 in the form of Theorem 11.1 from Chapter 11.

Theorem 25.3 *Given a function $f(x_2, x_1, t)$, suppose that f, $\partial f/\partial x_1$, and $\partial f/\partial x_2$ are continuous functions for $a_1 < x_1 < a_2$, $b_1 < x_2 < b_2$ and $t_1 < t < t_2$. Then*

---

[1] In the text we use the notation (x_1, x_2) to denote the column vector $\begin{pmatrix} x_1 \\ x_2 \end{pmatrix}$.

for all initial conditions

$$x(t_0) = x_0 \quad \text{and} \quad \dot{x}(t_0) = y_0 \tag{25.5}$$

with $a_1 < x_0 < a_2$, $b_1 < y_0 < b_2$, and $t_1 < t_0 < t_2$ there exists a unique solution of

$$\ddot{x} = f(\dot{x}, x, t) \tag{25.6}$$

on some interval I containing t_0, i.e. a continuous function with two continuous derivatives that satisfies (25.5) and the equation (25.6) on I.

A similar trick to this can be applied to deal with higher order equations, see Exercise 25.1.

Exercises

25.1 By choosing an appropriate collection of new variables $x_1, \ldots, x_n$ rewrite the nth order differential equation

$$\frac{d^n x}{dt^n} = f\left(\frac{d^{n-1} x}{dt^{n-1}}, \ldots, \frac{dx}{dt}, x, t\right)$$

as a set of n coupled linear first order equations. Find the conditions on the function $f(x_n, \ldots, x_1, t)$ for the original differential equation to have a unique solution.

25.2 Suppose that $\mathbf{f}$ is a Lipschitz function of $\mathbf{x}$, i.e. that for some $L > 0$

$$|\mathbf{f}(\mathbf{x}) - \mathbf{f}(\mathbf{y})| \le L|\mathbf{x} - \mathbf{y}|.$$

Use an argument similar to that of Exercise 6.3 to show that if $\mathbf{x}(t)$ and $\mathbf{y}(t)$ are two solutions of

$$d\mathbf{x}/dt = \mathbf{f}(\mathbf{x}) \quad \text{with} \quad \mathbf{x}(0) = \mathbf{x}_0 \tag{E25.1}$$

and $\mathbf{z}(t) = \mathbf{x}(t) - \mathbf{y}(t)$ then

$$\frac{d}{dt}|\mathbf{z}|^2 \le 2L|\mathbf{z}|^2,$$

and hence that the solution of (E25.1) is unique. (You might find the Cauchy–Schwarz inequality $|\mathbf{a} \cdot \mathbf{b}| \le |\mathbf{a}||\mathbf{b}|$ useful.)

26

Explicit solutions of coupled linear systems

In the following chapters we will consider what happens when we have two dependent variables, $x(t)$ and $y(t)$. In a general pair of coupled first order equations the derivative of x can depend not only on x and t, but also on y, and vice versa,

$$\begin{cases} \dot{x} = f(x, y, t) \\ \dot{y} = g(x, y, t). \end{cases}$$

The solutions $x(t)$ and $y(t)$ are, at least in principle, inextricably entangled with each other. Such systems are often referred to as 'two-dimensional', since the state of the system can be completely specified by the two variables x and y. The best way to understand the solutions of such coupled equations is graphically, using a two-dimensional version of the phase diagrams we drew in Chapter 7. The remainder of the book concentrates on this graphical approach.

However, in this chapter we first discuss a method for finding explicit solutions for a pair of coupled linear equations,

$$\begin{cases} \dot{x} = ax + by + f(t) \\ \dot{y} = cx + dy + g(t). \end{cases} \tag{26.1}$$

We saw in the previous chapter that by introducing extra variables we can rewrite a second order equation as two coupled first order equations. Given two coupled *linear* equations we can reverse this process. This method is the most reliable if you want to find an explicit solution of the equation, and in particular is good for dealing with the inhomogeneous problem (when $f(t)$ or $g(t)$ is non-zero).

We will assume here that $b \neq 0$. If $b = c = 0$ then the equations are not coupled and we could solve for x and y separately using the integrating factor method, while if $b = 0$ but $c \neq 0$ we can follow the method below swapping the rôles of x and y.

253

Since $b \neq 0$ we can rearrange the first equation of the pair in (26.1) to give y in terms of x and $\dot{x}$,

$$y = \frac{\dot{x} - ax - f(t)}{b}. \tag{26.2}$$

Differentiating this will give us $\dot{y}$ in terms of $\dot{x}$ and $\ddot{x}$,

$$\dot{y} = \frac{\ddot{x} - a\dot{x} - \dot{f}(t)}{b},$$

and we can now substitute these into the second equation from (26.1) to give

$$\frac{\ddot{x} - a\dot{x} - \dot{f}(t)}{b} = cx + d\frac{\dot{x} - ax - f(t)}{b} + g(t).$$

Rearranging this gives a second order equation for x,

$$\ddot{x} - (a + d)\dot{x} + (ad - bc)x = \dot{f}(t) + bg(t) - df(t). \tag{26.3}$$

We should be able to solve this to find $x(t)$, using the techniques for second order linear equations we have already covered. Once we know $x(t)$ we can then use the formula in (26.2) to work out $y(t)$.

Notice that it is not necessary to learn any new techniques to solve (26.1), apart from the 'trick' of substitution that is used here to turn the pair of equations into one second order equation.

Example 26.1 *By deriving the second order differential equation solved by x, find the general solution of the coupled equations*

$$\dot{x} = x + y$$
$$\dot{y} = 4x - 2y + 4e^{-2t}.$$

Find also the solution that satisfies the initial conditions $x(0) = 0$ *and* $y(0) = -1$.

Rearranging the first equation we can find y in terms of x,

$$y = \dot{x} - x, \tag{26.4}$$

and so, differentiating this,

$$\dot{y} = \ddot{x} - \dot{x}.$$

Substituting these into the second equation gives

$$\underbrace{\ddot{x} - \dot{x}}_{\dot{y}} = 4x - 2\underbrace{(\dot{x} - x)}_{y} + 4e^{-2t}$$

which simplifies to

$$\ddot{x} + \dot{x} - 6x = 4e^{-2t}. \tag{26.5}$$

To solve this we first find the complementary function, i.e. the solution of $\ddot{z} + \dot{z} - 6z = 0$. If we try $z(t) = e^{kt}$ the resulting auxiliary equation is $k^2 + k - 6 = 0$ with roots $k = 2$ and $k = -3$, yielding the complementary function

$$z(t) = Ae^{2t} + Be^{-3t}.$$

Since e^{-2t} is not a solution of the homogeneous equation we can try $x_p(t) = ce^{-2t}$ for a particular integral of (26.5). To obtain $4e^{-2t}$ on the right-hand side we need

$$4ce^{-2t} - 2ce^{-2t} - 6ce^{-2t} = 4e^{-2t},$$

so $c = -1$, and $x_p(t) = -e^{-2t}$ is a particular integral. The general solution for $x(t)$ is therefore

$$x(t) = Ae^{2t} + Be^{-3t} - e^{-2t}.$$

Having found $x(t)$ we can now find $y(t)$ using (26.4),

$$y(t) = \underbrace{[2Ae^{2t} - 3Be^{-3t} + 2e^{-2t}]}_{\dot{x}} - \underbrace{[Ae^{2t} + Be^{-3t} - e^{-2t}]}_{x}$$

$$= Ae^{2t} - 4Be^{-3t} + 3e^{-2t}.$$

The general solution is therefore

$$\begin{cases} x(t) = Ae^{2t} + Be^{-3t} - e^{-2t} \\ y(t) = Ae^{2t} - 4Be^{-3t} + 3e^{-2t}. \end{cases}$$

Note that the constants in the solutions for $x(t)$ and $y(t)$ are the same A and B, and the solutions are tied together by the choice of these constants.

The solution that satisfies the initial conditions $x(0) = 0$ and $y(0) = -1$ must have

$$A + B - 1 = 0 \quad \text{and} \quad A - 4B + 3 = -1,$$

i.e. $A = 0$ and $B = 1$; it is

$$x(t) = e^{-3t} - e^{-2t} \qquad y(t) = 3e^{-2t} - 4e^{-3t}. \qquad \square$$

Example 26.2 *By finding the second order differential equation solved by x, find the general solution of the coupled equations*

$$\dot{x} = 2x + 5y$$
$$\dot{y} = -2x.$$

Find also the solution satisfying the initial conditions $x(0) = 5$ and $y(0) = -4$.

Rearranging the first equation we obtain $y = \frac{1}{5}(\dot{x} - 2x)$. Therefore

$$\dot{y} = \frac{1}{5}(\ddot{x} - 2\dot{x}) = -2x,$$

and so

$$\ddot{x} - 2\dot{x} + 10x = 0.$$

To solve this we try $x = e^{kt}$ and obtain the auxiliary equation for k,

$$k^2 - 2k + 10 = 0.$$

The solutions of this equation are

$$k = \frac{2 \pm \sqrt{4 - 40}}{2} = 1 \pm 3i,$$

and so

$$x(t) = e^t(A\cos 3t + B\sin 3t).$$

Since $y = \frac{1}{5}(\dot{x} - 2x)$ the solution for $y(t)$ is given by

$$y(t) = \frac{1}{5}\{e^t[(3B + A)\cos 3t + (B - 3A)\sin 3t] - 2e^t[A\cos 3t + B\sin 3t]\}$$

$$= \frac{e^t}{5}[(3B - A)\cos 3t - (3A + B)\sin 3t].$$

Since A and B are arbitrary constants, if we set $A = 5C$ and $B = 5D$ the general solution is

$$\begin{aligned} x(t) &= e^t(5C\cos 3t + 5D\sin 3t) \\ y(t) &= e^t[(3D - C)\cos 3t - (3C + D)\sin 3t]. \end{aligned} \tag{26.6}$$

To ensure that $x(0) = 5$ and $y(0) = -4$ we need

$$5C = 5 \quad \text{and} \quad 3D - C = -4,$$

so that $C = 1$ and $D = -1$. Thus the solution satisfying these initial conditions is

$$\begin{aligned} x(t) &= e^t(5\cos 3t - 5\sin 3t) \\ y(t) &= -e^t(4\cos 3t + 2\sin 3t). \end{aligned}$$

The graphs of $x(t)$ and $y(t)$ against t are shown in Figure 26.1. ☐

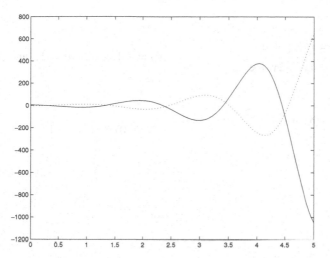

Fig. 26.1. Graphs of $x(t)$ [solid line] and $y(t)$ [dotted line] against t. Both are oscillations whose amplitude increases exponentially.

Exercises

26.1 Find the general solutions of the following differential equations by converting them into a single second-order equation. Also find the solution that satisfies the given initial conditions.

(i)

$$\dot{x} = 4x - y$$
$$\dot{y} = 2x + y + t^2, \qquad x(0) = 0 \text{ and } y(0) = 1;$$

(ii)

$$\dot{x} = x - 4y + \cos 2t$$
$$\dot{y} = x + y, \qquad x(0) = 1 \text{ and } y(0) = 1;$$

(iii)

$$\dot{x} = 2x + 2y$$
$$\dot{y} = 6x + 3y + e^t, \qquad x(0) = 0 \text{ and } \dot{x}(0) = 1;$$

(iv)

$$\dot{x} = 5x - 4y + e^{3t}$$
$$\dot{y} = x + y, \qquad x(0) = 1 \text{ and } y(0) = -1;$$

(v)

$$\dot{x} = 2x + 5y$$
$$\dot{y} = -2x + \cos 3t, \qquad x(0) = 2 \text{ and } y(0) = -1;$$

(vi)

$$\dot{x} = x + y + e^{-t}$$
$$\dot{y} = 4x - 2y + e^{2t},$$

$y(0) = -1$ and $x(0) = 1$; and

(vii)

$$\dot{x} = 8x + 14y$$
$$\dot{y} = 7x + y,$$

$x(0) = y(0) = 1.$

27

The matrix approach to linear equations: eigenvalues and eigenvectors

We now reconsider the coupled homogeneous linear system

$$\begin{cases} \dot{x} = ax + by \\ \dot{y} = cx + dy. \end{cases} \tag{27.1}$$

Our new approach may initially seem a little complicated, and it is not the best method if we want to find an explicit solution. However, it will enable us to draw the phase portrait (phase diagram) for this linear system after some fairly simple calculations. We will then use these linear phase portraits to analyse coupled nonlinear systems. This graphical approach forms the main subject matter of all that follows.

27.1 Rewriting the equation in matrix form

The starting point is the observation that there is a much more compact way to write (27.1) using vector and matrix notation. If we write

$$\mathbf{x}(t) = \begin{pmatrix} x(t) \\ y(t) \end{pmatrix}$$

and define a matrix $\mathbb{A}$ by

$$\mathbb{A} = \begin{pmatrix} a & b \\ c & d \end{pmatrix}$$

(see Appendix B for some background material on matrices) then we can rewrite (27.1) as

$$\dot{\mathbf{x}} = \mathbb{A}\mathbf{x}. \tag{27.2}$$

Notice that since this equation is linear, we have a superposition principle; linear combinations of solutions will still satisfy the equation, i.e. if $\mathbf{x}_1(t)$ and $\mathbf{x}_2(t)$ solve

259

(27.2) then so does

$$x(t) = \alpha x_1(t) + \beta x_2(t), \tag{27.3}$$

since

$$\frac{dx}{dt} = \frac{d}{dt}[\alpha x_1(t) + \beta x_2(t)] = \alpha \dot{x}_1 + \beta \dot{x}_2$$
$$= \alpha A x_1 + \beta A x_2 = A[\alpha x_1 + \beta x_2]$$
$$= Ax.$$

Now, observe that the way equation (27.2) is written makes it look like the simple linear equation $\dot{x} = ax$, whose solution we know is $x(t) = Ce^{at}$. We will try to find a solution of (27.2) by guessing that it has the same type of exponential dependence on time ($e^{\lambda t}$ for some λ); however, since x has two components the 'coefficient' (C in the solution $x(t) = Ce^{at}$) will have to be a constant vector v. Our trial solution is therefore

$$x(t) = e^{\lambda t} v. \tag{27.4}$$

If we substitute this guess into (27.2) then we obtain

$$\frac{d}{dt}[e^{\lambda t} v] = A[e^{\lambda t} v].$$

Since v is a constant vector the d/dt on the left-hand side only affects the exponential term:

$$\lambda e^{\lambda t} v = e^{\lambda t} A v.$$

Dividing by the non-zero factor $e^{\lambda t}$ this becomes

$$Av = \lambda v. \tag{27.5}$$

This is known as an **eigenvalue** equation. Any values of λ for which this equation has a non-zero solution for v are called the *eigenvalues* of A, and the corresponding v are called the *eigenvectors* of A. The remainder of this chapter discusses how to find the eigenvalues and eigenvectors of A, and we return to their relevance to the differential equation $dx/dt = Ax$ in the next chapter.

27.2 Eigenvalues and eigenvectors

For what values of λ does the eigenvalue equation

$$Av = \lambda v \tag{27.6}$$

have a non-zero solution for $\mathbf{v}$? If we rewrite the equation as

$$(\mathbb{A} - \lambda\mathbb{I})\mathbf{v} = \mathbf{0}, \qquad (27.7)$$

where $\mathbb{I}$ is the 2×2 identity matrix,

$$\mathbb{I} = \begin{pmatrix} 1 & 0 \\ 0 & 1 \end{pmatrix},$$

and $\mathbf{0} = (0, \; 0)$, it is clear that if $\mathbb{A} - \lambda\mathbb{I}$ is invertible then the only solution is $\mathbf{v} = \mathbf{0}$. So to have a non-trivial solution ($\mathbf{v} \neq \mathbf{0}$) we have to ensure that $\mathbb{A} - \lambda\mathbb{I}$ is singular, i.e. that

$$\det(\mathbb{A} - \lambda\mathbb{I}) = |\mathbb{A} - \lambda\mathbb{I}| = 0.$$

Rewriting this in full it becomes

$$\left| \begin{pmatrix} a & b \\ c & d \end{pmatrix} - \begin{pmatrix} \lambda & 0 \\ 0 & \lambda \end{pmatrix} \right| = \begin{vmatrix} a - \lambda & b \\ c & d - \lambda \end{vmatrix} = (a - \lambda)(d - \lambda) - bc = 0$$

which gives a quadratic equation for λ ('the characteristic equation')

$$\lambda^2 - (a + d)\lambda + (ad - bc) = 0. \qquad (27.8)$$

The solutions of this equation are the *eigenvalues* of the matrix $\mathbb{A}$, and for these eigenvalues we can expect to be able to find non-zero vectors $\mathbf{v}$ that have $\mathbb{A}\mathbf{v} = \lambda\mathbf{v}$; these are the corresponding *eigenvectors*. Since the eigenvalues are the solutions of a quadratic equation, they could be real and distinct, there could be only one, or they could be a complex conjugate pair. These different possibilities will have different implications for the solutions of the original differential equation, and we examine these three possibilities in turn in the following three chapters.

Given an eigenvalue λ the simplest way to find the associated eigenvector $\mathbf{v} = (v_1, \; v_2)$ is usually to solve the rearranged version (27.7) of the eigenvalue equation,

$$(\mathbb{A} - \lambda\mathbb{I}) \begin{pmatrix} v_1 \\ v_2 \end{pmatrix} = \mathbf{0}.$$

Note that if $\mathbf{v}$ is an eigenvector with

$$\mathbb{A}\mathbf{v} = \lambda\mathbf{v}$$

then any constant multiple of $\mathbf{v}$ is also an eigenvector, since

$$\mathbb{A}(c\mathbf{v}) = c\mathbb{A}\mathbf{v} = c\lambda\mathbf{v} = \lambda(c\mathbf{v}).$$

A canonical (standardised) choice of eigenvector is the vector with unit length (there are in fact two of these in the direction of **v**), but often it is easier to work with an eigenvector of integer values (if there is one).

This is all most easily illustrated by considering an example.

Example 27.1 *Find the eigenvalues and eigenvectors of the matrix*

$$\mathbb{A} = \begin{pmatrix} 2 & 2 \\ 6 & 3 \end{pmatrix}.$$

To find the eigenvalues of $\mathbb{A}$ we solve the equation $\det(\mathbb{A} - \lambda\mathbb{I}) = 0$; this is

$$\left| \begin{pmatrix} 2 & 2 \\ 6 & 3 \end{pmatrix} - \begin{pmatrix} \lambda & 0 \\ 0 & \lambda \end{pmatrix} \right| = \begin{vmatrix} 2-\lambda & 2 \\ 6 & 3-\lambda \end{vmatrix} = 0,$$

which gives the quadratic equation

$$(2 - \lambda)(3 - \lambda) - 12 = 0.$$

Multiplying this out and simplifying we obtain

$$\lambda^2 - 5\lambda - 6 = 0 \qquad \text{or} \qquad (\lambda - 6)(\lambda + 1) = 0.$$

The solutions of this, $\lambda = 6$ and $\lambda = -1$, are the eigenvalues of $\mathbb{A}$.

Now that we know the eigenvalues we can find the corresponding eigenvectors. If we write

$$\mathbf{v} = \begin{pmatrix} v_1 \\ v_2 \end{pmatrix}$$

then when $\lambda = 6$ we need $(\mathbb{A} - 6\mathbb{I})\mathbf{v} = \mathbf{0}$, i.e.

$$\left[\begin{pmatrix} 2 & 2 \\ 6 & 3 \end{pmatrix} - \begin{pmatrix} 6 & 0 \\ 0 & 6 \end{pmatrix} \right] \mathbf{v} = \begin{pmatrix} -4 & 2 \\ 6 & -3 \end{pmatrix} \begin{pmatrix} v_1 \\ v_2 \end{pmatrix} = \mathbf{0},$$

This matrix equation gives us two equations relating v_1 and v_2:

$$-4v_1 + 2v_2 = 0 \qquad \text{and} \qquad 6v_1 - 3v_2 = 0.$$

At this stage we will always obtain two equations, one of which is a multiple of the other. We need only consider one equation, then, and this tells us that $v_2 = 2v_1$; the general eigenvector corresponding to $\lambda_1 = 6$ is

$$\begin{pmatrix} v_1 \\ 2v_1 \end{pmatrix} = v_1 \begin{pmatrix} 1 \\ 2 \end{pmatrix}.$$

We usually drop the arbitrary constant, and talk about (for example) 'the eigenvector'

$$\mathbf{v}_1 = \begin{pmatrix} 1 \\ 2 \end{pmatrix},$$

remembering that in fact any multiple of $(1, 2)$ is an eigenvector.[1]

To check that $\mathbf{v}_1$ really is an eigenvector, simply multiply it by $\mathbb{A}$:

$$\mathbb{A}\mathbf{v}_1 = \begin{pmatrix} 2 & 2 \\ 6 & 3 \end{pmatrix}\begin{pmatrix} 1 \\ 2 \end{pmatrix} = \begin{pmatrix} 2+4 \\ 6+6 \end{pmatrix} = \begin{pmatrix} 6 \\ 12 \end{pmatrix} = 6\mathbf{v}_1.$$

So $\mathbf{v}_1$ is indeed an eigenvector corresponding to the eigenvalue 6.

For $\lambda = -1$ we want the eigenvector $\mathbf{v}_2 = (v_1, v_2)$ to solve $[\mathbb{A} - (-\mathbb{I})]\mathbf{v} = \mathbf{0}$,

$$\left[\begin{pmatrix} 2 & 2 \\ 6 & 3 \end{pmatrix} - \begin{pmatrix} -1 & 0 \\ 0 & -1 \end{pmatrix}\right]\mathbf{v} = \begin{pmatrix} 3 & 2 \\ 6 & 4 \end{pmatrix}\begin{pmatrix} v_1 \\ v_2 \end{pmatrix} = \mathbf{0}.$$

Again, there are two equations, with one a multiple of the other,

$$3v_1 + 2v_2 = 0 \qquad \text{and} \qquad 6v_1 + 4v_2 = 0;$$

so v_1 and v_2 are related by $v_2 = -3v_1/2$: the general eigenvector corresponding to the eigenvalue $\lambda = -1$ is

$$\begin{pmatrix} v_1 \\ -3v_1/2 \end{pmatrix} = v_1\begin{pmatrix} 1 \\ -3/2 \end{pmatrix}.$$

In this case we could choose

$$\mathbf{v}_2 = \begin{pmatrix} 2 \\ -3 \end{pmatrix}$$

as our representative eigenvector. Checking once more that this really is an eigenvector, we have

$$\mathbb{A}\mathbf{v}_2 = \begin{pmatrix} 2 & 2 \\ 6 & 3 \end{pmatrix}\begin{pmatrix} 2 \\ -3 \end{pmatrix} = \begin{pmatrix} 4-6 \\ 12-9 \end{pmatrix} = \begin{pmatrix} -2 \\ 3 \end{pmatrix} = -\mathbf{v}_2$$

as required.

We have now found the two eigenvalues and their corresponding eigenvectors,

$$\lambda_1 = 6 \quad \text{with} \quad \mathbf{v}_1 = \begin{pmatrix} 1 \\ 2 \end{pmatrix} \qquad \text{and} \qquad \lambda_2 = -1 \quad \text{with} \quad \mathbf{v}_2 = \begin{pmatrix} 2 \\ -3 \end{pmatrix}.$$

$\square$

---

[1] In some respects it is therefore more sensible to speak of an 'eigendirection', in that any vector $\mathbf{v}$ in the same direction as the vector $(1, 2)$ satisfies $\mathbb{A}\mathbf{v} = 6\mathbf{v}$.

Since we will be calculating many eigenvalues and eigenvectors in what follows, any work that we can save ourselves later on will be useful. To this end, we note here that there are some special cases in which it is possible to 'read off' the eigenvalues and eigenvectors from the matrix itself.

Example 27.2 *The eigenvalues of a diagonal matrix*

$$\begin{pmatrix} \lambda_1 & 0 \\ 0 & \lambda_2 \end{pmatrix}$$

are λ_1 and λ_2, with corresponding eigenvectors $\mathbf{v}_1 = (1,\ 0)$ and $\mathbf{v}_2 = (0,\ 1)$, respectively.

This is easy to check, either directly by matrix multiplication, or by applying the general method. □

Slightly less obvious, and therefore more useful, is the following.

Example 27.3 *The eigenvalues of both the matrices*

$$\begin{pmatrix} \lambda_1 & b \\ 0 & \lambda_2 \end{pmatrix} \quad and \quad \begin{pmatrix} \lambda_1 & 0 \\ c & \lambda_2 \end{pmatrix}$$

are λ_1 and λ_2. For the first, the eigenvector corresponding to λ_1 is $(1,\ 0)$; while for the second, the eigenvector corresponding to λ_2 is $(0,\ 1)$. In each case the other eigenvector needs to be found using the standard method.

The fact that λ_1 and λ_2 are the eigenvalues becomes obvious when you write down the characteristic equation; for example, if $\mathbb{A}$ is the matrix on the right then the eigenvalues λ are the solutions of

$$|\mathbb{A} - \lambda\mathbb{I}| = \begin{vmatrix} \lambda_1 - \lambda & 0 \\ c & \lambda_2 - \lambda \end{vmatrix} = (\lambda_1 - \lambda)(\lambda_2 - \lambda) = 0,$$

and hence $\lambda = \lambda_1$ or $\lambda = \lambda_2$. That the eigenvector corresponding to λ_2 is $(0,\ 1)$ essentially follows 'by inspection',

$$\begin{pmatrix} \lambda_1 & 0 \\ c & \lambda_2 \end{pmatrix} \begin{pmatrix} 0 \\ 1 \end{pmatrix} = \begin{pmatrix} 0 \\ \lambda_2 \end{pmatrix} = \lambda_2 \begin{pmatrix} 0 \\ 1 \end{pmatrix}.$$

The other eigenvector will need to be found via the standard calculation (find $\mathbf{v}$ with $(\mathbb{A} - \lambda_1\mathbb{I})\mathbf{v} = \mathbf{0}$). □

All the above examples have distinct real eigenvalues. We now look at an example in which there is only one eigenvalue.

Example 27.4 *Find the eigenvalues and eigenvectors of the matrix*

$$A = \begin{pmatrix} 5 & -4 \\ 1 & 1 \end{pmatrix}. \qquad (27.9)$$

The eigenvalues λ are the solutions of

$$\begin{vmatrix} 5 - \lambda & -4 \\ 1 & 1 - \lambda \end{vmatrix} = \lambda^2 - 6\lambda + 9 = (\lambda - 3)^2 = 0,$$

thus $\lambda = 3$ is a repeated eigenvalue. Since we have only one eigenvalue we can only expect to find one eigenvector $\mathbf{v}$; if $(A - \lambda I)\mathbf{v} = \mathbf{0}$ then

$$\begin{pmatrix} 2 & -4 \\ 1 & -2 \end{pmatrix}\begin{pmatrix} v_1 \\ v_2 \end{pmatrix} = \mathbf{0},$$

and so $\mathbf{v} = (2, 1)$. $\qquad\qquad\square$

It is also possible for the eigenvalues to be complex. Since they are the solutions of a quadratic equation, they will be a complex conjugate pair. The eigenvectors will now be complex too, and also come as a complex conjugate pair (so it will only be necessary to calculate one of them).

Example 27.5 *Find the eigenvalues and eigenvectors of the matrix*

$$A = \begin{pmatrix} 2 & 5 \\ -2 & 0 \end{pmatrix}.$$

The eigenvalues of A are given by the solutions of the quadratic equation

$$\begin{vmatrix} 2 - \lambda & 5 \\ -2 & -\lambda \end{vmatrix} = (2 - \lambda)(-\lambda) + 10 = \lambda^2 - 2\lambda + 10 = 0.$$

The roots of this equation are the complex conjugate pair

$$\lambda = \frac{2 \pm \sqrt{4 - 40}}{2} = 1 \pm 3i.$$

To find the eigenvector associated with $1 + 3i$ we have to solve $(A - \lambda I)\mathbf{v} = \mathbf{0}$, i.e.

$$\begin{pmatrix} 1 - 3i & 5 \\ -2 & -1 - 3i \end{pmatrix}\begin{pmatrix} v_1 \\ v_2 \end{pmatrix} = \mathbf{0},$$

or

$$\begin{pmatrix} (1 - 3i)v_1 + 5v_2 \\ -2v_1 - (1 + 3i)v_2 \end{pmatrix} = \mathbf{0}.$$

(Although not entirely clear, the second equation is, as usual, a multiple of the first.) Using the first equation we must have $5v_2 = (3i - 1)v_1$, and so we can take

$$\begin{pmatrix} 5 \\ 3i - 1 \end{pmatrix}$$

as a representative eigenvector with eigenvalue $1 + 3i$. Since

$$\begin{pmatrix} 2 & 5 \\ -2 & 0 \end{pmatrix} \begin{pmatrix} 5 \\ 3i - 1 \end{pmatrix} = \begin{pmatrix} 10 + 15i - 5 \\ -10 \end{pmatrix} = \begin{pmatrix} 5 + 15i \\ -5 \end{pmatrix}$$

$$= (1 + 3i) \begin{pmatrix} 5 \\ 3i - 1 \end{pmatrix},$$

$(5, 3i - 1)$ is indeed an eigenvector corresponding to the eigenvalue $1 + 3i$.

The eigenvector corresponding to $1 - 3i$, the complex conjugate of $1 + 3i$, will be the complex conjugate of the eigenvector we have already found,

$$\begin{pmatrix} 5 \\ -3i - 1 \end{pmatrix}.$$

To check that taking the complex conjugate works, we have

$$\begin{pmatrix} 2 & 5 \\ -2 & 0 \end{pmatrix} \begin{pmatrix} 5 \\ -3i - 1 \end{pmatrix} = \begin{pmatrix} 10 - 15i - 5 \\ -10 \end{pmatrix} = \begin{pmatrix} 5 - 15i \\ -5 \end{pmatrix}$$

$$= (1 - 3i) \begin{pmatrix} 5 \\ -3i - 1 \end{pmatrix},$$

as we should.   □

27.3 *Eigenvalues and eigenvectors with MATLAB

It is easy to find eigenvalues and eigenvectors using MATLAB. If A is a square matrix then the command eig(A) will return the eigenvalues of A.

```
>> A=[2 2; 6 3]

A = 2     2
    6     3

>> eig(A)

ans = -1
       6
```

If you also want MATLAB to find the eigenvectors for you then you can obtain them by typing [V D]=eig(A). This will return two matrices, V and D, where the two columns of V are the eigenvectors, chosen so that they have length one, and D is a diagonal matrix consisting of the eigenvalues.

```
>> [V D]=eig(A)

V = -0.5547    -0.4472
     0.8321    -0.8944

D = -1      0
     0      6
```

Note that, as you would expect, the first column of V is the eigenvector corresponding to the eigenvalue in the first position in the diagonal matrix D. However, since the eigenvectors have been chosen so that they have unit length they can appear to be quite complicated (in our analytical calculations we found eigenvectors $(2, -3)$ and $(1, 2)$). You can also use MATLAB to find the ratio of the components of the eigenvectors to each other if you want to write the eigenvectors as a pair of integers. First type format rat, which tells MATLAB to display its answers as fractions in their lowest terms, and then calculate the ratio of the components of the eigenvectors,

```
>> format rat
>> V(2,1)/V(1,1)

ans = -3/2

>> V(2,2)/V(1,2)

ans = 2
```

The first calculation shows that $(1, -3/2)$ is a possible choice for the first eigenvector (which gives our $(2, -3)$ after multiplying by two), while the second gives $(1, 2)$ immediately.

Exercises

27.1 Find the eigenvectors and eigenvalues of the following matrices:
    (i)

$$\begin{pmatrix} 1 & 2 \\ 1 & 0 \end{pmatrix},$$

(ii)

$$\begin{pmatrix} 2 & 2 \\ 0 & -4 \end{pmatrix},$$

(iii)

$$\begin{pmatrix} 7 & -2 \\ 26 & -1 \end{pmatrix},$$

(iv)

$$\begin{pmatrix} 9 & 2 \\ 2 & 6 \end{pmatrix},$$

(v)

$$\begin{pmatrix} 7 & 1 \\ -4 & 11 \end{pmatrix},$$

(vi)

$$\begin{pmatrix} 2 & -3 \\ 3 & 2 \end{pmatrix},$$

(vii)

$$\begin{pmatrix} 6 & 0 \\ 0 & -13 \end{pmatrix},$$

(viii)

$$\begin{pmatrix} 4 & -2 \\ 1 & 2 \end{pmatrix},$$

(ix)

$$\begin{pmatrix} 3 & -1 \\ 1 & 1 \end{pmatrix},$$

(x)

$$\begin{pmatrix} -7 & 6 \\ 12 & -1 \end{pmatrix}.$$

28

Distinct real eigenvalues

In this chapter, and the following two, our main aim is to show how to use a knowledge of the eigenvalues and eigenvectors of $\mathbb{A}$ in order to draw the phase diagram for the equation $\dot{\mathbf{x}} = \mathbb{A}\mathbf{x}$. As in Chapter 7, this phase diagram will illustrate the qualitative behaviour of the solutions by showing a representative choice of the curves traced out by the solutions $(x(t), y(t))$, labelled with an arrow to indicate in which direction the solution moves as t increases.

In each chapter we will examine one of the three possibilities (two distinct real eigenvalues, a complex conjugate pair of eigenvalues, or a repeated eigenvalue) and for each case we will show

(i) how an appropriate change of coordinates, based on the eigenvectors of the matrix $\mathbb{A}$, can be used to transform the differential equation into a standard, simpler (canonical) form;
(ii) how to find the explicit solution of this simple form of the equation;
(iii) how to draw the phase portrait for the simple equation;

and hence

(iv) how to find the explicit solution of the original equation; and
(v) how to draw its phase portrait.

Since we already have a reliable method for solving coupled linear equations, (ii) and (iv) will be much less important than (iii) and (v). Besides giving the mathematical justification of step (v), the coordinate transformations required to simplify the equation also provide a very natural illustration of the notion of the Jordan Canonical Form of a matrix, an important topic in the theory of linear algebra.

In the first of these three chapters we assume that $\mathbb{A}$ has two distinct real eigenvalues λ_1 and λ_2, with corresponding eigenvectors $\mathbf{v}_1$ and $\mathbf{v}_2$.

28.1 The explicit solution

In this, the simplest case, we do not need to make a change of coordinates in order to find the form of the explicit solution given the eigenvalues and eigenvectors.

In the previous chapter we found that when we tried $\mathbf{x}(t) = e^{\lambda t}\mathbf{v}$ as a solution of the linear equation $\dot{\mathbf{x}} = \mathbb{A}\mathbf{x}$ (see (27.4)), this gave rise to

$$\mathbb{A}\mathbf{v} = \lambda\mathbf{v}.$$

Given the eigenvalues and eigenvectors of $\mathbb{A}$ this means that we have obtained two possible solutions of the differential equation,

$$e^{\lambda_1 t}\mathbf{v}_1 \quad \text{and} \quad e^{\lambda_2 t}\mathbf{v}_2.$$

We saw in equation (27.3) in the previous chapter that a linear combination of two solutions is still a solution; thus the general solution of $d\mathbf{x}/dt = \mathbb{A}\mathbf{x}$ can be written as

$$\mathbf{x}(t) = Ae^{\lambda_1 t}\mathbf{v}_1 + Be^{\lambda_2 t}\mathbf{v}_2. \tag{28.1}$$

We now apply this method to a linear equation that is the homogeneous version of the equations in Example 26.1.

Example 28.1 *By finding the eigenvalues and eigenvectors of an appropriate matrix, find the general solution of the coupled system*

$$\begin{aligned} \dot{x} &= x + y \\ \dot{y} &= 4x - 2y. \end{aligned} \tag{28.2}$$

Rewritten as a matrix equation the problem becomes

$$\frac{d}{dt}\begin{pmatrix} x \\ y \end{pmatrix} = \begin{pmatrix} 1 & 1 \\ 4 & -2 \end{pmatrix}\begin{pmatrix} x \\ y \end{pmatrix}.$$

We will denote the matrix on the right-hand side by $\mathbb{A}$.

If we try a solution for the equation $\dot{\mathbf{x}} = \mathbb{A}\mathbf{x}$ of the form $\mathbf{x}(t) = e^{\lambda t}\mathbf{v}$, then as we saw above this leads to this eigenvalue problem

$$\lambda\mathbf{v} = \mathbb{A}\mathbf{v}.$$

To find the eigenvalues of $\mathbb{A}$ we solve the equation $\det(\mathbb{A} - \lambda\mathbb{I}) = 0$; this is

$$\begin{vmatrix} 1 - \lambda & 1 \\ 4 & -2 - \lambda \end{vmatrix} = 0,$$

which gives the quadratic equation

$$(1 - \lambda)(-2 - \lambda) - 4 = 0.$$

Multiplying this out and simplifying we obtain

$$\lambda^2 + \lambda - 6 = 0.$$

The solutions of this, $\lambda = 2$ or $\lambda = -3$, are the eigenvalues of $\mathbb{A}$.

Now that we know the eigenvalues we can find the corresponding eigenvectors. If we write

$$\mathbf{v} = \begin{pmatrix} v_1 \\ v_2 \end{pmatrix}$$

then when $\lambda = 2$ we need $(\mathbb{A} - 2\mathbb{I})\mathbf{v} = \mathbf{0}$, i.e.

$$\left[\begin{pmatrix} 1 & 1 \\ 4 & -2 \end{pmatrix} - \begin{pmatrix} 2 & 0 \\ 0 & 2 \end{pmatrix} \right] \mathbf{v} = \begin{pmatrix} -1 & 1 \\ 4 & -4 \end{pmatrix} \begin{pmatrix} v_1 \\ v_2 \end{pmatrix} = \mathbf{0},$$

We obtain two equations relating v_1 and v_2,

$$v_1 - v_2 = 0 \quad \text{and} \quad 4v_1 - 4v_2 = 0,$$

and so $v_1 = v_2$. A representative eigenvector is therefore

$$\mathbf{v}_1 = \begin{pmatrix} 1 \\ 1 \end{pmatrix}.$$

For $\lambda = -3$ we want the eigenvector $\mathbf{v}_2 = (v_1, \ v_2)$ to solve $(\mathbb{A} + 3\mathbb{I})\mathbf{v} = \mathbf{0}$,

$$\left[\begin{pmatrix} 1 & 1 \\ 4 & -2 \end{pmatrix} - \begin{pmatrix} -3 & 0 \\ 0 & -3 \end{pmatrix} \right] \mathbf{v} = \begin{pmatrix} 4 & 1 \\ 4 & 1 \end{pmatrix} \begin{pmatrix} v_1 \\ v_2 \end{pmatrix} = \mathbf{0}.$$

This provides just one equation relating v_1 and v_2,

$$4v_1 + v_2 = 0,$$

so $v_2 = -4v_1$, and

$$\mathbf{v}_2 = \begin{pmatrix} 1 \\ -4 \end{pmatrix}$$

is an eigenvector.

The general solution is therefore (cf. (28.1))

$$\mathbf{x}(t) = A e^{2t} \begin{pmatrix} 1 \\ 1 \end{pmatrix} + B e^{-3t} \begin{pmatrix} 1 \\ -4 \end{pmatrix}. \qquad \square$$

28.2 Changing coordinates

When we have two distinct real eigenvalues we will change to a coordinate system that uses the eigenvectors as axes. It is a general result that eigenvectors

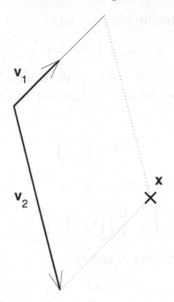

Fig. 28.1. Writing $\mathbf{x}$ as a combination of $\mathbf{v}_1$ and $\mathbf{v}_2$; here $\mathbf{x} = 2\mathbf{v}_1 + \mathbf{v}_2$.

corresponding to different eigenvalues are linearly independent (see Appendix B), which means that any vector can be written as an appropriate linear combination of $\mathbf{v}_1$ and $\mathbf{v}_2$,

$$\mathbf{x} = \begin{pmatrix} x \\ y \end{pmatrix} = \tilde{x}\mathbf{v}_1 + \tilde{y}\mathbf{v}_2, \tag{28.3}$$

see Figure 28.1.

Writing (28.3) in full gives

$$\begin{pmatrix} x \\ y \end{pmatrix} = \begin{pmatrix} \tilde{x}v_{11} + \tilde{y}v_{21} \\ \tilde{x}v_{12} + \tilde{y}v_{22} \end{pmatrix}$$

(where v_{ij} is the jth component of $\mathbf{v}_i$). We could also write this as

$$\begin{pmatrix} x \\ y \end{pmatrix} = \begin{pmatrix} v_{11} & v_{21} \\ v_{12} & v_{22} \end{pmatrix} \begin{pmatrix} \tilde{x} \\ \tilde{y} \end{pmatrix},$$

or more conveniently

$$\mathbf{x} = [\mathbf{v}_1 \ \mathbf{v}_2]\,\tilde{\mathbf{x}}, \tag{28.4}$$

where $\tilde{\mathbf{x}} = (\tilde{x}, \ \tilde{y})$ and we are using the notation $[\mathbf{v}_1 \ \mathbf{v}_2]$ to mean the 2×2 matrix with columns $\mathbf{v}_1$ and $\mathbf{v}_2$. To further simplify notation, we will write $\mathbb{P} = [\mathbf{v}_1 \ \mathbf{v}_2]$, so (28.4) is just $\mathbf{x} = \mathbb{P}\tilde{\mathbf{x}}$. To find $\tilde{\mathbf{x}}$ given $\mathbf{x}$, we have to multiply both sides by the inverse of $\mathbb{P}$,

$$\tilde{x} = \mathbb{P}^{-1}x = [\mathbf{v}_1 \ \mathbf{v}_2]^{-1}x. \tag{28.5}$$

In order to write the differential equation

$$\frac{dx}{dt} = \mathbb{A}x \tag{28.6}$$

in these new coordinates, we need to calculate $d\tilde{x}/dt$ in terms of $\tilde{x}$. Using (28.5), the original equation (28.6), and (28.4), we have

$$\begin{aligned}
\frac{d\tilde{x}}{dt} &= \frac{d}{dt}(\mathbb{P}^{-1}x) \\
&= \mathbb{P}^{-1}\frac{dx}{dt} \\
&= \mathbb{P}^{-1}\mathbb{A}x \\
&= \mathbb{P}^{-1}\mathbb{A}\mathbb{P}\tilde{x},
\end{aligned}$$

where we have used $x = \mathbb{P}\tilde{x}$ in the last line. We have obtained a new linear equation for $\tilde{x}$,

$$\frac{d\tilde{x}}{dt} = \mathbb{P}^{-1}\mathbb{A}\mathbb{P}\tilde{x}. \tag{28.7}$$

Although this looks more complicated, if we now substitute for $\mathbb{P}$ we have

$$\begin{aligned}
\mathbb{P}^{-1}\mathbb{A}\mathbb{P} &= [\mathbf{v}_1 \ \mathbf{v}_2]^{-1}\mathbb{A}[\mathbf{v}_1 \ \mathbf{v}_2] \\
&= [\mathbf{v}_1 \ \mathbf{v}_2]^{-1}[\mathbb{A}\mathbf{v}_1 \ \mathbb{A}\mathbf{v}_2] \\
&= [\mathbf{v}_1 \ \mathbf{v}_2]^{-1}[\lambda_1\mathbf{v}_1 \ \lambda_2\mathbf{v}_2].
\end{aligned}$$

That $\mathbb{A}[\mathbf{v}_1 \ \mathbf{v}_2] = [\mathbb{A}\mathbf{v}_1 \ \mathbb{A}\mathbf{v}_2]$ follows from the definition of matrix multiplication (see Appendix B), and then since $\mathbb{A}\mathbf{v}_j = \lambda_j\mathbf{v}_j$ we have $[\mathbb{A}\mathbf{v}_1 \ \mathbb{A}\mathbf{v}_2] = [\lambda_1\mathbf{v}_1 \ \lambda_2\mathbf{v}_2]$.

We can rewrite

$$\begin{aligned}
[\lambda_1\mathbf{v}_1 \ \lambda_2\mathbf{v}_2] &= \begin{pmatrix} \lambda_1 v_{11} & \lambda_2 v_{21} \\ \lambda_1 v_{12} & \lambda_2 v_{22} \end{pmatrix} = \begin{pmatrix} v_{11} & v_{21} \\ v_{12} & v_{22} \end{pmatrix}\begin{pmatrix} \lambda_1 & 0 \\ 0 & \lambda_2 \end{pmatrix} \\
&= [\mathbf{v}_1 \ \mathbf{v}_2]\begin{pmatrix} \lambda_1 & 0 \\ 0 & \lambda_2 \end{pmatrix},
\end{aligned}$$

and so we have

$$\mathbb{P}^{-1}\mathbb{A}\mathbb{P} = [\mathbf{v}_1 \ \mathbf{v}_2]^{-1}[\mathbf{v}_1 \ \mathbf{v}_2]\begin{pmatrix} \lambda_1 & 0 \\ 0 & \lambda_2 \end{pmatrix} = \begin{pmatrix} \lambda_1 & 0 \\ 0 & \lambda_2 \end{pmatrix}.$$

Therefore the equation for $\tilde{\mathbf{x}}$ is

$$\frac{d\tilde{\mathbf{x}}}{dt} = \begin{pmatrix} \lambda_1 & 0 \\ 0 & \lambda_2 \end{pmatrix} \tilde{\mathbf{x}}, \tag{28.8}$$

which in terms of the coordinates $\tilde{x}$ and $\tilde{y}$ is

$$\begin{aligned} d\tilde{x}/dt &= \lambda_1 \tilde{x} \\ d\tilde{y}/dt &= \lambda_2 \tilde{y}. \end{aligned} \tag{28.9}$$

We have obtained equations for $\tilde{x}$ and $\tilde{y}$ that are no longer coupled. We can solve these equations easily to find the solutions

$$\tilde{x}(t) = A e^{\lambda_1 t} \quad \text{and} \quad \tilde{y}(t) = B e^{\lambda t}.$$

Note that if we use the expression $\mathbf{x} = \tilde{x}\mathbf{v}_1 + \tilde{y}\mathbf{v}_2$ for $\mathbf{x}$ in terms of $\tilde{x}$ and $\tilde{y}$ (this was equation (28.3)) then we recover the general solution of the original problem as given in (28.1),

$$\mathbf{x}(t) = A e^{\lambda_1 t}\mathbf{v}_1 + B e^{\lambda_2 t}\mathbf{v}_2.$$

We now apply these ideas to the differential equation we considered as Example 28.1. Note that this is *not* a sensible way to find the explicit solution of the coupled equations in the example, since we can write down this solution as soon as we know the eigenvalues and eigenvectors of $\mathbb{A}$.

Example 28.2 *By means of an appropriate coordinate transformation decouple the equations*

$$\begin{aligned} \dot{x} &= x + y \\ \dot{y} &= 4x - 2y \end{aligned}$$

and hence write down their general solution.

Rewriting the equation in matrix form we have

$$\dot{\mathbf{x}} = \underbrace{\begin{pmatrix} 1 & 1 \\ 4 & -2 \end{pmatrix}}_{\mathbb{A}} \mathbf{x}.$$

We found the eigenvalues and eigenvectors of the matrix $\mathbb{A}$ for Example 28.1: they are

$$\lambda_1 = 2 \quad \text{with} \quad \mathbf{v}_1 = \begin{pmatrix} 1 \\ 1 \end{pmatrix} \quad \text{and} \quad \lambda_2 = -3 \quad \text{with} \quad \mathbf{v}_2 = \begin{pmatrix} 1 \\ -4 \end{pmatrix}.$$

Note that $\mathbf{v}_1$ and $\mathbf{v}_2$ are linearly independent; if

$$\alpha \begin{pmatrix} 1 \\ 1 \end{pmatrix} + \beta \begin{pmatrix} 1 \\ -4 \end{pmatrix} = 0$$

then $\alpha + \beta = 0$ and $\alpha - 4\beta = 0$. The only solution of these two equations is $\alpha = \beta = 0$, and so the two vectors are linearly independent as claimed.

In order to write a vector $\mathbf{x}$ in terms of these two eigenvectors,

$$\begin{pmatrix} x \\ y \end{pmatrix} = \tilde{x}\mathbf{v}_1 + \tilde{y}\mathbf{v}_2, \tag{28.10}$$

we need

$$\begin{pmatrix} x \\ y \end{pmatrix} = \tilde{x}\begin{pmatrix} 1 \\ 1 \end{pmatrix} + \tilde{y}\begin{pmatrix} 1 \\ -4 \end{pmatrix} = \begin{pmatrix} \tilde{x}+\tilde{y} \\ \tilde{x}-4\tilde{y} \end{pmatrix} = \underbrace{\begin{pmatrix} 1 & 1 \\ 1 & -4 \end{pmatrix}}_{\mathbb{P}}\begin{pmatrix} \tilde{x} \\ \tilde{y} \end{pmatrix}$$

('$\mathbf{x} = \mathbb{P}\tilde{\mathbf{x}}$').

Multiplying the extreme left- and right-hand sides of this equation by $\mathbb{P}^{-1}$ will give $\tilde{\mathbf{x}}$ in terms of $\mathbf{x}$,

$$\begin{pmatrix} \tilde{x} \\ \tilde{y} \end{pmatrix} = \begin{pmatrix} 1 & 1 \\ 1 & -4 \end{pmatrix}^{-1}\begin{pmatrix} x \\ y \end{pmatrix}.$$

Computing the inverse we arrive at

$$\begin{pmatrix} \tilde{x} \\ \tilde{y} \end{pmatrix} = \frac{1}{5}\begin{pmatrix} 4 & 1 \\ 1 & -1 \end{pmatrix}\begin{pmatrix} x \\ y \end{pmatrix},$$

i.e.

$$\tilde{x} = \frac{1}{5}(4x + y) \qquad \text{and} \qquad \tilde{y} = \frac{1}{5}(x - y).$$

Referred to these new coordinate axes the equation becomes

$$\frac{d\tilde{\mathbf{x}}}{dt} = \mathbb{P}^{-1}A\mathbb{P}\tilde{\mathbf{x}}$$

$$= \frac{1}{5}\begin{pmatrix} 4 & 1 \\ 1 & -1 \end{pmatrix}\begin{pmatrix} 1 & 1 \\ 4 & -2 \end{pmatrix}\begin{pmatrix} 1 & 1 \\ 1 & -4 \end{pmatrix}\tilde{\mathbf{x}}$$

$$= \frac{1}{5}\begin{pmatrix} 4 & 1 \\ 1 & -1 \end{pmatrix}\begin{pmatrix} 2 & -3 \\ 2 & 12 \end{pmatrix}\tilde{\mathbf{x}}$$

$$= \frac{1}{5}\begin{pmatrix} 10 & 0 \\ 0 & -15 \end{pmatrix}\tilde{\mathbf{x}}$$

$$= \begin{pmatrix} 2 & 0 \\ 0 & -3 \end{pmatrix}\tilde{\mathbf{x}}.$$

So in our new variables we obtain the decoupled equations

$$\frac{d\tilde{x}}{dt} = 2\tilde{x} \qquad \frac{d\tilde{y}}{dt} = -3\tilde{y}.$$

The solutions of these can easily be seen to be $\tilde{x}(t) = Ae^{2t}$ and $\tilde{y}(t) = Be^{-3t}$, and so the solution of the original equation can be recovered from (28.10),

$$\mathbf{x}(t) = Ae^{2t}\begin{pmatrix} 1 \\ 1 \end{pmatrix} + Be^{-3t}\begin{pmatrix} 1 \\ -4 \end{pmatrix}. \qquad \square$$

28.3 Phase diagrams for uncoupled equations

We have just seen that in a new coordinate system $\tilde{\mathbf{x}} = (\tilde{x}, \tilde{y})$ that uses the eigenvectors as axes the original equation becomes

$$\frac{d\tilde{\mathbf{x}}}{dt} = \begin{pmatrix} \lambda_1 & 0 \\ 0 & \lambda_2 \end{pmatrix}\tilde{\mathbf{x}}. \qquad (28.11)$$

This gives the two decoupled equations $d\tilde{x}/dt = \lambda_1\tilde{x}$ and $d\tilde{y}/dt = \lambda_2\tilde{y}$, whose solutions are

$$\tilde{x}(t) = \tilde{x}_0 e^{\lambda_1 t} \qquad \text{and} \qquad \tilde{y}(t) = \tilde{y}_0 e^{\lambda_2 t}. \qquad (28.12)$$

We now draw the phase diagram for (28.11), assuming that λ_1 and λ_2 are both non-zero (for the exceptional case when one of them is zero see Exercise 28.6). We can use our explicit form for $\tilde{x}(t)$ and $\tilde{y}(t)$ in order to find the equation of the curves traced out by these solutions as t increases.[1] Since

$$\left(\frac{\tilde{x}(t)}{\tilde{x}_0}\right)^{\lambda_2/\lambda_1} = e^{\lambda_2 t}$$

we have

$$\left(\frac{\tilde{y}(t)}{\tilde{y}_0}\right) = \left(\frac{\tilde{x}(t)}{\tilde{x}_0}\right)^{\lambda_2/\lambda_1}.$$

So the trajectories will trace out the curves

$$\tilde{y} = K\tilde{x}^{\lambda_2/\lambda_1},$$

where the constant K depends on the initial conditions (in fact we have $K = \tilde{y}_0\tilde{x}_0^{-\lambda_2/\lambda_1}$). Note that it is also easy to see from (28.12) that given an initial condition on one of the axes ($\tilde{x}(0) = 0$ or $\tilde{y}(0) = 0$) then the solution remains on the axis; we say that the axes are *invariant*.

If both eigenvalues are negative then $\tilde{x}$ and $\tilde{y}$ decrease to zero as t increases. Suppose that $\lambda_1 < \lambda_2 < 0$; then $\lambda_1/\lambda_2 > 1$, and a typical phase portrait is shown

---

[1] It is also possible to derive this equation from (28.9), without finding the solutions first, by solving the equation $d\tilde{y}/d\tilde{x} = \lambda_2\tilde{y}/\lambda_1\tilde{x}$, see Exercise 28.4.

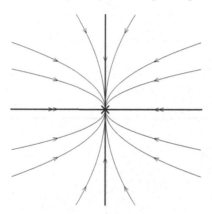

Fig. 28.2. A stable node.

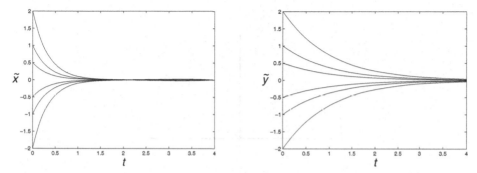

Fig. 28.3. Plots of $\tilde{x}(t)$ [solid line] and $\tilde{y}(t)$ [dotted line] against t. Both $\tilde{x}$ and $\tilde{y}$ decay to zero, but note that $\tilde{x}$ decays to zero much faster than $\tilde{y}$. (In the pictures $\mathrm{d}\tilde{x}/\mathrm{d}t = -3\tilde{x}$ and $\mathrm{d}\tilde{y}/\mathrm{d}t = -\tilde{y}$.)

in Figure 28.2 (like the graph of $x = Ky^r$ with $r > 1$; a 'rotated parabola' when $r = 2$). In this case the origin is called a **stable node**. Note that the axes are invariant as remarked above.

The axis associated with the faster rate of decay has been marked with two arrows on the figure; the solutions approach the origin, tangential to the direction corresponding to the value of λ with smaller modulus. You can see the reason for this by imagining the x component of the solution decaying very fast, so that the trajectory gets close to the y-axis quickly, and then tends towards the origin 'almost vertically'. You can see this in Figure 28.3, which shows the graphs of $\tilde{x}(t)$ and $\tilde{y}(t)$ against t for some sample initial conditions.

If both eigenvalues are positive then both $\tilde{x}$ and $\tilde{y}$ tend to $\pm\infty$ depending on the signs of the initial conditions. Suppose that $\lambda_2 > \lambda_1 > 0$; then $\lambda_2/\lambda_1 > 1$ and solutions move on curves like those shown in Figure 28.4: the graph is similar to $y = Kx^r$ with $r > 1$ (e.g. a parabola when $r = 2$).

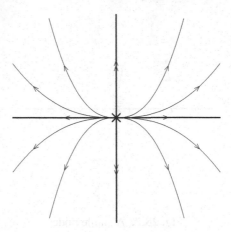

Fig. 28.4. An unstable node.

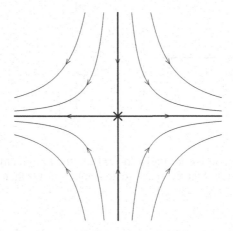

Fig. 28.5. A saddle point.

The direction in which growth is faster has been marked with a double arrow; note that solutions move away from the origin tangent to the direction that corresponds to slower growth (the smaller value of λ). In this situation the origin is known as an **unstable node**.

If λ_1 and λ_2 are of opposite signs then one of $\tilde{x}$ and $\tilde{y}$ increases, while the other decreases; there is one 'stable direction' and one 'unstable direction'. For example, if $\lambda_2 < 0 < \lambda_1$ then $\tilde{x}(t)$ tends to $\pm\infty$ while $\tilde{y}(t)$ tends to zero. The phase diagram is shown in Figure 28.5; the curves have equations like $xy^r = K$ (e.g. when $r = 1$ they are hyperbolae). The origin is called a **saddle point**, or just a **saddle**. (The behaviour here is like a ball rolling on a horse's saddle, see Figure C.1, for example.)

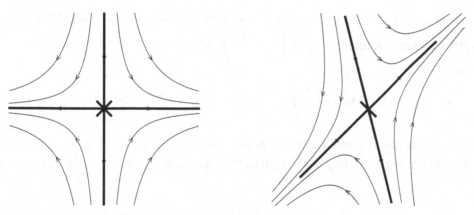

Fig. 28.6. In the left-hand picture ($\tilde{\mathbf{x}}$ coordinates) the eigenvectors appear as orthogonal axes; in the right-hand picture the axes have 'twisted' back to their true positions, moving all the other trajectories too.

28.4 Phase diagrams for coupled equations

Once we have drawn the phase diagram for the ($\tilde{x}$, $\tilde{y}$) variables it is fairly straightforward to draw the phase diagram for the original (x, y) variables.

The transformation we used to change coordinates had the effect of making the eigenvectors of the matrix $\mathbb{A}$ the coordinates axes for the new variables $\tilde{\mathbf{x}}$. This means that the $\tilde{x}$ and $\tilde{y}$ axes, at right-angles in the phase diagrams we drew in Figures 28.2, 28.4 and 28.5, actually correspond to the eigenvectors in the (x, y) plane.

When we transform a picture in the ($\tilde{x}$, $\tilde{y}$) plane back to the (x, y) plane the axes will 'twist' and line up with the eigenvectors. To draw the phase diagram in the (x, y) plane we also have to 'twist' all the other trajectories accordingly; this is illustrated in Figure 28.6 for the case of a saddle point.

In order to draw the phase diagram for a particular example it is not necessary to make the transformation to the new coordinate system, draw the phase portrait there, and then transform it back. Instead, once you have found the eigenvalues and eigenvectors first draw the eigenvectors (in the (x, y) plane), label the eigenvectors with an arrow in the appropriate direction (away from the origin if the eigenvalue is positive and so corresponds to an unstable direction, towards the origin if the eigenvalue is negative and thus corresponds to a stable direction), and then 'fill in' between the two eigenvectors in a consistent way. This idea is illustrated in the following example.

Example 28.3 *By finding the eigenvectors and eigenvalues of an appropriate matrix draw the phase portrait for the equation*

$$\dot{x} = x + y$$
$$\dot{y} = 4x - 2y.$$

This is the same example we looked at above; we have already found the eigen-values and eigenvectors of $\mathbb{A} = \begin{pmatrix} 1 & 1 \\ 4 & -2 \end{pmatrix}$, which are

$$\lambda_1 = 2 \quad \text{with} \quad \mathbf{v}_1 = \begin{pmatrix} 1 \\ 1 \end{pmatrix} \qquad \text{and} \qquad \lambda_2 = -3 \quad \text{with} \quad \mathbf{v}_2 = \begin{pmatrix} 1 \\ -4 \end{pmatrix}.$$

To draw the phase portrait, first we draw lines in the direction of the eigenvectors, and add arrows according to the sign of the eigenvalue, as in Figure 28.7. It is then

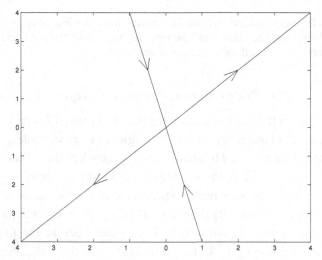

Fig. 28.7. First draw the eigenvectors and add arrows; away from the origin if $\lambda > 0$, towards the origin if $\lambda < 0$.

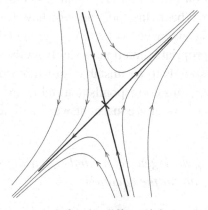

Fig. 28.8. The phase portrait for the differential equation of Example 28.3.

  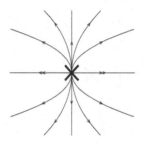

Fig. 28.9. Stable manifolds for a stable node (the whole plane), a saddle point (the eigenvector corresponding to the negative eigenvalue) and an unstable node (just the origin).

simple to add more illustrative trajectories to give the complete phase diagram of Figure 28.8.

28.5 Stable and unstable manifolds

We now introduce some terminology that will prove extremely useful when we come to study nonlinear systems. At present it may seem overly complicated, but it is easier to understand in these simple linear examples.

The **stable manifold**[2] of the origin, written $W^s(0)$, is all those points lying on trajectories that approach the origin as $t \to \infty$; for a stable node this stable manifold is all of $\mathbb{R}^2$; for a saddle this stable manifold is just the eigenvector corresponding to the negative eigenvalue; and for an unstable node the stable manifold is just the origin. These alternatives are illustrated in Figure 28.9.

Note that when the origin is a saddle point its stable manifold separates the plane into two parts; in our picture all the points to the right of the stable manifold move away 'to the right' (so to $x = +\infty$), while all the points to its left move away 'to the left' (to $x = -\infty$). In this case the stable manifold plays a particularly significantly rôle and is often called the 'separatrix'.

The **unstable manifold** of the origin, written as $W^u(0)$, is not quite 'all those points that move away from the origin'; it consists of all those points lying on trajectories that would approach the origin if we reversed the sense of time, i.e. reversed their arrows. When the origin is a stable node its unstable manifold is just $\{0\}$; for a saddle point the eigenvector corresponding to the positive eigenvalue is the unstable manifold; and for an unstable node the unstable manifold is the whole of $\mathbb{R}^2$, see Figure 28.10.

---

[2] A *manifold* is an abstract mathematical object; the easiest way of imagining it is as a generalisation of a 'surface'. In the relatively simple systems we are looking at here this generality will not be needed, although we will see some of the possible complexity in the final chapter when we consider the Lorenz equations. It is probably more helpful for now to think of $W^s(0)$ as the 'stable set' of the origin.

  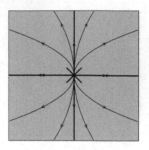

Fig. 28.10. Unstable manifolds for a stable node (just the origin), a saddle point (the eigenvector corresponding to the positive eigenvalue) and an unstable node (the whole plane).

Exercises

28.1 Write down the general solution and draw the phase portrait for the equation $\dot{x} = A x$, when the eigenvalues and eigenvectors of A are as follows. You should take particular care with stable (or unstable) nodes to ensure that the trajectories approach (or move away from) the origin tangent to the correct eigenvector.

   (i) $\lambda_1 = 1$, $v_1 = (1, \ 1)$ and $\lambda_2 = 2$, $v_2 = (1, \ -1)$;

   (ii) $\lambda_1 = 1$, $v_1 = (1, \ 0)$ and $\lambda_2 = -2$, $v_2 = (1, \ 1)$;

   (iii) $\lambda_1 = -2$, $v_1 = (1, \ 2)$ and $\lambda_2 = -3$, $v_2 = (2, \ -3)$;

   (iv) $\lambda_1 = 3$, $v_1 = (2, \ 3)$ and $\lambda_2 = -5$, $v_2 = (0, \ 1)$;

   (v) $\lambda_1 = 3$, $v_1 = (1, \ 2)$ and $\lambda_2 = 1$, $v_2 = (1, \ -3)$;

   (vi) $\lambda_1 = 2$, $v_1 = (0, \ 1)$ and $\lambda_2 = -3$, $v_2 = (1, \ 5)$;

   (vii) $\lambda_1 = 1$, $v_1 = (1, \ 1)$ and $\lambda_2 = 2$, $v_2 = (2, \ 1)$; and

   (viii) $\lambda_1 = -3$, $v_1 = (1, \ 3)$ and $\lambda_2 = -1$, $v_2 = (-3, \ 2)$.

28.2 For the following equations find the eigenvalues and eigenvectors of the matrix on the right-hand side, and hence find the coordinate transformation that will decouple the equations. Show that this transformation has the desired effect. (You can also write down the general solution and draw the phase portrait for the equation if you wish.)

   (i)

$$\frac{dx}{dt} = \begin{pmatrix} 8 & 14 \\ 7 & 1 \end{pmatrix} x$$

   (ii)

$$\frac{dx}{dt} = \begin{pmatrix} 2 & 0 \\ -5 & -3 \end{pmatrix} x$$

   (iii)

$$\frac{dx}{dt} = \begin{pmatrix} 11 & -2 \\ 3 & 4 \end{pmatrix} x$$

and

(iv)

$$\frac{d\mathbf{x}}{dt} = \begin{pmatrix} 1 & 20 \\ 40 & -19 \end{pmatrix} \mathbf{x}.$$

28.3 (C) Given a matrix

$$\mathbb{A} = \begin{pmatrix} a & b \\ c & d \end{pmatrix},$$

the M-file lportrait.m will draw the phase portrait for the linear equation $\dot{\mathbf{x}} = \mathbb{A}\mathbf{x}$. The program draws the trajectory forwards and backwards from a given initial condition, placing an arrow there indicating the direction the solution moves as t increases. Draw the phase portraits for the equations in the previous exercise using this program.

28.4 (T) Using the chain rule, if $y = y(x(t))$ then

$$\frac{dy}{dt} = \frac{dy}{dx}\frac{dx}{dt},$$

from which it follows that

$$\frac{dy}{dx} = \frac{dy}{dt} \bigg/ \frac{dx}{dt}.$$

Therefore if

$$\frac{dx}{dt} = \lambda_1 x \qquad \text{and} \qquad \frac{dy}{dt} = \lambda_2 y \tag{E28.1}$$

we have

$$\frac{dy}{dx} = \frac{\lambda_2 y}{\lambda_1 x}.$$

Solve this to find the equation of the curves traced out by trajectories of (E28.1).

28.5 (T) We have seen in this chapter that if $\mathbb{A}$ has distinct real eigenvalues λ_1 and λ_2, with corresponding eigenvectors $\mathbf{v}_1$ and $\mathbf{v}_2$, then

$$\mathbb{P}^{-1}\mathbb{A}\mathbb{P} = \begin{pmatrix} \lambda_1 & 0 \\ 0 & \lambda_2 \end{pmatrix},$$

where $\mathbb{P} = [\mathbf{v}_1 \ \mathbf{v}_2]$. It follows, conversely, that the matrix with these eigenvalues and eigenvector is

$$\mathbb{A} = \mathbb{P} \begin{pmatrix} \lambda_1 & 0 \\ 0 & \lambda_2 \end{pmatrix} \mathbb{P}^{-1}.$$

(This is how the M-file makematrix.m constructs matrices with specified eigenvalues and eigenvectors.) Find the matrices whose eigenvalues and eigenvectors are as follows:

(i) $\lambda_1 = 3$, $\mathbf{v}_1 = (1, 2)$ and $\lambda_2 = 6$, $\mathbf{v}_2 = (1, -1)$;
(ii) $\lambda_1 = 3$, $\mathbf{v}_1 = (1, 0)$ and $\lambda_2 = -1$, $\mathbf{v}_2 = (2, 1)$; and
(iii) $\lambda_1 = 5$, $\mathbf{v}_1 = (1, 1)$ and $\lambda_2 = 1$, $\mathbf{v}_2 = (1 - 1,)$.

(You could now check your phase portraits for Exercise 28.1, using the M-file makematrix.m to find the matrix with the specified eigenvalues and eigenvectors, and then lportrait.m to draw the phase portraits.)

28.6 Suppose that A has two eigenvalues, $\lambda_1 = 0$ with eigenvector v_1 and $\lambda_2 \neq 0$ with eigenvector v_2.

(i) Write down the general solution of the equation $\dot{x} = Ax$.

(ii) After changing to a coordinate system referred to the eigenvectors the equation will become

$$\frac{d\tilde{x}}{dt} = \begin{pmatrix} 0 & 0 \\ 0 & \lambda_2 \end{pmatrix} \tilde{x},$$

i.e

$$\frac{d\tilde{x}}{dt} = 0 \quad \text{and} \quad \frac{d\tilde{y}}{dt} = \lambda\tilde{y}.$$

By solving these equations draw the phase portrait in the $(\tilde{x}, \tilde{y})$ system, and hence sketch the phase portrait for the original coordinates.

(iii) Draw the phase portrait for the equation

$$\frac{dx}{dt} = \begin{pmatrix} -2 & 2 \\ 1 & -1 \end{pmatrix} x.$$

29

More phase portraits: complex eigenvalues

We have seen how to find explicit solutions of the equation $\dot{\mathbf{x}} = \mathbb{A}\mathbf{x}$, and how to draw the corresponding phase portrait, when the matrix $\mathbb{A}$ has a pair of distinct real eigenvalues. In this chapter we treat the first of the two remaining possibilities; a complex conjugate pair of eigenvalues $\lambda_{\pm} = \rho \pm i\omega$, with a corresponding pair of complex conjugate eigenvectors $\boldsymbol{\eta}_{\pm} = \mathbf{v}_1 \pm i\mathbf{v}_2$.

29.1 The explicit solution

If all you want is an explicit solution then it is unlikely that you would want to find the eigenvalues and eigenvectors and use the method here, since the method of Chapter 26 is much simpler. However, if you have the eigenvalues and eigenvectors of the matrix then this method will provide you with the explicit solution after only a little work.

If we try to construct the general solution of $\dot{\mathbf{x}} = \mathbb{A}\mathbf{x}$ as we did in the previous chapter, by setting $\mathbf{x}(t) = Ce^{\lambda_+ t}\boldsymbol{\eta}_+ + De^{\lambda_- t}\boldsymbol{\eta}_-$, we obtain

$$\mathbf{x}(t) = Ce^{(\rho+i\omega)t}[\mathbf{v}_1 + i\mathbf{v}_2] + C^*e^{(\rho-i\omega)t}[\mathbf{v}_1 - i\mathbf{v}_2].$$

Here we have taken the first coefficient to be a complex number $C = \alpha + i\beta$, and the second to be C^* (its complex conjugate), which ensures that $\mathbf{x}(t)$ is real.

We therefore have (since $z^* + z = 2\,\mathrm{Re}[z]$)

$$\mathbf{x}(t) = 2\,\mathrm{Re}\,[Ce^{(\rho+i\omega)t}(\mathbf{v}_1 + i\mathbf{v}_2)]. \tag{29.1}$$

Since $C = \alpha + i\beta$ this is

$$\begin{aligned}
\mathbf{x}(t) &= 2e^{\rho t}\,\mathrm{Re}\,[(\alpha + i\beta)(\cos\omega t + i\sin\omega t)(\mathbf{v}_1 + i\mathbf{v}_2)]\\
&= 2e^{\rho t}\,\mathrm{Re}\,[((\alpha\cos\omega t - \beta\sin\omega t) + i(\beta\cos\omega t + \alpha\sin\omega t))(\mathbf{v}_1 + i\mathbf{v}_2)]\\
&= 2e^{\rho t}[(\alpha\cos\omega t - \beta\sin\omega t)\mathbf{v}_1 - (\beta\cos\omega t + \alpha\sin\omega t)\mathbf{v}_2].
\end{aligned}$$

Since α and β are arbitrary constants we can set $A = 2\alpha$ and $B = -2\beta$, so that the general solution is finally

$$\mathbf{x}(t) = e^{\rho t}[(A \cos \omega t + B \sin \omega t)\mathbf{v}_1 + (B \cos \omega t - A \sin \omega t)\mathbf{v}_2]. \qquad (29.2)$$

We illustrate this method to find, once again, the solution of Example 26.2. It is probably better to go through the algebra each time than try to remember (29.2).

Example 29.1 *Find the general solution of the equation*

$$\dot{x} = 2x + 5y$$
$$\dot{y} = -2x.$$

We found the eigenvalues and eigenvectors of the matrix $\begin{pmatrix} 2 & 5 \\ -2 & 0 \end{pmatrix}$ in Example 27.5,

$$\rho \pm i\omega = 1 \pm 3i \quad \text{with eigenvectors} \quad \mathbf{v}_1 \pm i\mathbf{v}_2 = \begin{pmatrix} 5 \\ -1 \end{pmatrix} \pm i \begin{pmatrix} 0 \\ 3 \end{pmatrix}.$$

Therefore, using (29.1),

$$\mathbf{x}(t) = 2\text{Re}\left\{(\alpha + i\beta)e^{(1+3i)t}\left[\begin{pmatrix} 5 \\ -1 \end{pmatrix} + i \begin{pmatrix} 0 \\ 3 \end{pmatrix}\right]\right\}$$

$$= 2e^t\text{Re}\left\{(\alpha + i\beta)(\cos 3t + i \sin 3t)\left[\begin{pmatrix} 5 \\ -1 \end{pmatrix} + i \begin{pmatrix} 0 \\ 3 \end{pmatrix}\right]\right\}$$

$$= 2e^t\text{Re}\left\{\left[(\alpha \cos 3t - \beta \sin 3t) + i(\beta \cos 3t - \alpha \sin 3t)\right]\right.$$

$$\times \left.\left[\begin{pmatrix} 5 \\ -1 \end{pmatrix} + i \begin{pmatrix} 0 \\ 3 \end{pmatrix}\right]\right\}$$

$$= 2e^t\left\{(\alpha \cos 3t - \beta \sin 3t)\begin{pmatrix} 5 \\ -1 \end{pmatrix} - (\beta \cos 3t - \alpha \sin 3t)\begin{pmatrix} 0 \\ 3 \end{pmatrix}\right\}$$

With $C = 2\alpha$ and $D = -2\beta$ we obtain

$$\begin{pmatrix} x(t) \\ y(t) \end{pmatrix} = e^t \begin{pmatrix} 5C \cos 3t + 5D \sin 3t \\ (3D - C) \cos 3t - (3C + D) \sin 3t \end{pmatrix},$$

which agrees with the solution (26.6) that we obtained earlier.   □

Although the formula in (29.2) does provide us with an explicit solution, it is still not obvious how the solution behaves. Drawing the phase portrait will make things much clearer.

29.2 Changing coordinates and the phase portrait

We now make a coordinate transformation to put the equation into a standard form, choosing the real and imaginary parts of the eigenvectors, v_1 and v_2, as our new axes. If you want to avoid the details, you should go to equation (29.6) which gives the form of the differential equation in these new coordinates.

Before we make our change of coordinates, it will be useful to note that since $\eta_+ = v_1 + iv_2$ is an eigenvector of $\mathbb{A}$ with eigenvalue $\lambda_+ = \rho + i\omega$, we have

$$\mathbb{A}[v_1 + iv_2] = (\rho + i\omega)[v_1 + iv_2] = (\rho v_1 - \omega v_2) + i(\omega v_1 + \rho v_2).$$

Taking real and imaginary parts of this equation gives

$$\mathbb{A}v_1 = \rho v_1 - \omega v_2 \qquad \text{and} \qquad \mathbb{A}v_2 = \omega v_1 + \rho v_2. \qquad (29.3)$$

It is relatively straightforward to check that v_1 and v_2 are linearly independent, see Exercise 29.4, so we can write any vector as a linear combination

$$x = \tilde{x}v_1 + \tilde{y}v_2. \qquad (29.4)$$

We saw above in (28.7) that if we make this coordinate transformation then we can write

$$x = [v_1 \quad v_2]\tilde{x}$$

and the differential equation satisfied by $\tilde{x}$ is

$$\frac{d\tilde{x}}{dt} = [v_1 \quad v_2]^{-1}\mathbb{A}[v_1 \quad v_2]\tilde{x}.$$

In the current case we have

$$\begin{aligned}
[v_1 \quad v_2]^{-1}\mathbb{A}[v_1 \quad v_2] &= [v_1 \quad v_2]^{-1}[\mathbb{A}v_1 \quad \mathbb{A}v_2] \\
&= [v_1 \quad v_2]^{-1}[\rho v_1 - \omega v_2 \quad \omega v_1 + \rho v_2] \\
&= [v_1 \quad v_2]^{-1}[v_1 \quad v_2]\begin{pmatrix} \rho & \omega \\ -\omega & \rho \end{pmatrix} \\
&= \begin{pmatrix} \rho & \omega \\ -\omega & \rho \end{pmatrix}.
\end{aligned}$$

So we have obtained an equation for $\tilde{\mathbf{x}}$ which is in the standard, simpler form

$$\frac{d\tilde{\mathbf{x}}}{dt} = \begin{pmatrix} \rho & \omega \\ -\omega & \rho \end{pmatrix} \tilde{\mathbf{x}}. \qquad (29.5)$$

Example 29.2 *By means of an appropriate choice of coordinates transform the equations*

$$\dot{x} = 2x + 5y$$
$$\dot{y} = -2x$$

into the standard form

$$\frac{d\tilde{\mathbf{x}}}{dt} = \begin{pmatrix} 1 & 3 \\ -3 & 1 \end{pmatrix} \tilde{\mathbf{x}}.$$

The eigenvectors were found in Example 27.5. They are

$$\mathbf{v}_1 \pm i\mathbf{v}_2 = \begin{pmatrix} 5 \\ -1 \end{pmatrix} \pm i \begin{pmatrix} 0 \\ 3 \end{pmatrix},$$

corresponding to the eigenvalues $1 \pm 3i$. We therefore set

$$\mathbf{x} = \tilde{x} \begin{pmatrix} 5 \\ -1 \end{pmatrix} + \tilde{y} \begin{pmatrix} 0 \\ 3 \end{pmatrix} = \underbrace{\begin{pmatrix} 5 & 0 \\ -1 & 3 \end{pmatrix}}_{\mathbb{P}} \tilde{\mathbf{x}},$$

and so

$$\tilde{\mathbf{x}} = \frac{1}{15} \begin{pmatrix} 3 & 0 \\ 1 & 5 \end{pmatrix} \mathbf{x}.$$

The equation satisfied by $\tilde{\mathbf{x}}$ is

$$\frac{d\tilde{\mathbf{x}}}{dt} = \mathbb{P}^{-1} \mathbb{A} \mathbb{P} \tilde{\mathbf{x}}$$

$$= \frac{1}{15} \begin{pmatrix} 3 & 0 \\ 1 & 5 \end{pmatrix} \begin{pmatrix} 2 & 5 \\ -2 & 0 \end{pmatrix} \begin{pmatrix} 5 & 0 \\ -1 & 3 \end{pmatrix} \tilde{\mathbf{x}}$$

$$= \frac{1}{15} \begin{pmatrix} 6 & 15 \\ -8 & 5 \end{pmatrix} \begin{pmatrix} 5 & 0 \\ -1 & 3 \end{pmatrix} \tilde{\mathbf{x}}$$

$$= \frac{1}{15} \begin{pmatrix} 15 & 45 \\ -45 & 15 \end{pmatrix} \tilde{\mathbf{x}}$$

$$= \begin{pmatrix} 1 & 3 \\ -3 & 1 \end{pmatrix} \tilde{\mathbf{x}},$$

as required. Note that, since $\lambda_\pm = 1 \pm 3i$, this agrees with (29.5).   □

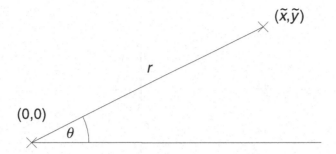

Fig. 29.1. Plane polar coordinates; $\tilde{x} = r\cos\theta$ and $\tilde{y} = r\sin\theta$.

Since the equation in the new coordinates is

$$\frac{d\tilde{\mathbf{x}}}{dt} = \begin{pmatrix} \rho & \omega \\ -\omega & \rho \end{pmatrix} \tilde{\mathbf{x}}, \tag{29.6}$$

the equations for the components $\tilde{x}$ and $\tilde{y}$ are

$$\frac{d\tilde{x}}{dt} = \rho\tilde{x} + \omega\tilde{y} \qquad \frac{d\tilde{y}}{dt} = -\omega\tilde{x} + \rho\tilde{y}. \tag{29.7}$$

The easiest way to understand how the solutions of this equation behave is to change coordinates yet again, this time into plane polar coordinates. For $\tilde{\mathbf{x}} = (\tilde{x}, \tilde{y})$ we set

$$\tilde{x} = r\cos\theta \qquad \text{and} \qquad \tilde{y} = r\sin\theta, \tag{29.8}$$

see Figure 29.1. Then

$$r^2 = \tilde{x}^2 + \tilde{y}^2 \qquad \text{and} \qquad \theta = \tan^{-1}(\tilde{y}/\tilde{x}).$$

In order to find the equations satisfied by r and θ we differentiate using the chain rule (see Appendix C),

$$2r\dot{r} = 2\tilde{x}\dot{\tilde{x}} + 2\tilde{y}\dot{\tilde{y}} \qquad \Rightarrow \qquad \dot{r} = \frac{\tilde{x}\dot{\tilde{x}} + \tilde{y}\dot{\tilde{y}}}{r}$$

and

$$\dot{\theta} = \frac{1}{1 + (\tilde{y}/\tilde{x})^2} \frac{\tilde{x}\dot{\tilde{y}} - \tilde{y}\dot{\tilde{x}}}{\tilde{y}^2} = \frac{\tilde{x}\dot{\tilde{y}} - \tilde{y}\dot{\tilde{x}}}{r^2}.$$

Substituting in for $\dot{\tilde{x}}$ and $\dot{\tilde{y}}$ from (29.7) we get the simple system

$$\dot{r} = \rho r$$
$$\dot{\theta} = -\omega.$$

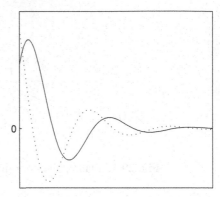

Fig. 29.2. The phase portrait of a stable spiral point, and the graphs of a sample solution with $\tilde{x}(t)$ (solid line) and $\tilde{y}(t)$ (dotted line) plotted against t. Solutions tend to zero, oscillating as they go.

These equations we can solve easily,[1]

$$r(t) = r(0)e^{\rho t} \qquad \text{and} \qquad \theta(t) = \theta(0) - \omega t.$$

The trajectories spiral round the origin with angular velocity $-\omega$, and when $\rho \neq 0$ the distance from the origin either increases exponentially to infinity if $\rho > 0$, or decreases to zero exponentially if $\rho < 0$.

Note that once again the stability of the origin is determined by the eigenvalues. If the real part of the eigenvalues is negative the origin is stable, and called a **stable spiral**. This case is illustrated in Figure 29.2, with the graphs of $\tilde{x}(t)$ and $\tilde{y}(t)$ for some sample solutions shown alongside.

If the real part of the eigenvalues is positive then the origin is an **unstable spiral**, as illustrated in Figure 29.3.

When the eigenvalues are purely imaginary then $\rho = 0$ and so $\dot{r} = 0$; the distance r from the origin is constant. In this case the orbits are circles centred at the origin, as in Figure 29.4. Since the motion repeats itself periodically as it goes round and round the same closed curve, the solutions $\tilde{x}(t)$ and $\tilde{y}(t)$ repeated themselves periodically, and such curves are called *periodic orbits*. In this case the origin is referred to as a **centre**. Note that a centre is stable (if you start close to it you stay close to it) but it is not attracting (trajectories do not tend to it).

---

[1] Given the solutions in this form we can write down the general solution of our original equation in a more memorable way than we did in Section 29.1. It follows from (29.8) that

$$\tilde{x}(t) = Re^{\rho t}\cos(T - \omega t) \qquad \text{and} \qquad \tilde{y}(t) = Re^{\rho t}\sin(T - \omega t),$$

where $R = r(0) > 0$ and $T = \theta(0)$ will be the arbitrary constants in our solution; using (29.4) we have

$$\mathbf{x}(t) = Re^{\rho t}\left[\cos(T - \omega t)\mathbf{v}_1 + \sin(T - \omega t)\mathbf{v}_2\right].$$

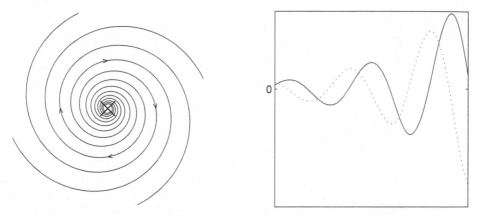

Fig. 29.3. An unstable spiral, along with the graph of a sample solution against t ($\tilde{x}(t)$ solid, $\tilde{y}(t)$ dotted). Solutions oscillate, but the amplitude of the oscillations tends exponentially to infinity.

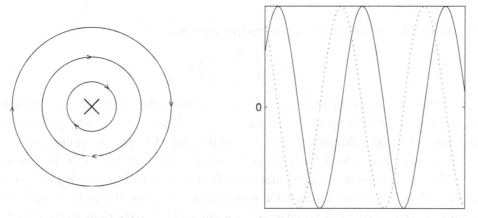

Fig. 29.4. A centre; the phase portrait consists of closed curves around the origin. The solutions repeat periodically with period $2\pi/\omega$, where ω is the imaginary part of the eigenvalues, as shown on the right, where a sample solution ($\tilde{x}(t)$ solid, $\tilde{y}(t)$ dotted) is plotted against t.

29.3 The phase portrait for the original equation

Note that in all three of these cases the rôle played by the eigenvectors (rather than the eigenvalues) is not as clear as it was in the case when the eigenvalues were distinct real numbers. Indeed, 'twisting' any of these pictures will still leave them qualitatively the same (although generally circles will be deformed into ellipses). You can therefore base your phase portrait for the original equation entirely on the eigenvalues and still obtain an accurate picture. We illustrate this for the example (27.5) that started this section.

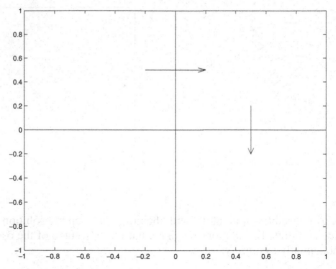

Fig. 29.5. How to find out which way orbits rotate around the origin.

Example 29.3 *Draw the phase portrait for the equation*

$$\dot{\mathbf{x}} = \begin{pmatrix} 2 & 5 \\ -2 & 0 \end{pmatrix} \mathbf{x}. \qquad (29.9)$$

We have already found the eigenvalues of the matrix that occurs in this equation: they are $1 \pm 3i$. So the origin is an unstable spiral. Since the imaginary part (which is what makes the trajectories 'spin' around the origin) is $\pm i$, you cannot just read off the direction in which the trajectories 'rotate'. The best way is to concentrate on a line level with the origin, either $x = 0$ or $y = 0$, and by looking at $\dot{y}$ or $\dot{x}$ work out the direction in which the trajectories are going. If you look on $x = 0$ with $y = 0$ then $\dot{x} = 5y > 0$ and trajectories are moving to the right, so the rotation is clockwise; similarly on $y = 0$ with $x > 0$ we have $\dot{y} = -2x < 0$, so the motion is downwards, which once again shows that trajectories are rotating clockwise. See Figure 29.5.

It is now simple to draw the phase portrait, which is shown in Figure 29.6, along with the graphs of $x(t)$ and $y(t)$ for a sample solution.

Exercises

29.1 Draw the phase portrait for the equation $d\mathbf{x}/dt = \mathbb{A}\mathbf{x}$, when the eigenvalues ($\lambda_\pm$) and eigenvectors ($\boldsymbol{\eta}_\pm$) of $\mathbb{A}$ are as follows. Also given is the sign of $\dot{x}$ when $x = 0$ and $y > 0$.

    (i) $\lambda_\pm = 1 \pm 3i$ with $\boldsymbol{\eta}_\pm = (1, \ 2 \mp i)$, $\dot{x} < 0$;

    (ii) $\lambda_\pm = \pm 3i$ with $\boldsymbol{\eta}_\pm = (1 \pm 2i, \ 1 \mp 3i)$, $\dot{x} < 0$;

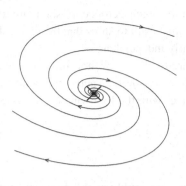

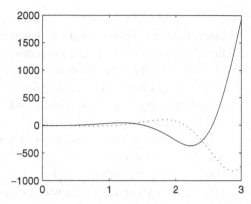

Fig. 29.6. The phase portrait for equation (29.9), and the graph of a sample solution plotted against t, with $x(t)$ solid and $y(t)$ dotted.

   (iii) $\lambda_\pm = -2 \pm i$ with $\eta_\pm = (1 \mp i, \; 3 \pm i)$, $\dot{x} > 0$;

   (iv) $\lambda_\pm = -1 \pm i$ with $\eta_\pm = (1, \; \pm i)$, $\dot{x} > 0$;

    (v) $\lambda_\pm = 2 \pm 2i$ with $\eta_\pm = (\pm 3i, \; 5 \mp 4i)$, $\dot{x} < 0$;

   (vi) $\lambda_\pm = 5 \pm 3i$ with $\eta_\pm = (2 \pm 5i, \; \mp i)$, $\dot{x} < 0$;

  (vii) $\lambda_\pm = \pm 7i$ with $\eta_\pm = (1 \pm i, \; -1 \pm 2i)$, $\dot{x} > 0$; and

 (viii) $\lambda_\pm = -13 \pm 17i$ with $\eta_\pm = (\pm 6i - 8, \; 4 \mp 5i)$, $\dot{x} > 0$.

29.2 Write down the general solution of the equation $d\mathbf{x}/dt = \mathbb{A}$ when the eigenvalues $(\lambda_\pm)$ and eigenvectors $(\eta_\pm)$ of $\mathbb{A}$ are those in the previous exercise.

29.3 For the following equations find the eigenvalues and eigenvectors of the matrix on the right-hand side, and hence find the coordinate transformation that will put the equations into their standard simple (canonical) form. Show that this transformation has the desired effect.

   (i)

$$\frac{d\mathbf{x}}{dt} = \begin{pmatrix} 0 & -1 \\ 1 & -1 \end{pmatrix} \mathbf{x};$$

  (ii)

$$\frac{d\mathbf{x}}{dt} = \begin{pmatrix} -2 & 3 \\ -6 & 4 \end{pmatrix} \mathbf{x};$$

  (iii)

$$\frac{d\mathbf{x}}{dt} = \begin{pmatrix} -11 & -2 \\ 13 & -9 \end{pmatrix} \mathbf{x};$$

   and

  (iv)

$$\frac{d\mathbf{x}}{dt} = \begin{pmatrix} 7 & -5 \\ 10 & -3 \end{pmatrix} \mathbf{x}.$$

29.4 (T) In the previous chapter we used the result that the eigenvectors corresponding to distinct eigenvalues are linearly independent. Use this result to show that the real and imaginary parts of complex eigenvectors are linearly independent.

29.5 (T) Following the same line of reasoning as in Exercise 28.5, show how to construct a matrix with a complex conjugate pair of eigenvalues $\lambda_\pm = \rho \pm i\omega$ and corresponding eigenvectors $\eta_\pm = v_1 \pm iv_2$. Hence find the matrices with the following eigenvalues and eigenvectors:

   (i) $\lambda_\pm = 3 \pm 3i$ with $\eta_\pm = (2 \pm i, \ 1 \mp i)$;

   (ii) $\lambda_\pm = \pm 3i$ with $\eta_\pm = (\pm i, \ 3 \pm 2i)$; and

   (iii) $\lambda_\pm = -2 \pm i$ with $\eta_\pm = (1 \pm i, \ 1 \mp i)$.

   (The M-file `makematrix.m` will do this for you. You could use this to check that the signs of $\dot{x}$ given in Exercise 29.1 are correct by finding the appropriate matrix $\mathbb{A}$ and then looking at $\dot{x}$ when $x = 0$ and $y > 0$.)

30

Yet more phase portraits: a repeated real eigenvalue

We now treat the final case, in which the matrix $\mathbb{A}$ has a non-zero repeated real eigenvalue. There are two very different situations in which this can happen, and we will treat them separately. The case of a repeated eigenvalue zero is the subject of Exercise 30.5.

30.1 $\mathbb{A}$ is a multiple of the identity: stars

The first possibility is that $\mathbb{A}$ is a multiple of the identity,

$$\mathbb{A} = \begin{pmatrix} \lambda & 0 \\ 0 & \lambda \end{pmatrix}.$$

In this case the equation $\dot{\mathbf{x}} = \mathbb{A}\mathbf{x}$ decouples with no extra work required on our part,

$$\dot{x} = \lambda x \qquad \dot{y} = \lambda y.$$

(In fact the equation would decouple in any coordinate system.) For such a matrix, for any vector $\mathbf{v}$ we have $\mathbb{A}\mathbf{v} = \lambda\mathbf{v}$, and so all vectors are eigenvectors.

The phase portrait is particularly simple. Since any vector is an eigenvector, $\mathbf{x}(t) = A e^{\lambda t}\mathbf{v}$ is a solution for any $\mathbf{v}$; in particular $\mathbf{x}(t) = e^{\lambda t}\mathbf{x}(0)$ is a solution, and so solutions move on lines emanating from the origin. Depending on the sign of λ we have a stable or unstable **star**; the phase portrait for the stable case is shown in Figure 30.1.

30.2 $\mathbb{A}$ is not a multiple of the identity: improper nodes

When there is just one eigenvalue and the matrix is not a multiple of the identity then things are more difficult. In this case we will only be able to find one eigenvector. Even though we only have one eigenvector it is still possible to find a new

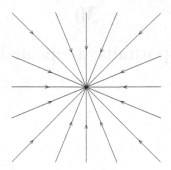

Fig. 30.1. A stable star.

coordinate system in which the matrix takes a 'canonical form', although the argument is a little subtle.

We will suppose that the eigenvalue is λ and that the corresponding eigenvector is $\mathbf{v}$. To begin, choose any vector $\mathbf{v}_2$ that is not in the same direction as $\mathbf{v}$. Since these two vectors are linearly independent they form a basis for $\mathbb{R}^2$, and we can write any vector $\mathbf{x}$ as a linear combination of $\mathbf{v}_2$ and $\mathbf{v}$.

We now show that $\mathbf{v}_1 = (\mathbb{A} - \lambda\mathbb{I})\mathbf{v}_2$ is in the same direction as $\mathbf{v}$, i.e. is an eigenvector. To see this,[1] we write $\mathbf{v}_1$ as a combination of the vectors $\mathbf{v}_2$ and $\mathbf{v}$,

$$(\mathbb{A} - \lambda\mathbb{I})\mathbf{v}_2 = \alpha\mathbf{v}_2 + \beta\mathbf{v},$$

and apply $\mathbb{A} - \lambda\mathbb{I}$ to both sides. The term on the right-hand side involving $\mathbf{v}$ vanishes (since $\mathbf{v}$ is an eigenvector with eigenvalue λ) and we obtain

$$(\mathbb{A} - \lambda\mathbb{I})[(\mathbb{A} - \lambda\mathbb{I})\mathbf{v}_2] = \alpha[(\mathbb{A} - \lambda\mathbb{I})\mathbf{v}_2],$$

or

$$(\mathbb{A} - \lambda\mathbb{I})\mathbf{v}_1 = \alpha\mathbf{v}_1.$$

But this says that $\mathbb{A}\mathbf{v}_1 = (\lambda + \alpha)\mathbf{v}_1$, i.e. that $\mathbf{v}_1$ is an eigenvector with eigenvalue $\lambda + \alpha$. Since there is only one eigenvalue we must have $\alpha = 0$ and $\mathbf{v}_1$ must be an eigenvector lying in the direction of $\mathbf{v}$. Note that since $\mathbf{v}_1 = (\mathbb{A} - \lambda\mathbb{I})\mathbf{v}_2$ we have

$$\mathbb{A}\mathbf{v}_2 = \mathbf{v}_1 + \lambda\mathbf{v}_2. \tag{30.1}$$

---

[1] A much more elegant, but less elementary, approach is to use the Cayley–Hamilton Theorem, see Exercises 30.2 and 30.3.

We now refer our coordinates to the axes $\mathbf{v}_1$ and $\mathbf{v}_2$,

$$\mathbf{x} = \tilde{x}\mathbf{v}_1 + \tilde{y}\mathbf{v}_2,$$

and then, using (30.1) in the second line,

$$\frac{d\tilde{\mathbf{x}}}{dt} = [\mathbf{v}_1 \ \mathbf{v}_2]^{-1}\mathbb{A}[\mathbf{v}_1 \ \mathbf{v}_2]\tilde{\mathbf{x}}$$

$$= [\mathbf{v}_1 \ \mathbf{v}_2]^{-1}[\lambda\mathbf{v}_1 \ \mathbf{v}_1 + \lambda\mathbf{v}_2]\tilde{\mathbf{x}}$$

$$= [\mathbf{v}_1 \ \mathbf{v}_2]^{-1}[\mathbf{v}_1 \ \mathbf{v}_2]\begin{pmatrix} \lambda & 1 \\ 0 & \lambda \end{pmatrix}\tilde{\mathbf{x}}.$$

In the new coordinates our original equation becomes

$$\frac{d\tilde{\mathbf{x}}}{dt} = \begin{pmatrix} \lambda & 1 \\ 0 & \lambda \end{pmatrix}\tilde{\mathbf{x}}. \tag{30.2}$$

Writing this as a coupled system gives

$$\dot{\tilde{x}} = \lambda\tilde{x} + \tilde{y}$$
$$\dot{\tilde{y}} = \lambda\tilde{y}. \tag{30.3}$$

Note that although the equations have not completely decoupled, we can solve the $\tilde{y}$ equation on its own; its solution is $\tilde{y}(t) = Ae^{\lambda t}$. Substituting this for $\tilde{y}(t)$ in the equation for $\tilde{x}$,

$$\frac{d\tilde{x}}{dt} = \lambda\tilde{x} + Ae^{\lambda t}.$$

This is a linear equation, and using the integrating factor $e^{-\lambda t}$ we have

$$\frac{d}{dt}[e^{-\lambda t}\tilde{x}(t)] = A.$$

Integrating between 0 and t we obtain $e^{-\lambda t}\tilde{x}(t) = At + B$, and finally

$$\tilde{x}(t) = Be^{\lambda t} + Ate^{\lambda t}.$$

This means that the solution in the original variables is

$$\mathbf{x}(t) = [Ate^{\lambda t} + Be^{\lambda t}]\mathbf{v}_1 + Ae^{\lambda t}\mathbf{v}_2, \tag{30.4}$$

where to use this formula you need to remember that $\mathbf{v}_1 = (\mathbb{A} - \lambda\mathbb{I})\mathbf{v}_2$. You are probably better off using the 'second order equation' method of Chapter 26 if you want to find an explicit solution.

Drawing the phase diagram in this case is a little more difficult than before, even with the exact expressions for the solutions. You can get some idea of how

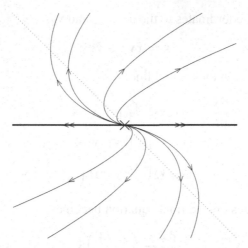

Fig. 30.2. The phase portrait for an improper node for $\lambda > 0$. On the dotted line $d\tilde{x}/dt = 0$.

the picture should look as follows; we assume here that λ is positive. Note first that $d\tilde{y}/dt$ is positive when $\tilde{y}$ is positive, and negative when $\tilde{y}$ is negative; also, $d\tilde{x}/dt$ is positive while $\tilde{y} > -\lambda\tilde{x}$, and is negative when $\tilde{y} < -\lambda\tilde{x}$. Trajectories cross the line $\tilde{y} = -\lambda\tilde{x}$ vertically. You should be able to put these ingredients together to get something like the collection of 'S' shaped trajectories shown in the phase portrait of Figure 30.2. The stability of the origin, referred to in this case as an **improper node**, depends on whether λ is positive or negative.

Example 30.1 *Draw the phase portrait for the equation*

$$\frac{d\mathbf{x}}{dt} = \begin{pmatrix} 5 & -4 \\ 1 & 1 \end{pmatrix} \mathbf{x}. \tag{30.5}$$

We found the (repeated) eigenvalue $\lambda = 3$ of the matrix $\begin{pmatrix} 5 & -4 \\ 1 & 1 \end{pmatrix}$ above (see Example 27.4), and its solitary eigenvector $(2, 1)$. To draw the phase portrait we first draw the eigenvector and label it with arrows moving away from the origin (since the eigenvalue is positive). To work out whether the trajectories move on a collection of 'forwards' or 'backwards' S shapes we can use the same method as we did for the rotating cases; above the stationary point ($x = 0$ and $y > 0$) we have $\dot{x} = 5x - 4y = -4y < 0$ and so the trajectories are moving left, see Figure 30.3. $\qquad\square$

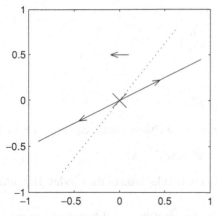 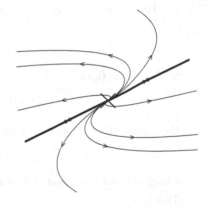

Fig. 30.3. On the left are the steps for drawing the phase portrait for equation (30.5). On the dotted line $\dot{x} = 5x - 4y = 0$; when $y > 0$ and $x = 0$ we have $\dot{x} < 0$. The phase portrait is on the right.

Exercises

30.1 Find the eigenvalue and eigenvector of the matrices occurring in the following equations, and hence draw the phase portrait. Find also the coordinate transformation that will put the equation into canonical form and show that this works. Write down the general solution.

(i)

$$\frac{d\mathbf{x}}{dt} = \begin{pmatrix} 5 & -4 \\ 1 & 1 \end{pmatrix} \mathbf{x};$$

(ii)

$$\frac{d\mathbf{x}}{dt} = \begin{pmatrix} -6 & 2 \\ -2 & -2 \end{pmatrix} \mathbf{x};$$

(iii)

$$\frac{d\mathbf{x}}{dt} = \begin{pmatrix} -3 & -1 \\ 1 & -5 \end{pmatrix} \mathbf{x};$$

(iv)

$$\frac{d\mathbf{x}}{dt} = \begin{pmatrix} 13 & 0 \\ 0 & 13 \end{pmatrix} \mathbf{x};$$

and

(v)

$$\frac{d\mathbf{x}}{dt} = \begin{pmatrix} 7 & -4 \\ 1 & 3 \end{pmatrix} \mathbf{x}.$$

30.2 (T) The characteristic equation for a 2×2 matrix

$$\mathbb{A} = \begin{pmatrix} a & b \\ c & d \end{pmatrix}$$

is $|\mathbb{A} - k\mathbb{I}| = 0$, i.e.

$$k^2 - (a+d)k + (ad - bc) = 0.$$

By explicit calculation show that $\mathbb{A}$ satisfies its own characteristic equation, i.e. that

$$\mathbb{A}^2 - (a+d)\mathbb{A} + (ad - bc)\mathbb{I} = \mathbb{O},$$

where $\mathbb{O}$ is the 2×2 matrix of zeros. This is a particular case of the Cayley-Hamilton Theorem.

30.3 (T) If $\mathbb{A}$ has a repeated eigenvalue λ with eigenvector $\mathbf{v}$ then its characteristic equation can be written

$$(k - \lambda)^2 = 0.$$

Use the Cayley-Hamilton Theorem from the previous exercise to deduce that

$$(\mathbb{A} - \lambda\mathbb{I})^2 = \mathbb{O},$$

and hence that $(\mathbb{A} - \lambda\mathbb{I})\mathbf{x}$ is an eigenvector of $\mathbb{A}$ for any choice of vector $\mathbf{x}$ such that $\mathbb{A}\mathbf{x} \neq \lambda\mathbf{x}$.

30.4 (T) By following the ideas of Exercise 28.5, show how to construct a matrix with a single eigenvalue λ and corresponding eigenvector $\mathbf{v}$. (There will be many such matrices.) Find two matrices with eigenvalue -1 and eigenvector $(1, 1)$.

30.5 (T) Suppose that the matrix $\mathbb{A}$ has zero as a repeated eigenvalue, with eigenvector $\mathbf{v}$. Then we can change to coordinates referred to $\mathbf{v}_2$ and $\mathbf{v}_1 = \mathbb{A}\mathbf{v}_2$, where $\mathbf{v}_2$ is any vector in a different direction to $\mathbf{v}$, so that $\tilde{\mathbf{x}} = \tilde{x}\mathbf{v}_1 + \tilde{y}\mathbf{v}_2$. The equation becomes

$$\frac{d\tilde{\mathbf{x}}}{dt} = \begin{pmatrix} 0 & 1 \\ 0 & 0 \end{pmatrix}\tilde{\mathbf{x}},$$

and so

$$\frac{d\tilde{x}}{dt} = \tilde{y} \qquad \text{and} \qquad \tilde{y} = 0.$$

(i) Solve the equations for $\tilde{x}(t)$ and $\tilde{y}(t)$, and hence write down the general solution for $\mathbf{x}(t)$.

(ii) Draw the phase diagram in the $(\tilde{x}, \tilde{y})$ plane, and hence in the (x, y) plane.

(iii) Draw the phase diagram for the equation

$$\frac{d\mathbf{x}}{dt} = \begin{pmatrix} -1 & 1 \\ -1 & 1 \end{pmatrix}\mathbf{x}.$$

31

Summary of phase portraits for linear equations

When confronted with an example $d\mathbf{x}/dt = \mathbb{A}\mathbf{x}$ you should first calculate the eigenvalues of $\mathbb{A}$, and then if they are real calculate the eigenvectors. With this information you can draw the phase portrait. The various possibilities are summarised below, and illustrated in Figure 31.1.

(i) Distinct real eigenvalues
- $\lambda_1 < \lambda_2 < 0$ gives a *stable node*: all trajectories approach the origin, tangent to the eigenvector corresponding to λ_2.
- $\lambda_1 > \lambda_2 > 0$ gives an *unstable node*: all trajectories move away from the origin, tangent to the eigenvector corresponding to λ_2.
- $\lambda_1 < 0 < \lambda_2$ gives a *saddle*: the only trajectories to approach the origin are those starting on the 'stable eigenvector', while all other trajectories move away.
(ii) Complex conjugate eigenvalues $\rho \pm i\omega$ (to find the direction of rotation you need to check, for example, the sign of $\dot{x}$ on the line $x = 0$)
- $\rho < 0$ gives a *stable spiral*: all trajectories spiral into the origin.
- $\rho > 0$ gives an *unstable spiral*: all trajectories spiral out from the origin.
- $\rho = 0$, i.e. $\lambda = \pm i\omega$, gives a *centre*: trajectories close and we have a family of *periodic orbits*.
(iii) A repeated real eigenvalue
- The matrix is a multiple of the identity: we have a stable or unstable *star* depending on the sign of the eigenvalue ($\lambda < 0$ gives stability, which is the case shown in Figure 31.1).
- The matrix is not a multiple of the identity: we get the S-shaped phase portrait of an *improper node*, whose stability depends on the sign of λ (stable for $\lambda < 0$).

31.1 *Jordan canonical form

The coordinate transformations that we have used in the previous three chapters are those that put the matrix $\mathbb{A}$ into its *Jordan canonical form*. In order to explain this, we first need to discuss the relationship between matrices and linear

301

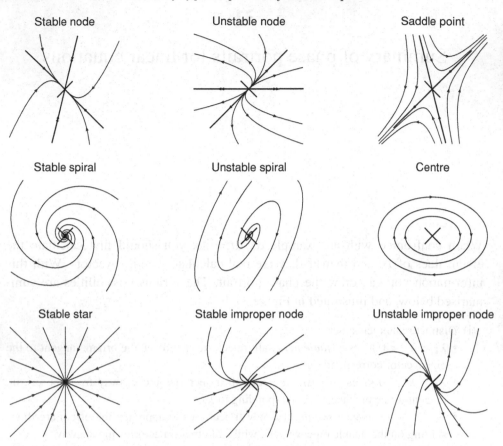

Fig. 31.1. Sample phase portraits listed by the stability type of the stationary point: (i) distinct real eigenvalues on the first row, (ii) complex eigenvalues on the second row, (iii) equal real eigenvalues of the third row.

transformations. We will only do this for the case of linear transformations of the plane.

31.1.1 Representation of vectors in different coordinate systems

A vector $\mathbf{x}$ lying in the plane can be specified without reference to a coordinate system. For each particular choice of coordinate axes, $\mathbf{x}$ will be represented by a different pair of coordinates. For example, if $\mathbf{x}$ has coordinates (x, y) with respect to the coordinate axes $\mathbf{e}_1$ and $\mathbf{e}_2$ this means that

$$\mathbf{x} = x\mathbf{e}_1 + y\mathbf{e}_2,$$

while with a different choice of axes, $\mathbf{v}_1$ and $\mathbf{v}_2$, say, the same vector would have coordinates $(\tilde{x}, \tilde{y})$, i.e.

$$\mathbf{x} = \tilde{x}\mathbf{v}_1 + \tilde{y}\mathbf{v}_2.$$

Writing both expressions for $\mathbf{x}$ in matrix form we have

$$[\mathbf{e}_1 \; \mathbf{e}_2]\begin{pmatrix} x \\ y \end{pmatrix} = [\mathbf{v}_1 \; \mathbf{v}_2]\begin{pmatrix} \tilde{x} \\ \tilde{y} \end{pmatrix},$$

which enables us to transform between the two different coordinate systems,

$$\begin{pmatrix} x \\ y \end{pmatrix} = \mathbb{S}\begin{pmatrix} \tilde{x} \\ \tilde{y} \end{pmatrix} \quad \text{and} \quad \begin{pmatrix} \tilde{x} \\ \tilde{y} \end{pmatrix} = \mathbb{S}^{-1}\begin{pmatrix} x \\ y \end{pmatrix}, \tag{31.1}$$

where $\mathbb{S} = [\mathbf{e}_1 \; \mathbf{e}_2]^{-1}[\mathbf{v}_1 \; \mathbf{v}_2]$.

31.1.2 Linear transformations of the plane and 2×2 matrices

A transformation of the plane $L : \mathbf{x} \mapsto L[\mathbf{x}]$ is linear if

$$L[\alpha\mathbf{x} + \beta\mathbf{y}] = \alpha L[\mathbf{x}] + \beta L[\mathbf{y}]$$

for all $\mathbf{x}, \mathbf{y} \in \mathbb{R}^2$ and all $\alpha, \beta \in \mathbb{R}$. In particular, this means that

$$L[\mathbf{x}] = L[x\mathbf{e}_1 + y\mathbf{e}_2] = xL[\mathbf{e}_1] + yL[\mathbf{e}_2],$$

and so in order to work out $L[\mathbf{x}]$ we only need to know $L[\mathbf{e}_1]$ and $L[\mathbf{e}_2]$.

For each choice of coordinates axes we can find a matrix $\mathbb{A}$ such that if $\mathbf{x}$ has coordinates (x, y), then $L[\mathbf{x}]$ has coordinates

$$\mathbb{A}\begin{pmatrix} x \\ y \end{pmatrix}.$$

For example, if we work in the $\mathbf{e}_1$–$\mathbf{e}_2$ coordinate system, in which (x, y) represents the point $\mathbf{x} = x\mathbf{e}_1 + y\mathbf{e}_2$, then if

$$L[\mathbf{e}_1] = a_{11}\mathbf{e}_1 + a_{21}\mathbf{e}_2, \qquad L[\mathbf{e}_2] = a_{12}\mathbf{e}_1 + a_{22}\mathbf{e}_2,$$

and we set

$$\mathbb{A} = \begin{pmatrix} a_{11} & a_{12} \\ a_{21} & a_{22} \end{pmatrix},$$

we have

$$A \begin{pmatrix} x \\ y \end{pmatrix} = \begin{pmatrix} a_{11}x + a_{12}y \\ a_{21}x + a_{22}y \end{pmatrix}$$

$$= x \begin{pmatrix} a_{11} \\ a_{21} \end{pmatrix} + y \begin{pmatrix} a_{12} \\ a_{22} \end{pmatrix},$$

which is how we write $xL[\mathbf{e}_1] + yL[\mathbf{e}_2] = L[\mathbf{x}]$ in the $\mathbf{e}_1$–$\mathbf{e}_2$ coordinate system.

If we were to change the coordinate axes then the matrix representing the transformation L would have to change too. To find the new matrix, suppose that the point $\mathbf{x}$ has coordinates $(\tilde{x}, \tilde{y})$ with respect to the axes $\mathbf{v}_1$ and $\mathbf{v}_2$, so that $\mathbf{x} = \tilde{x}\mathbf{v}_1 + \tilde{y}\mathbf{v}_2$. Then its coordinates (x, y) with respect to the axes $\mathbf{e}_1$ and $\mathbf{e}_2$ are given by

$$\begin{pmatrix} x \\ y \end{pmatrix} = S \begin{pmatrix} \tilde{x} \\ \tilde{y} \end{pmatrix}$$

(see (31.1)). We know that in the $\mathbf{e}_1$–$\mathbf{e}_2$ coordinate system, the coordinates of $L[\mathbf{x}]$ can be found by multiplying by A, and so are

$$AS \begin{pmatrix} \tilde{x} \\ \tilde{y} \end{pmatrix}.$$

Using (31.1) again the coordinates of the vector $L[\mathbf{x}]$ in the $\mathbf{v}_1$–$\mathbf{v}_2$ coordinate system are

$$S^{-1}AS \begin{pmatrix} \tilde{x} \\ \tilde{y} \end{pmatrix}.$$

The upshot of this is that the transformation L is represented in the new coordinate system by the matrix

$$B = S^{-1}AS.$$

31.1.3 Similar matrices and the Jordan canonical form

We have seen that if two matrices A and B represent the same linear transformation with respect to different coordinate axes then for some non-singular matrix S we have

$$B = S^{-1}AS.$$

Such matrices are said to be *similar*.

It is a natural question whether there is a particular choice of matrix S that will find the 'simplest' (in some way) matrix that is similar to a given matrix A. In other words, whether there is a 'natural' coordinate system in which the linear transformation L can be easily expressed.

There is such a form, known as the Jordan canonical form. All similar matrices have the same eigenvalues, and the simplest matrix depends on these eigenvalues. It is always one of the following possibilities, which should by now be familiar. If A has distinct real eigenvalues λ_1 and λ_2 then the Jordan canonical form is

$$\begin{pmatrix} \lambda_1 & 0 \\ 0 & \lambda_2 \end{pmatrix};$$

if A has a complex conjugate pair of eigenvalues $\rho \pm i\omega$ then the canonical form is

$$\begin{pmatrix} \rho & -\omega \\ \omega & \rho \end{pmatrix};$$

and if A has only one eigenvalue λ then the canonical form is either

$$\begin{pmatrix} \lambda & 0 \\ 0 & \lambda \end{pmatrix} \quad \text{or} \quad \begin{pmatrix} \lambda & 1 \\ 0 & \lambda \end{pmatrix}.$$

These possibilities follow from an analysis that closely parallels that of the previous three chapters.

Exercises

31.1 Draw the phase portrait for the equation $dx/dt = Ax$ when the eigenvalues and eigenvectors of A are the following:
  (i) $\lambda_1 = 3$ with $v_1 = (1, 1)$ and $\lambda_2 = -2$ with $v_2 = (1, -2)$;
  (ii) complex conjugate eigenvalues $\lambda_\pm = -1 \pm 3i$, with $\dot{x} < 0$ when $x = 0$ and $y > 0$;
  (iii) a single eigenvalue $\lambda = 13$ with eigenvector $(3, 2)$, and $\dot{x} > 0$ when $x = 0$ and $y > 0$;
  (iv) $\lambda_1 = -2$ with $v_1 = (2, 1)$ and $\lambda_2 = -3$ with $v_2 = (1, -1)$;
  (v) a single eigenvalue $\lambda = -3$ with eigenvector $(1, -1)$, and $\dot{x} > 0$ when $x = 0$ and $y > 0$;
  (vi) $\lambda = \pm 2i$, where $\dot{y} < 0$ when $y = 0$ and $x > 0$;
  (vii) $\lambda_1 = 1$ with $v_1 = (3, 2)$ and $\lambda_2 = 5$ with $v_2 = (1, -4)$;
  (viii) $\lambda = 5 \pm i$, and $\dot{y} > 0$ when $y = 0$ and $x > 0$; and
  (ix) a single eigenvalue $\lambda = -7$, with the matrix A a multiple of the identity.

Part VI

Coupled nonlinear equations

32

Coupled nonlinear equations

We now turn our attention to coupled nonlinear systems. We will concentrate on autonomous systems in which the right-hand side does not depend explicitly on time,

$$
\begin{aligned}
dx/dt &= f(x, y) \\
dy/dt &= g(x, y).
\end{aligned}
\tag{32.1}
$$

Using the vector notation $\mathbf{x} = (x, \ y)$ and $\mathbf{f}(\mathbf{x}) = (f(x, y), \ g(x, y))$, this equation can be rewritten

$$
\frac{d\mathbf{x}}{dt} = \mathbf{f}(\mathbf{x}).
$$

Our approach will be to try to understand the dynamics of these equations (the behaviour of their solutions) in a qualitative way by drawing the phase diagram in the (x, y) plane ('the phase plane'), just as we have done for linear equations in the past three chapters. We will find that we can piece together the phase portrait for nonlinear systems from a collection of phase portraits for linear (or nearly linear) systems near the stationary points.

32.1 Some comments on phase portraits

A stationary point is a point (x^*, y^*) at which $\dot{x} = \dot{y} = 0$, i.e. where

$$
f(x^*, y^*) = 0 \qquad \text{and} \qquad g(x^*, y^*) = 0.
$$

Because solutions are unique, it follows that if (x^*, y^*) is a stationary point then solutions starting at (x^*, y^*) remain there for all time. The phase portraits we drew in the previous chapters were fairly simple, since we only ever had a single stationary point at the origin.

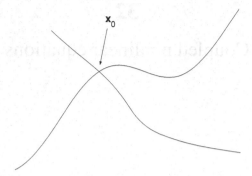

Fig. 32.1. Trajectories cannot cross; x_0 must be a stationary point.

Our phase portraits will show the stationary points marked by crosses, and include a representative collection of trajectories (the curves traced out by solutions as they change in time) with the direction of increasing time indicated by an arrow. As with the phase diagrams we drew in Chapter 7 we lose any information about the rates at which the solutions change, but may be able to understand their behaviour much more easily from the diagram than we would from an explicit solution.

An important point to notice is that curves in the phase diagram cannot cross one another; this is a consequence of the uniqueness of solutions. If we had a situation like that illustrated in Figure 32.1, with two curves emanating from a point x_0, then there would be two solutions starting at x_0; this cannot happen, since solutions are unique. The only way that we can get the kind of situation pictured is if x_0 is a stationary point. If this is the case there is no contradiction, since although the trajectories meet at x_0, they do not actually pass through it. (If $x(s) = x_0$ for some s, where x_0 is a stationary point, then we must have $x(t) = x_0$ for all t.) In all our phase diagrams for linear equations (see Figure 31.1) there are apparent 'crossings'; but they all occur at the origin, which is a stationary point.

32.2 Competition of species

We will illustrate the general method by considering a simple ecological model for two species that are competing for the same resources, e.g. a herd of sheep and cows grazing over the same fields. If we denote the numbers of the two species (measured, let us suppose, in hundreds) as $x(t)$ and $y(t)$ then *in isolation* we might expect the size of both populations to obey the logistic equations

$$\dot{x} = x(A - ax) \qquad \text{and} \qquad \dot{y} = y(B - dy) \qquad (32.2)$$

Fig. 32.2. The phase portrait for either species in isolation; the left stationary point represents a population of zero, the right-hand one the equilibrium value ($x = A/a$ or $y = B/d$).

(with A, B, a, and d positive), cf. Section 8.5. In particular these equations predict that left to themselves each species would settle down to a constant population (for the first, $x = A/a$, and for the second, $y = B/d$ as can be seen from the phase diagram in Figure 32.2).

However, since there are limited resources both species will be disadvantaged by the presence of the other. So we would expect the model

$$\dot{x} = x(A \quad - ax \quad \underbrace{-by}_{y \text{ inhibits } x})$$

$$\dot{y} = y(B \quad \underbrace{-cx}_{x \text{ inhibits } y} \quad -dy)$$

with b and c positive to reflect this.

In this chapter we will consider these equation with a particular choice of the parameters A, B, a, b, c, and d,

$$\begin{aligned} \dot{x} &= x(8 - 4x - y) \\ \dot{y} &= y(3 - 3x - y). \end{aligned} \tag{32.3}$$

For these models we are only interested in the behaviour of the solutions for $x, y \geq 0$, since we want our populations to be positive. (It should be clear, however, that the equations are mathematically sensible for any values of x and y.)

32.3 Direction fields

Given a coupled pair of nonlinear equations (similar to (32.3)) we can get a very good indication of what the phase portrait should look like by drawing the *direction field*. This is a set of arrows pointing in the direction of the vector $\dot{x}$ and whose length is proportional to the magnitude of $\dot{x}$. The direction field shows the direction in which solutions move, and how fast. The curves traced out by solutions will be everywhere tangential to the direction field, since the field shows how they are moving instantaneously, see Figure 32.3.

The direction field for equation (32.3) is shown in Figure 32.4, which was produced using MATLAB's quiver command.

```
>> [x, y] = meshgrid(0:1/3:2.5, 0:1/3:3.5);
>> xd = x.*(8-4*x-y); yd= y.*(3-3*x-y);
>> quiver(x,y,xd,yd)
```

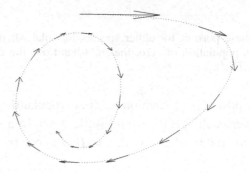

Fig. 32.3. A fanciful trajectory shown as a dotted line, and some arrows from the direction field tangential to it. The solution moves faster along the trajectory where the arrows are larger.

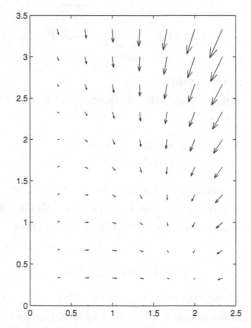

Fig. 32.4. The direction field for equation (32.3).

Having drawn the direction field, it should be relatively easy to 'join the dots' and draw the phase portrait as in Figure 32.5.

Of course, in practice it is not really convenient to draw so many arrows of the direction field unless you have access to a computer package. However, a related approach that can be useful more generally is to draw the 'nullclines'. These are the lines (or curves) on which $\dot{x}$ or $\dot{y}$ is equal to zero. For our example $\dot{x} = 0$ when $x = 0$ or when $8 - 4x - y = 0$, and $\dot{y}$ is zero when $y = 0$ or $3 - 3x - y = 0$. These nullclines are shown in Figure 32.6, along with some sample trajectories.

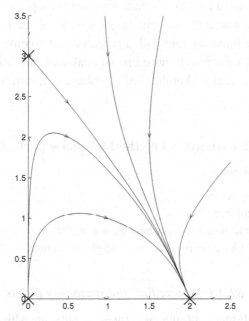

Fig. 32.5. The phase portrait for equation (32.3).

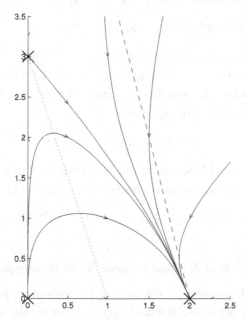

Fig. 32.6. The 'nullclines' for equation (32.3). The dashed line is $8 - 4x - y = 0$, on which $\dot{x} = 0$, and the dotted line is $3 - 3x - y = 0$, on which $\dot{y} = 0$. So, from left to right, we have regions where x and y are both increasing; where x increases but y decreases; and where x and y are both decreasing.

With some thought (and a lot of imagination) it might be possible to draw the phase portrait if you know where the nullclines are. (We used the nullclines to draw the phase portrait for the linear system with a repeated real eigenvalue in Section 30.2.)

However, we will concentrate here on the analytical method that enables us to draw qualitatively accurate 'sketches' of the phase diagram using our knowledge of linear systems.

32.4 Analytical method for phase portraits

The method has four stages:

- find all the stationary points,
- linearise near the stationary points,
- draw the phase portrait near the stationary points, and then
- join up these 'local' phase portraits to give the global picture.

32.4.1 Step 1: find the stationary points

Recall that the stationary points are those points at which x and y do not change, i.e. those points (x^*, y^*) such that $f(x^*, y^*) = g(x^*, y^*) = 0$. For equations (32.3) we therefore need (x^*, y^*) to satisfy

$$x^*(8 - 4x^* - y^*) = 0 \qquad \text{and} \qquad y^*(3 - 3x^* - y^*) = 0.$$

We can satisfy the first equation if we choose $x^* = 0$. The second equation is then satisfied if either $y^* = 0$ or $y^* = 3$. The choice $y^* = 0$ for the second equation also allows $x^* = 2$ as a solution of the first equation. A final possibility is provided by the solution of the simultaneous equations

$$8 - 4x^* - y^* = 0 \qquad \text{and} \qquad 3 - 3x^* - y^* = 0$$

gives $x^* = 5$ and $y^* = -12$. Since we are only interested in stationary points with x^* and y^* non-negative, we can concentrate on the three stationary points

$$(0, 0), \qquad (2, 0) \qquad \text{and} \qquad (0, 3).$$

32.4.2 Step 2: linearise near the stationary points

The next step is to determine the stability of the stationary points by linearising. The idea, essentially, is to look at what happens 'near to' the stationary points.

To do this we suppose that $\mathbf{x}(t)$ is close to a stationary point $\mathbf{x}^* = (x^*, y^*)$, and write

$$x(t) = x^* + \xi(t) \qquad \text{and} \qquad y(t) = y^* + \eta(t), \tag{32.4}$$

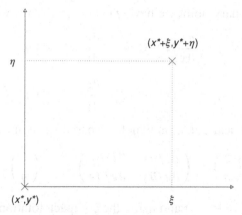

Fig. 32.7. Choosing new coordinates $\xi = x - x^*$ and $\eta = y - y^*$ near a stationary point (x^*, y^*).

where $\xi(t)$ and $\eta(t)$ are small, see Figure 32.7. The new coordinates (ξ, η) treat (x^*, y^*) as the origin.

We now write down the equations satisfied by $\xi(t)$ and $\eta(t)$. Since x^* and y^* are constants, it follows from differentiating the two equations in (32.4) that

$$\dot{x} = \dot{\xi} \quad \text{and} \quad \dot{y} = \dot{\eta}.$$

Therefore

$$\dot{\xi} = \dot{x} = f(x, y) = f(x^* + \xi, y^* + \eta)$$
$$\dot{\eta} = \dot{y} = g(x, y) = g(x^* + \xi, y^* + \eta).$$

Now we use the Taylor expansion of f and g about (x^*, y^*). Recall (or see Appendix C) that the Taylor expansion of a function f of two variables is given by

$$f(x + \xi, y + \eta) = f + f_x\xi + f_y\eta + \tfrac{1}{2}f_{xx}\xi^2 + f_{xy}\xi\eta + \tfrac{1}{2}f_{yy}\eta^2 + \cdots$$

where $f_x = \partial f/\partial x$, etc., and all the partial derivatives on the right-hand side are evaluated at the point (x, y). Then

$$\dot{\xi} = f(x^*, y^*) + \frac{\partial f}{\partial x}(x^*, y^*)\xi + \frac{\partial f}{\partial y}(x^*, y^*)\eta + \cdots$$

$$\dot{\eta} = g(x^*, y^*) + \frac{\partial g}{\partial x}(x^*, y^*)\xi + \frac{\partial g}{\partial y}(x^*, y^*)\eta + \cdots,$$

where the '$\cdots$' are terms of higher order in ξ and η. If ξ and η are sufficiently small we would expect that we can ignore the higher order terms and still have a good approximation of the rate of change of $\xi(t)$ and $\eta(t)$.

Since $\mathbf{x}^*$ is a stationary point we have $f(x^*, y^*) = g(x^*, y^*) = 0$, and so

$$\dot{\xi} = \frac{\partial f}{\partial x}(x^*, y^*)\,\xi + \frac{\partial f}{\partial y}(x^*, y^*)\,\eta$$

$$\dot{\eta} = \frac{\partial g}{\partial x}(x^*, y^*)\,\xi + \frac{\partial g}{\partial y}(x^*, y^*)\,\eta.$$

Notice that this is a linear equation which we can rewrite in matrix form as

$$\begin{pmatrix} \dot{\xi} \\ \dot{\eta} \end{pmatrix} = \begin{pmatrix} \partial f/\partial x & \partial f/\partial y \\ \partial g/\partial x & \partial g/\partial y \end{pmatrix}_{|(x^*, y^*)} \begin{pmatrix} \xi \\ \eta \end{pmatrix}. \tag{32.5}$$

Alternatively, with $\boldsymbol{\xi} = (\xi, \eta)$ and using the compact notation

$$\mathbf{Df}(\mathbf{x}^*) = \begin{pmatrix} \partial f/\partial x & \partial f/\partial y \\ \partial g/\partial x & \partial g/\partial y \end{pmatrix}_{|(x^*, y^*)},$$

this becomes

$$\frac{d\boldsymbol{\xi}}{dt} = \mathbf{Df}(\mathbf{x}^*)\,\boldsymbol{\xi}.$$

This equation is known as the 'linearisation' of (32.1) about the stationary point (x^*, y^*), since by considering only solutions 'sufficiently close' to (x^*, y^*) we have approximated the original nonlinear equation by a linear equation. From the work we did in earlier chapters on linear equations we can understand the behaviour of this equation using only the eigenvalues and eigenvectors of the matrix

$$\mathbf{Df}(\mathbf{x}^*) = \begin{pmatrix} \partial f/\partial x & \partial f/\partial y \\ \partial g/\partial x & \partial g/\partial y \end{pmatrix}_{|(x^*, y^*)}.$$

For our example, where

$$f(x, y) = x(8 - 4x - y) \qquad \text{and} \qquad g(x, y) = y(3 - 3x - y),$$

we have

$$\mathbf{Df}(x, y) = \begin{pmatrix} 8 - 8x - y & -x \\ -3y & 3 - 3x - 2y \end{pmatrix}. \tag{32.6}$$

32.4.3 The Hartman–Grobman Theorem

It can be proved (but is well beyond the scope of this book) that the phase portrait of the original nonlinear problem 'sufficiently close' to a stationary point looks 'essentially the same' as that of the linear equation in (32.5) *provided that the eigenvalues have non-zero real part*. This result, known as the Hartman–Grobman

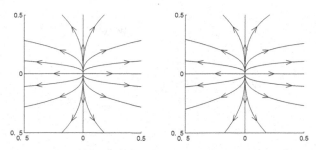

Fig. 32.8. The left-hand picture shows the phase portrait for the linearised equation close to the origin; the right-hand picture shows the phase portrait of the nonlinear equation close to the origin. The two pictures are indistinguishable. (Despite appearances this is an honest figure, with one picture produced using the linearisation and one using the full nonlinear equation!)

Theorem, gives the mathematical foundation for the whole technique that we will be using.

Figure 32.8 shows some solutions of

$$\dot{x} = x(8 - 4x - y)$$
$$\dot{y} = y(3 - 3x - y)$$

near the origin, along with some solutions of the linearised equations

$$\dot{x} = 8x \qquad \dot{y} = 3y.$$

The two pictures appear to be identical.

The condition that the eigenvalues have non-zero real part means that the stationary point cannot be a centre. Indeed, Exercise 32.1 shows that nonlinear terms can turn a linearised centre into a stable or unstable focus. However, examiners and problem setters have a habit of finding nonlinear systems in which stationary points that are centres for the linearised equation do indeed sit in the centre of a family of periodic orbits in the nonlinear problem. In all the examples below we will check carefully that our linearised centres correspond to centres of the nonlinear equation.

The concepts of the stable and unstable manifolds, which perhaps seemed somewhat artificial when they were introduced in Section 28.5, are much more useful when dealing with nonlinear equations. With essentially the same definitions as before, the stable manifold of a stationary point $\mathbf{x}^*$, $W^s(\mathbf{x}^*)$, consists of all those points lying on trajectories that tend to $\mathbf{x}^*$ as $t \to \infty$; while the unstable manifold $W^u(\mathbf{x}^*)$ consists of all those points lying on trajectories that would tend to $\mathbf{x}^*$ were the direction of time reversed. The stable and unstable manifolds of a saddle point in a nonlinear system are tangential to the eigenvectors corresponding to the

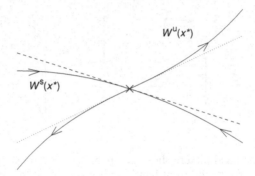

Fig. 32.9. The stable and unstable manifolds of a stationary point are tangential to the eigenvectors: $W^u(x^*)$ is tangent to the unstable eigenvector (dotted line), and $W^s(x^*)$ is tangent to the stable eigenvector (dashed line).

negative and positive eigenvalues of the linearisation; once again a proof is beyond the scope of this book. This is illustrated in Figure 32.9.

32.4.4 Step 3: find the stability type of each stationary point

The next step is to work out the 'stability type' (stable node, saddle, unstable focus, etc.) of each stationary point. Since the phase portrait 'close' to each stationary point looks like the phase portrait of the linearised system, we can use this to draw the phase portrait near each stationary point.

For our example, the matrix of partial derivatives was given in (32.6),

$$\mathbf{Df}(x, y) = \begin{pmatrix} 8 - 8x - y & -x \\ -3y & 3 - 3x - 2y \end{pmatrix}.$$

Near $(0, 0)$ we set $\xi = x$ and $\eta = y$, and so the linearisation is

$$\frac{d\xi}{dt} = \begin{pmatrix} 8 & 0 \\ 0 & 3 \end{pmatrix} \xi.$$

We can just read off the eigenvalues of this matrix (see Example 27.2); they are $\lambda = 8$ and $\lambda = 3$, and so this is an unstable node. The eigenvector corresponding to $\lambda = 8$ is $(1, 0)$ (the x-axis), and that corresponding to $\lambda = 3$ is $(0, 1)$ (the y-axis). So the linearised phase portrait near the origin looks like Figure 32.10, where the trajectories for negative values of x and y (which are not of interest given the application of the model) are shown as dotted lines.

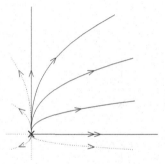

Fig. 32.10. The linearised phase portrait near $(0, 0)$.

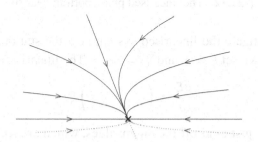

Fig. 32.11. The linearised phase portrait near $(2, 0)$.

Near the stationary point at $(2, 0)$ we set $\xi = x - 2$ and $\eta = y$ and the linearised system is

$$\frac{d\xi}{dt} = \begin{pmatrix} -8 & -2 \\ 0 & -3 \end{pmatrix} \xi.$$

The eigenvalues of the matrix we can read off as $\lambda_1 = -8$ and $\lambda_2 = -3$ (see Example 27.3), and so this point is a stable node. The eigenvector corresponding to $\lambda_1 = -8$ is just $\mathbf{v}_1 = (1, \ 0)$ (the x-axis), while that corresponding to $\lambda_2 = -3$ has to be found from

$$\begin{pmatrix} -5 & -2 \\ 0 & 0 \end{pmatrix} \begin{pmatrix} v_1 \\ v_2 \end{pmatrix} = \mathbf{0},$$

and so is $\mathbf{v}_2 = (2, \ -5)$. The linearised phase portrait near $(2, 0)$ is shown in Figure 32.11, where once again the 'uninteresting' orbits with $y < 0$ are shown as dotted lines.

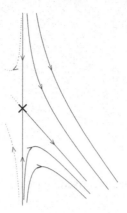

Fig. 32.12. The linearised phase portrait near $(0, 3)$.

Finally we investigate the linearised system near the stationary point on the y-axis, $(0, 3)$, where we set $\xi = x$ and $\eta = y - 3$. The linearisation is given by

$$\frac{d\xi}{dt} = \begin{pmatrix} 5 & 0 \\ -9 & -3 \end{pmatrix} \xi.$$

Once again we can just read off the eigenvalues, which are $\lambda_1 = 5$ and $\lambda_2 = -3$; this stationary point is a saddle. The eigenvector corresponding to $\lambda_1 = 5$ ('the unstable direction'), $\mathbf{v}_1 = (v_1, \ v_2)$, is determined by

$$\begin{pmatrix} 0 & 0 \\ -9 & -8 \end{pmatrix} \begin{pmatrix} v_1 \\ v_2 \end{pmatrix} = \mathbf{0};$$

one choice would be $\mathbf{v}_1 = (8, \ -9)$. The 'stable eigenvector' corresponding to $\lambda_2 = -3$ is $\mathbf{v}_2 = (0, \ 1)$ (the y-axis). This stationary point is a saddle, and the linearised phase portrait nearby is shown in Figure 32.12.

Plotting these three 'local' phase portraits on the region $x, y \geq 0$ gives the partial phase portrait shown in Figure 32.13.

32.4.5 Step 4: 'join the dots'

Now we want to join up the local portraits to give the global picture; the principle is to join up the local phase portraits in a consistent way. The full phase portrait is shown in Figure 32.14. (We saw this before at the beginning of the chapter, but now we can draw it without recourse to a computer-generated plot of the direction field.)

It should be easy to read the fate of our two species from the phase portrait. The point at $(2, 0)$ is globally attracting; no matter where a trajectory begins (unless it is on one of the axes) it tends towards this point. So whatever the initial balance

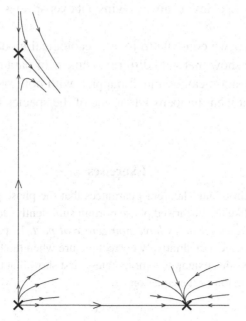

Fig. 32.13. The local phase portraits plotted together.

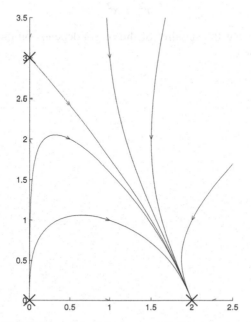

Fig. 32.14. The global phase portrait for equation (32.3).

between the species, species 'x' always wins out over species 'y', which eventually dies out.

In the next chapter we continue to look at ecological models using these phase plane methods, and show that with different choices of parameters the outcome in this competitive situation can be much happier, with both species able to coexist. We will also look at what happens when one of the species forms the prey of the other.

Exercises

32.1 The Hartman–Grobman Theorem guarantees that the phase portrait for a nonlinear equation looks like the linearised phase portrait sufficiently close to a stationary point *provided that the eigenvalues have non-zero real part*. In particular, the linearised system may not give a qualitatively correct picture when the linearised equation produces a centre, as this example demonstrates. First show that the origin is a centre for the linearised version of the equation

$$\dot{x} = -y + \lambda x(x^2 + y^2)$$
$$\dot{y} = x + \lambda y(x^2 + y^2).$$

Now write down the equation satisfied by r, where

$$r^2 = x^2 + y^2,$$

and hence show that the stability of the origin depends on the sign of λ. Draw the phase portrait for $\lambda < 0$.

33

Ecological models

In this chapter we first investigate what other types of behaviour can arise in models of competitive species, and then we consider the more aggressive situation in which one species preys on the other. The simple models that we treat here are known as Lotka–Volterra systems.

33.1 Competing species

It is possible to treat the general model for competing species

$$\dot{x} = x(A - ax - by)$$
$$\dot{y} = y(B - cx - dy),$$

(33.1)

see Exercise 33.3. However, the general treatment is much less illuminating than considering particular examples, and here we deal with two cases that have behaviour which is significantly different from that of the previous chapter.

33.1.1 Weak competition

First we consider the example,

$$\dot{x} = x(4 - 2x - 2y)$$
$$\dot{y} = y(9 - 6x - 3y),$$

(33.2)

for which there are four non-negative stationary points (where the right-hand sides are zero): if $x = 0$ then we could have $y = 0$ or $y = 3$; if $y = 0$ then we could have the additional stationary point that arises when $x = 2$; and finally there is an interior stationary point when $x = y = 1$, corresponding to a coexistent state in which there are an equal number of both species. So the four possibilities are

$$(0, 0), \quad (2, 0), \quad (0, 3) \quad \text{and} \quad (1, 1).$$

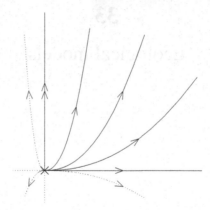

Fig. 33.1. The linearised phase portrait near $(0, 0)$.

With $\mathbf{f}(x, y) = (x(4 - 2x - 2y), \; y(9 - 6x - 3y))$ we have

$$\mathrm{D}\mathbf{f}(x, y) = \begin{pmatrix} 4 - 4x - 2y & -2x \\ -6y & 9 - 6x - 6y \end{pmatrix}.$$

We now look at the linearisation about the four stationary points above. Near the origin we have

$$\frac{d\boldsymbol{\xi}}{dt} = \begin{pmatrix} 4 & 0 \\ 0 & 9 \end{pmatrix} \boldsymbol{\xi},$$

so the eigenvalues are 4 and 9, corresponding to the x- and y-axes respectively. As before, the origin is an unstable node, as shown in Figure 33.1.

Near the stationary point on the x-axis, $(2, 0)$, we have

$$\frac{d\boldsymbol{\xi}}{dt} = \begin{pmatrix} -4 & -4 \\ 0 & -3 \end{pmatrix} \boldsymbol{\xi},$$

so the eigenvalues are $\lambda_1 = -4$ and $\lambda_2 = -3$. The point $(2, 0)$ is a stable node. While the eigenvector corresponding to $\lambda = -4$ is easily seen to be $\mathbf{v}_1 = (1, \; 0)$ (it lies along the x-axis), for the other we need to find $\mathbf{v}_2 = (v_1, \; v_2)$ that satisfies

$$\begin{pmatrix} -1 & -4 \\ 0 & 0 \end{pmatrix} \begin{pmatrix} v_1 \\ v_2 \end{pmatrix} = \mathbf{0},$$

and so $\mathbf{v}_2 = (-4, \; 1)$. The local phase portrait is shown in Figure 33.2.

The stationary point $(0, 3)$ on the y-axis will also turn out to be a stable node. Indeed, the linearised system about this point is

$$\frac{d\boldsymbol{\xi}}{dt} = \begin{pmatrix} -2 & 0 \\ -18 & -9 \end{pmatrix} \boldsymbol{\xi},$$

Fig. 33.2. The linearised phase portrait near $(2, 0)$.

Fig. 33.3. The linearised phase portrait near $(0, 3)$.

and so the eigenvalues are -2 and -9, both negative, and this is another stable node. The eigenvector $\mathbf{v}_1 = (v_1, \ v_2)$ corresponding to $\lambda_1 = -2$ is determined by

$$\begin{pmatrix} 0 & 0 \\ -18 & -7 \end{pmatrix} \begin{pmatrix} v_1 \\ v_2 \end{pmatrix},$$

and so one choice is $(7, \ -18)$; the eigenvector corresponding to $\lambda_2 = -9$ can easily be seen to be $\mathbf{v}_2 = (0, \ 1)$, i.e. along the y-axis. Figure 33.3 shows the local phase portrait.

Finally, near the interior stationary point $(1, 1)$ the linearisation is

$$\frac{d\boldsymbol{\xi}}{dt} = \begin{pmatrix} -2 & -2 \\ -6 & -3 \end{pmatrix} \boldsymbol{\xi}.$$

The eigenvalues of the matrix are the solutions λ of the characteristic equation

$$\begin{vmatrix} -2 - \lambda & -2 \\ -6 & -3 - \lambda \end{vmatrix} = (-2 - \lambda)(-3 - \lambda) + 12 = \lambda^2 + 5\lambda - 6$$

$$= (\lambda + 6)(\lambda - 1) = 0,$$

Fig. 33.4. The linearised phase portrait near (1, 1).

and so are $\lambda_1 = 1$ and $\lambda_2 = -6$; this stationary point is a saddle. The eigenvector $\mathbf{v}_1 = (v_1, \; v_2)$ corresponding to the unstable direction ($\lambda_1 = 1$) is determined by

$$\begin{pmatrix} -3 & -2 \\ -6 & -4 \end{pmatrix} \begin{pmatrix} v_1 \\ v_2 \end{pmatrix} = \mathbf{0},$$

and so is $\mathbf{v}_1 = (2, \; -3)$; while that corresponding to the stable direction ($\lambda_2 = -6$) can be found from

$$\begin{pmatrix} 4 & -2 \\ -6 & 3 \end{pmatrix} \begin{pmatrix} v_1 \\ v_2 \end{pmatrix} = \mathbf{0},$$

and is $\mathbf{v}_2 = (1, \; 2)$. The local saddle point behaviour is shown in Figure 33.4.

Figure 33.5 shows the local phase portraits near the stationary points plotted together. The full phase portrait is shown in Figure 33.6, and is consistent with the local patches we drew in Figure 33.5.

What you can see here is that the stable manifold of the interior stationary point (the bold line in the figure) separates two regions of behaviour. Above this 'separatrix' all trajectories are attracted to the stationary point (0, 3) which lies on the y-axis: species x dies out, and species y settles down to a steady population. Below the stable manifold all trajectories tend to (2, 0); species y dies out, and species x stabilises at a constant value. Only for very special initial conditions, those that lie precisely on the stable manifold, will we end up with the two species coexisting in equal numbers at the interior stationary point (1, 1). Any small fluctuations away from this point will drive one of the species to extinction.

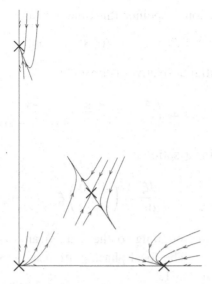

Fig. 33.5. The local phase portraits near the stationary points.

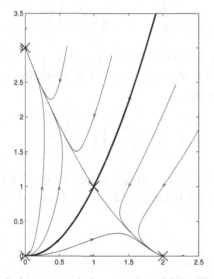

Fig. 33.6. The global phase portrait for equation (33.2). The 'separatrix' (the stable manifold of $(1, 2)$) is the bold curve.

33.1.2 Coexistence

As a final competitive model we will consider the system

$$
\begin{aligned}
\dot{x} &= x(4 - 2x - y) \\
\dot{y} &= y(9 - 3x - 3y).
\end{aligned}
$$

$$(33.3)$$

Again there are four stationary points; this time they are

$$(0, 0), \qquad (2, 0), \qquad (0, 3) \qquad \text{and} \qquad (1, 2).$$

The matrix **Df** of partial derivatives is given by

$$\mathbf{Df}(x, y) = \begin{pmatrix} 4 - 4x - y & -x \\ -3y & 9 - 3x - 6y \end{pmatrix}.$$

Near the origin the linearisation is

$$\frac{d\xi}{dt} = \begin{pmatrix} 4 & 0 \\ 0 & 9 \end{pmatrix} \xi,$$

with eigenvalues 4 (corresponding to the x-axis) and 9 (corresponding to the y-axis); an unstable node. The local phase portrait is the same as that in the last example, shown in Figure 33.1.

Near $(0, 3)$ we put $\xi = x$, $\eta = y - 3$, and linearise to obtain

$$\frac{d\xi}{dt} = \begin{pmatrix} 1 & 0 \\ -9 & -9 \end{pmatrix} \xi.$$

The eigenvalues of the matrix are $\lambda_1 = 1$ and $\lambda_2 = -9$; this stationary point is a saddle. The eigenvector $\mathbf{v}_1 = (v_1, v_2)$ corresponding to $\lambda_1 = 1$ (the unstable direction) is determined by

$$\begin{pmatrix} 0 & 0 \\ -9 & -10 \end{pmatrix} \begin{pmatrix} v_1 \\ v_2 \end{pmatrix} = \mathbf{0},$$

and so is $(10, -9)$; while that corresponding to $\lambda_2 = -9$ (the stable direction) is just $(0, 1)$. The local phase portrait is shown in Figure 33.7.

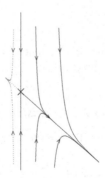

Fig. 33.7. The linearised phase portrait near $(0, 3)$.

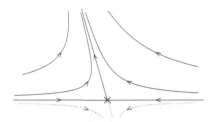

Fig. 33.8. The linearised phase portrait near $(2, 0)$.

The linearisation about $(2, 0)$ is

$$\frac{d\xi}{dt} = \begin{pmatrix} -4 & -2 \\ 0 & 3 \end{pmatrix} \xi;$$

the eigenvalues of this matrix are $\lambda_1 = -4$ with eigenvector $\mathbf{v}_1 = (1, 0)$, and $\lambda_2 = 3$ with corresponding eigenvector determined by

$$\begin{pmatrix} -7 & -2 \\ 0 & 0 \end{pmatrix} \begin{pmatrix} v_1 \\ v_2 \end{pmatrix} = \mathbf{0},$$

i.e. $\mathbf{v}_2 = (-2, 7)$. This is another saddle, with the local phase portrait shown in Figure 33.8.

Finally, the linearised system near the interior stationary point $(1, 2)$ is

$$\frac{d\xi}{dt} = \begin{pmatrix} -2 & -1 \\ -6 & -6 \end{pmatrix} \xi$$

[where $\xi = \mathbf{x} - (1, 2)$]. The eigenvalues of this matrix are the solutions λ of

$$\begin{vmatrix} -2 - \lambda & -1 \\ -6 & -6 - \lambda \end{vmatrix} = (-2 - \lambda)(-6 - \lambda) - 6 = \lambda^2 + 8\lambda + 6 = 0.$$

Hence

$$\lambda = \frac{-8 \pm \sqrt{64 - 24}}{2} = -4 \pm \sqrt{10}.$$

Note first that both $\lambda_1 = -4 - \sqrt{10}$ and $\lambda_2 = -4 + \sqrt{10}$ are negative, so this stationary point is a stable node. The eigenvector corresponding to λ_1 is given by

$$\begin{pmatrix} 2 + \sqrt{10} & -1 \\ -6 & -2 + \sqrt{10} \end{pmatrix} \begin{pmatrix} v_1 \\ v_2 \end{pmatrix} = \mathbf{0},$$

and so is $\mathbf{v}_1 = (1, 2 + \sqrt{10})$; while that corresponding to λ_2 is given by

$$\begin{pmatrix} 2 - \sqrt{10} & -1 \\ -6 & -2 - \sqrt{10} \end{pmatrix} \begin{pmatrix} v_1 \\ v_2 \end{pmatrix} = \mathbf{0},$$

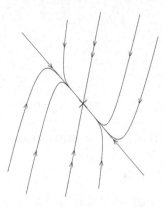

Fig. 33.9. The linearised phase portrait near $(1, 2)$.

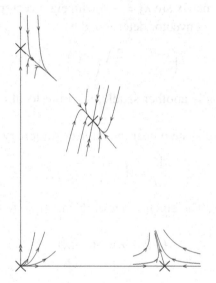

Fig. 33.10. The local phase portraits near the stationary points.

and so is $v_2 = (1, \ 2 - \sqrt{10})$. Since λ_2 has the smaller modulus, trajectories approach the interior fixed point tangent to the v_2 direction; the local phase portrait near $(1, 2)$ is shown in Figure 33.9.

The local phase portraits are all combined on Figure 33.10, and the global phase portrait is shown in Figure 33.11. For this choice of parameters the interior stationary point attracts all trajectories, and so any initial condition that includes some of both species will lead to a state of coexistence in which there are twice as many of species y as there are of species x. If there is only species x then it will settle to its own equilibrium $x = 2$ and if there is only species y then it will settle to $y = 3$.

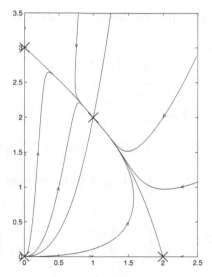

Fig. 33.11. The global phase portrait. From any initial mix of species the solution converges to $(1, 2)$.

33.2 Predator-prey models I

Another class of models governs two species when one is the prey of the other, e.g. hawks and rabbits. If x is the prey and y the predator, we would expect the prey to thrive happily without the predator, so in isolation it should obey the logistic equation

$$\dot{x} = rx(k - x).$$

In contrast, if left to itself with no prey, and hence no food, species y would die out,

$$\dot{y} = -sy.$$

Any interaction between the species now favours y (since they need x to eat) but clearly disadvantages x (who are eaten). So we end up with the model

$$\dot{x} = rx(k - x - ay)$$
$$\dot{y} = y(-s + bx),$$

where all the parameters are positive. When $b < ks$ there are only two stationary points with $x, y \geq 0$, and the predator will eventually die out, leaving the prey to settle down by itself to its natural equilibrium ($x = k$); this possibility is treated in Exercise 33.1 (vii).

We consider here an example that has $b > ks$ and so exhibits more interesting behaviour,

$$\dot{x} = x(1 - 2x - y)$$
$$\dot{y} = y(-2 + 6x).$$

(33.4)

The three stationary points here are

$$(0, 0), \qquad (\tfrac{1}{2}, 0) \qquad \text{and} \qquad (\tfrac{1}{3}, \tfrac{1}{3}).$$

With $\mathbf{f}(x, y) = (x(1 - 2x - y), \ y(-2 + 6x))$ the matrix $\mathbf{Df}$ of partial derivatives is given by

$$\mathbf{Df}(x, y) = \begin{pmatrix} 1 - 4x - y & -x \\ 6y & -2 + 6x \end{pmatrix}.$$

Near the origin the linearisation is simply

$$\frac{d\boldsymbol{\xi}}{dt} = \begin{pmatrix} 1 & 0 \\ 0 & -2 \end{pmatrix} \boldsymbol{\xi},$$

so the origin is a saddle; the eigenvalues are 1, corresponding to the x-axis, i.e. the eigenvector $(1, \ 0)$, and -2, corresponding to the y-axis, i.e. $(0, \ 1)$. See Figure 33.12.

Near $(\tfrac{1}{2}, 0)$ the linearisation gives

$$\frac{d\boldsymbol{\xi}}{dt} = \begin{pmatrix} -1 & -\tfrac{1}{2} \\ 0 & 1 \end{pmatrix} \boldsymbol{\xi},$$

and so the eigenvalues are $\lambda_1 = -1$ and $\lambda_2 = 1$ and this stationary point is a saddle. While the eigenvector corresponding to $\lambda_1 = -1$ lies along the x-axis (it is

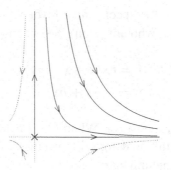

Fig. 33.12. The linearised phase portrait near $(0, 0)$.

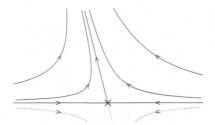

Fig. 33.13. The linearised phase portrait near $\left(\frac{1}{2}, 0\right)$.

$\mathbf{v}_1 = (1, \ 0))$, the eigenvector corresponding to $\lambda_2 = 1$ is determined by

$$\begin{pmatrix} -2 & -\frac{1}{2} \\ 0 & 0 \end{pmatrix} \begin{pmatrix} v_1 \\ v_2 \end{pmatrix} = \mathbf{0},$$

so is $\mathbf{v}_2 = (-1, \ 4)$. See Figure 33.13.

The linearisation about the interior stationary point is

$$\frac{d\xi}{dt} = \begin{pmatrix} -\frac{2}{3} & -\frac{1}{3} \\ 2 & 0 \end{pmatrix} \xi.$$

The eigenvalues of this matrix are given by the solutions λ of the characteristic equation

$$\begin{vmatrix} -\frac{2}{3} - \lambda & -\frac{1}{3} \\ 2 & -\lambda \end{vmatrix} = \lambda^2 + \frac{2\lambda}{3} + \frac{2}{3} = 0.$$

Using the quadratic formula these are

$$\lambda = -\frac{\frac{2}{3} \pm \sqrt{\frac{4}{9} - \frac{8}{3}}}{2} = \frac{-1 \pm i\sqrt{5}}{3},$$

and so the interior stationary point is a stable spiral. Trajectories therefore spiral in towards $\left(\frac{1}{3}, \frac{1}{3}\right)$; to find out whether they spiral in clockwise or anti-clockwise, we look at the linearised equations near $\left(\frac{1}{3}, \frac{1}{3}\right)$. The equation for η is $\dot{\eta} = 2\xi$, so that for $\xi > 0$ (to the right of the stationary point, since $x = \frac{1}{3} + \xi$) trajectories are moving up, while for $\xi < 0$ (to the left of the stationary point) trajectories are moving down. The spiralling motion is therefore anti-clockwise (see Figure 33.14).

The full phase portrait is shown in Figure 33.15. Since the interior stationary point is globally attracting, the predator and the prey settle down to a state in which there is enough prey to keep the predators alive and the predators never eat themselves to extinction.

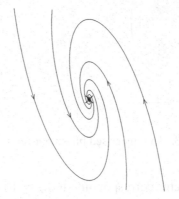

Fig. 33.14. Spiral motion near the interior stationary point $(\frac{1}{3}, \frac{1}{3})$.

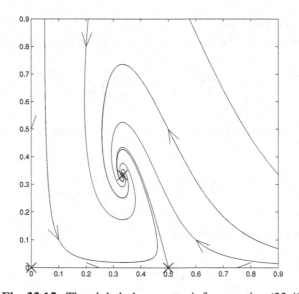

Fig. 33.15. The global phase portrait for equation (33.4).

33.3 Predator-prey models II

We now suppose that the prey, if left alone, would obey the exponential population model

$$\dot{x} = kx$$

rather than the logistic model we used previously; this would be the case for a population (perhaps unrealistically) not limited by local resources. An example might be whales and plankton. In this case, the equations become

$$\dot{x} = kx(1 - ay)$$
$$\dot{y} = y(bx - s),$$

(with all coefficients positive) and we will analyse these in general.

There are only two stationary points, at $(0, 0)$ and at $(s/b, 1/a)$. The matrix of partial derivatives of the right-hand side is

$$\mathrm{Df}(x, y) = \begin{pmatrix} k(1 - ay) & -akx \\ by & bx - s \end{pmatrix}.$$

Near the origin this gives the linearisation

$$\frac{d\xi}{dt} = \begin{pmatrix} k & 0 \\ 0 & -s \end{pmatrix} \xi,$$

and so the origin is a saddle, with the stable direction running along the y-axis and the unstable direction along the x-axis. About the interior stationary point the linearisation is

$$\frac{d\xi}{dt} = \begin{pmatrix} 0 & -aks/b \\ b/a & 0 \end{pmatrix} \xi.$$

The eigenvalues λ of the matrix satisfy

$$\begin{vmatrix} -\lambda & -aks/b \\ b/a & -\lambda \end{vmatrix} = \lambda^2 + ks = 0,$$

and so $\lambda = \pm i\sqrt{ks}$.

The linearised flow near the interior stationary point suggests that it might be a centre, but recall that we can only guarantee that the phase portrait for the nonlinear equation looks like the linearised phase portrait when the eigenvalues have *non-zero real part*. So we need to do a little more work to check that there really are closed orbits around $(s/b, 1/a)$. We can do this because it is possible to find the equation of the curves traced out by trajectories.

Using the chain rule, if $y(t) = y(x(t))$ then we have

$$\frac{dy}{dt} = \frac{dy}{dx}\frac{dx}{dt},$$

from which it follows that when $\dot{x} \neq 0$ we have

$$\frac{dy}{dx} = \frac{dy}{dt} \Big/ \frac{dx}{dt}.$$

(you can think of this heuristically as 'cancelling the dts'). So therefore along a trajectory

$$\frac{dy}{dx} = \frac{\dot{y}}{\dot{x}} = \frac{y(bx - s)}{kx(1 - ay)}$$

(cf. example in Section 8.6.2). This equation is separable,

$$\frac{1 - ay}{y} dy = \frac{bx - s}{kx} dx,$$

and so

$$\frac{1}{y} - a\, dy = \frac{b}{k} - \frac{s}{kx} dx.$$

Integrating both sides gives (since x and y are positive)

$$\ln y - ay = (bx/k) - \frac{s}{k} \ln x + C,$$

i.e.

$$C(x, y) = \ln y - ay + \frac{s}{k} \ln x - \frac{bx}{k}$$

is constant on trajectories.

In order to understand the forms of curves of constant C, we can find the turning points of the function $C(x, y)$ (see Appendix C for a brief discussion). These occur when $\partial C/\partial x = \partial C/\partial y = 0$. Because

$$\frac{\partial C}{\partial x} = \frac{s}{kx} - \frac{b}{k} \quad \text{and} \quad \frac{\partial C}{\partial y} = \frac{1}{y} - a,$$

there is only one turning point at $(s/b, 1/a)$, the interior stationary point. Calculating the matrix of second partial derivatives

$$\begin{pmatrix} \partial^2 C/\partial x^2 & \partial^2 C/\partial x \partial y \\ \partial^2 C/\partial y \partial x & \partial^2 C/\partial y^2 \end{pmatrix} = \begin{pmatrix} -s/kx^2 & 0 \\ 0 & -1/y^2 \end{pmatrix},$$

it is easy to see that both the eigenvalues of this matrix are negative (they are $-s/kx^2$ and $-1/y^2$), and so C is a maximum at this stationary point. It follows that the curves of constant C are closed curves near the stationary point, and hence the trajectories form a collection of periodic orbits.

Example 33.1 *Draw the phase portrait for the equations*

$$\dot{x} = x(1 - y)$$
$$\dot{y} = y(2x - 4)$$

(33.5)

and find the equations of the trajectories.

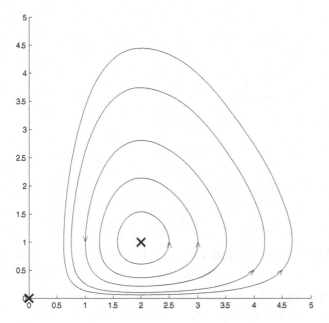

Fig. 33.16. The phase portrait for equation (33.5).

Using the analysis above, there are two fixed points, one at the origin (which is a saddle) and one at $(2, 1)$ which is a centre. The phase portrait is shown in Figure 33.16.

On the curves traced out by solutions we have

$$\frac{dy}{dx} = \frac{\dot{y}}{\dot{x}} = \frac{y(2x - 4)}{x(1 - y)}.$$

Separating the variables,

$$\frac{1 - y}{y} \, dy = \frac{2x - 4}{x} \, dx,$$

and integrating both sides gives

$$\ln y - y = 2x - 4 \ln x + C,$$

so that

$$C(x, y) = \ln y + 4 \ln x - y - 2x$$

is constant on trajectories. The level sets of this function C, produced by the MATLAB code

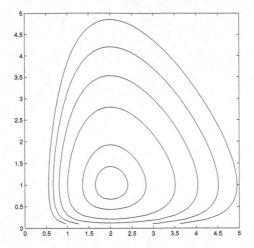

Fig. 33.17. Curves on which $C(x, y) = \ln y + 4 \ln x - y - 2x$ is constant.

```
>> [x, y]=meshgrid(.1:.1:4.5, .1:.1:4.5);
>> z=log(y)+4*log(x)-y-2*x;
>> contour(x,y,z,[-2.3 -2.5 -3.5 -4 -4.5]);
```

are shown in Figure 33.17 – note how they correspond to the trajectories in Figure 33.16.

Exercises

33.1 For each of the following models of two species, describe first the type of situation being modelled, then find the stationary points, determine their stability type and draw the phase portrait for $x, y \geq 0$. Finally, say what the phase portrait means for the two species.

(i)

$$\dot{x} = x(2 - x - y)$$
$$\dot{y} = y(2 - 2x - 2y)$$

(ii)

$$\dot{x} = x(2 - x - y)$$
$$\dot{y} = y(2 - x/4 - 2y)$$

(iii)

$$\dot{x} = x(2 - x - 3y)$$
$$\dot{y} = y(2 - 3x - 2y)$$

(iv)

$$\dot{x} = x(1 - 2y)$$
$$\dot{y} = y(-2 + 3x),$$

find also the equations of the curves along which the solutions move,

(v)

$$\dot{x} = x(4 - x/2 - 3y)$$
$$\dot{y} = y(-2 + x)$$

(vi)

$$\dot{x} = x(10 - x - 3y)$$
$$\dot{y} = y(1 + x - 10y)$$

(vii)

$$\dot{x} = x(3 - x - y)$$
$$\dot{y} = y(-2 + x).$$

(You could use the MATLAB program `lotkaplane.m` to help draw some of these phase portraits. It asks for the parameters that occur in the general form of the equations

$$\dot{x} = x(A + ax + by)$$
$$\dot{y} = y(B + cx + dy)$$

and then draws the trajectory forwards and backwards through specified initial conditions.)

33.2 The situation in which two species cooperate, so that the presence of one enhances the environment for the other, can be modelled by a coupled pair of equations of the form

$$\dot{x} = x(A - ax + by)$$
$$\dot{y} = y(B + cx - dy),$$

where all the parameters are positive. Draw the phase portraits for the following cooperative equations:

(i)

$$\dot{x} = x(1 - x + y)$$
$$\dot{y} = y(1 + x - 2y),$$

and

(ii)

$$\dot{x} = x(2 - x + y)$$
$$\dot{y} = y(4 + 2x - y).$$

33.3 (T) Consider the general model of two competing species,

$$\dot{x} = x(A - ax - by)$$
$$\dot{y} = y(B - cx - dy),$$

where all the parameters are positive. Assuming that the intercepts of the nullclines (lines on which $\dot{x} = 0$ and $\dot{y} = 0$) with the x- and y-axes do not coincide, by considering the relative positions of these intercepts show that there are four distinct possibilities for the behaviour of solutions, and find the parameter ranges over which they occur. Check that your results are consistent with what you found for the competitive examples in Exercise 33.1.

34

Newtonian dynamics

In this chapter we apply phase plane ideas to various one-dimensional systems that model a particle moving under Newton's laws of motion. First we consider systems in which the energy is constant, and then we consider systems in which there is some dissipation.

34.1 One-dimensional conservative systems

We consider a particle of mass m moving on a line in a potential force field, such that its potential energy at position x is given by $V(x)$. Then its kinetic energy is $\frac{1}{2}m\dot{x}^2$, and its total energy is

$$E = \tfrac{1}{2}m\dot{x}^2 + V(x). \tag{34.1}$$

If the energy is conserved then we can differentiate to give

$$m\dot{x}\ddot{x} + V'(x)\dot{x} = 0;$$

provided that $\dot{x} \neq 0$ we can cancel this term and obtain[1]

$$m\ddot{x} = -V'(x). \tag{34.2}$$

By setting $y = \dot{x}$, we can rewrite this as the coupled system

$$\begin{aligned} \dot{x} &= y \\ m\dot{y} &= -V'(x). \end{aligned} \tag{34.3}$$

In all that follows we will take $m = 1$ for simplicity.

If you think of these equations as describing the motion of a bead sliding on a wire whose height at coordinate x is given by $V(x)$ you will get the correct

---

[1] By assuming the continuity of the function $x(t)$ and its derivatives it is possible to justify this equation for all t, even at those times when $\dot{x}(t) = 0$.

qualitative idea of how the solutions should behave, although this interpretation is
not entirely accurate, as discussed in the next section.

Because of the relatively simple form of these equations the possible behaviour
of the solutions is restricted. First, note that at any stationary point (x^*, y^*) we
must have $y^* = 0$ (zero velocity), and x^* must be a turning point of $V(x)$ (since
we need $V'(x^*) = 0$). The linearisation near such a stationary point is simply

$$\frac{d\xi}{dt} = \begin{pmatrix} 0 & 1 \\ -V''(x^*) & 0 \end{pmatrix} \xi,$$

so that the eigenvalues of the matrix are the solutions of

$$\begin{vmatrix} -\lambda & 1 \\ -V''(x^*) & -\lambda \end{vmatrix} = \lambda^2 + V''(x^*) = 0,$$

i.e. $\lambda = \pm\sqrt{-V''(x^*)}$, giving either a pair of real eigenvalues of opposite sign if
$V''(x^*) < 0$ or a pair of purely imaginary eigenvalues if $V''(x^*) > 0$.

Thus any stationary points corresponding to maxima of V (where $V''(x^*) < 0$)
will be saddle points, while those corresponding to minima of V (where $V''(x^*)$
> 0) will be centres for the linearised equation.

Since trajectories move on the curves $\frac{1}{2}y^2 + V(x) = E$, these 'linearised centres' will in fact be centres for the full nonlinear equations, as we can see by considering the curves of constant E (you may find it useful to refer here to Appendix C).
Maxima and minima of E occur when $\partial E/\partial x = V'(x) = 0$ and $\partial E/\partial y = y = 0$,
precisely at the stationary points. The Hessian matrix of second derivatives of E is
simply

$$\begin{pmatrix} V''(x) & 0 \\ 0 & 1 \end{pmatrix}.$$

Since its eigenvalues are $V''(x)$ and 1, minima of V are also minima of E, and
maxima of V are saddle points of E. Thus when $V''(x^*) > 0$ the point $(x^*, 0)$ is a
local minimum of E, and so nearby curves of constant E are closed.

Example 34.1 *A particle of mass 1 moves on a line under the influence of a potential* $V(x) = x - \frac{1}{3}x^3$, *as illustrated in Figure 34.1. Sketch the phase portrait,
and describe the motion.*

The energy

$$E = \frac{1}{2}y^2 + x - \frac{1}{3}x^3$$

will be constant along any trajectory, and the curves of constant E are shown in
Figure 34.2.

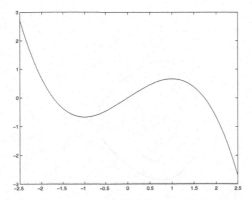

Fig. 34.1. The potential $V(x) = x - \frac{1}{3}x^3$ plotted against x.

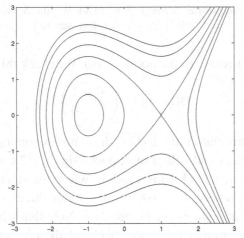

Fig. 34.2. Curves of constant $E(x, y) = \frac{1}{2}y^2 + x - \frac{1}{3}x^3$.

The equation of motion is $\ddot{x} = -V'(x)$,

$$\ddot{x} = -1 + x^2,$$

and setting $y = \dot{x}$ yields the coupled equations

$$\begin{aligned} \dot{x} &= y \\ \dot{y} &= -1 + x^2. \end{aligned} \tag{34.4}$$

For a stationary point we need $y = 0$ and $-1 + x^2 = 0$, so there are two stationary points, $(-1, 0)$ and $(1, 0)$. Looking at the potential, you might expect oscillations about $x = -1$ (these would be closed orbits around the point $(-1, 0)$), and

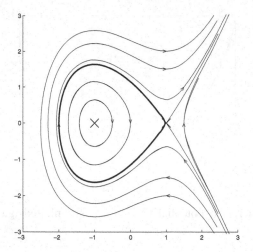

Fig. 34.3. The phase portrait for equation (34.4). Within the bold curve the particle oscillates about $x = 1$, while outside it rolls away to $x = +\infty$.

instability near $x = 1$ (close to the point $(1, 0)$). We check this by calculating

$$\mathbf{Df}(x, y) = \begin{pmatrix} 0 & 1 \\ 2x & 0 \end{pmatrix}.$$

The eigenvalues of this matrix are $\pm\sqrt{2x}$; the complex conjugate pair $\pm\sqrt{2}i$ when
$x = -1$ and two distinct real values of opposite sign $\pm\sqrt{2}$ when $x = 1$.

So $(-1, 0)$ is a centre and $(1, 0)$ is a saddle, where the eigenvectors are $(1, \sqrt{2})$
in the unstable direction and $(1, -\sqrt{2})$ in the stable direction. 'Joining the dots'
we get the phase portrait shown in Figure 34.3. Note that when $E < E_{\text{crit}} = 2/3$
the particle can move to and fro in the 'well' of the potential. When $E = E_{\text{crit}}$
there is one orbit that starts from the top of the rise, rolls to the left, and then rolls
back and comes to rest exactly where it started; this trajectory, shown as a bolder
line in the figure, forms both the stable and unstable manifold of the point $(1, 0)$.
When $E > E_{\text{crit}}$ the particle always rolls off to $x = +\infty$.

34.2 *A bead on a wire

Earlier in the chapter we said that it was possible to get an accurate qualitative idea
of the dynamics of the equation

$$\ddot{x} = -V'(x) \tag{34.5}$$

by imagining a bead sliding on a wire whose height is given by $h = V(x)$. However, (34.5) is not the right model for this situation, as we now see by considering
the 'bead on a wire' problem in more detail.

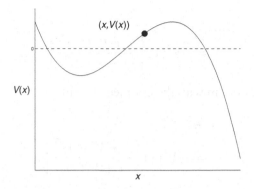

Fig. 34.4. A bead on a wire whose shape is given by $h = V(x)$.

We assume that the bead slides (without friction) on a wire bent into the shape given by $h = V(x)$ (where h is the height above some reference level), see Figure 34.4. In particular we want to relate this to the behaviour of a particle moving on a line in a potential forcefield $V(x)$. Again for convenience we take the mass of the bead to be 1.

Although $V(x)$ is the potential energy (taking units in which $g = 1$), the kinetic energy of the bead also has to include the vertical component of its motion. Since the position of the bead when its horizontal coordinate is x is

$$\mathbf{x} = (x, V(x)),$$

its velocity is

$$\dot{\mathbf{x}} = (\dot{x}, V'(x)\dot{x}).$$

So its kinetic energy, $\frac{1}{2}|\dot{\mathbf{x}}|^2$, is

$$\tfrac{1}{2}|\dot{\mathbf{x}}|^2 = \tfrac{1}{2}\dot{x}^2[1 + (V'(x))^2],$$

and the total energy is

$$E = \tfrac{1}{2}\dot{x}^2[1 + (V'(x))^2] + V(x).$$

Differentiating this gives

$$0 = \dot{x}\ddot{x}[1 + V'(x)^2] + \dot{x}^2 V'(x)V''(x)\dot{x} + V'(x)\dot{x};$$

dividing by $\dot{x}$ and rearranging we have

$$\ddot{x}[1 + V'(x)^2] = -V'(x)V''(x)\dot{x}^2 - V'(x),$$

or

$$\ddot{x} = -V'(x)\left[\frac{1 + V''(x)\dot{x}^2}{1 + V'(x)^2}\right].$$

Setting $y = \dot{x}$ we end up with the coupled system

$$\dot{x} = y \tag{34.6}$$

$$\dot{y} = -V'(x)\left[\frac{1 + V''(x)y^2}{1 + V'(x)^2}\right]. \tag{34.7}$$

We will now see that these complicated looking equations have the same stationary points as the system

$$\dot{x} = y \qquad \dot{y} = -V'(x)$$

and that these stationary points have the same stability properties for both sets of equations.

First, note that at any stationary point (x^*, y^*) we must still have $y^* = 0$, using equation (34.6). For $\dot{y} = 0$ equation (34.7) then requires

$$-V'(x)\left[\frac{1}{1 + V'(x)^2}\right] = 0$$

(the expression in the square brackets has simplified since $y^* = 0$). Since the denominator $1 + V'(x)^2$ is always strictly positive, we must have $V'(x^*) = 0$ for a stationary point. So, as before, the stationary points $(x^*, 0)$ occur when x^* is one of the turning points of $V(x)$.

With a little more algebra we can also show that the stationary points have the same stability type as their counterparts in the simpler problem, which depended only on the sign $V''(x^*)$. Here $\mathbf{Df}$ is given by the daunting expression

$$\mathbf{Df}(\mathbf{x}) = \begin{pmatrix} 0 & 1 \\ -V''\left[\frac{1+V''y^2}{1+V'^2}\right] - V'\left[\frac{(1+V'^2)V'''y^2 - 2V'V''(1+V''y^2)}{(1+V'^2)^2}\right] & \frac{-2yV'V''}{1+V'^2} \end{pmatrix},$$

where V', V'', and V''' are understood to depend on x. However, near any stationary point (x^*, y^*) the linearisation is simply

$$\frac{d\xi}{dt} = \begin{pmatrix} 0 & 1 \\ -V''(x^*) & 0 \end{pmatrix}\xi,$$

just as before, since $y^* = 0$ and $V'(x^*) = 0$ at any stationary point. Just as for the simpler model, if $V''(x^*) > 0$ then the stationary point is a centre, and if $V''(x^*) < 0$ then it is a saddle.

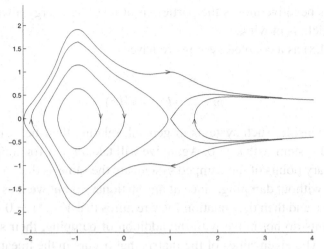

Fig. 34.5. The phase portrait for a bead moving on a wire bent in the shape $h = x - \frac{1}{3}x^3$.

For a bead sliding on a wire bent into the shape $h(x) = x - \frac{1}{3}x^3$ as in Example 34.1 the equations are $\dot{x} = y$ and

$$\dot{y} = -(1 - x^2)\left[\frac{1 - 2xy^2}{2 - 2x^2 + x^4}\right].$$

The phase portrait is shown in Figure 34.5. Although the individual trajectories have changed shape, the qualitative behaviour is the same as for the particle moving in the potential $V(x)$ (see Figure 34.3); if the energy is below a certain level the bead oscillates, while if the energy is too high the bead escapes to $x = +\infty$.

34.3 Dissipative systems

In Chapter 13 we looked at models of oscillating systems, both with and without damping. There we modelled the effect of damping by including in the equation for $\ddot{x}$ an additional term $-k\dot{x}$ (with $k > 0$), representing a force acting to oppose the motion.

If we include a similar factor in equation (34.2) then it becomes

$$m\ddot{x} = -k\dot{x} - V'(x). \tag{34.8}$$

The new damping term has the effect of dissipating the energy in the system. If we now calculate dE/dt we have

$$\frac{dE}{dt} = \frac{d}{dt}\left[\frac{1}{2}m\dot{x}^2 + V(x)\right]$$

$$= m\dot{x}\ddot{x} + V'(x)\dot{x} = -k\dot{x}^2 - V'(x)\dot{x} + V'(x)\dot{x}$$

$$= -k\dot{x}^2 < 0.$$

Since $-k\dot{x}^2$ is negative unless the particle is at rest, the energy always decreases while the particle is moving.

Writing (34.8) as a coupled system we have

$$\dot{x} = y$$
$$m\dot{y} = -ky - V'(x).$$

We will first consider such systems in general, relating their behaviour to that of the undamped system with $k = 0$. Again we will take $m = 1$ for simplicity.

The stationary points of the damped system are the same as the stationary points of the system without damping, since at any stationary point we must have $y^* = 0$ to make $\dot{x}$ zero, and then the equation for $\dot{y}$ requires that $V'(x^*) = 0$. Although the stationary points do not move with the addition of damping, their stability properties change. The eigenvalues of the matrix that occurs in the linearised equation near a stationary point $(x^*, 0)$,

$$\frac{d\boldsymbol{\xi}}{dt} = \begin{pmatrix} 0 & 1 \\ -V''(x^*) & -k \end{pmatrix} \boldsymbol{\xi}, \tag{34.9}$$

are the solutions of the characteristic equation

$$\begin{vmatrix} -\lambda & 1 \\ -V''(x^*) & -k - \lambda \end{vmatrix} = \lambda^2 + k\lambda + V''(x^*) = 0.$$

Using the quadratic formula,

$$\lambda = \frac{-k \pm \sqrt{k^2 - 4V''(x^*)}}{2}.$$

Without damping ($k = 0$) any maximum of V (where $V''(x^*) < 0$) was a saddle point. Even with damping this is still the case, since the expression within the square root is always positive and greater than k, giving one positive and one negative eigenvalue. However, the stability type of minima (where $V''(x^*) > 0$) changes. Where before we had a purely complex pair of eigenvalues, and so a centre, now the real part of both eigenvalues will be negative, and depending on whether $k^2 > 4V''(x^*)$ or $k^2 < 4V''(x^*)$ the stationary point will be a stable node or a stable spiral.

We now return to the system of Example 34.1 and investigate the effect of such a damping term.

Example 34.2 *Draw the phase portrait for the system*

$$\dot{x} = y$$
$$\dot{y} = -y - 1 + x^2.$$

As we noticed in general above, the stationary points are the same as they were for the undamped system, namely $(\pm 1, 0)$. However, their stability properties have changed. If we now calculate $\mathbf{Df}$ we have

$$\mathbf{Df}(x, y) = \begin{pmatrix} 0 & 1 \\ 2x & -1 \end{pmatrix}.$$

About $(1, 0)$ the linearisation is

$$\frac{d\xi}{dt} = \begin{pmatrix} 0 & 1 \\ 2 & -1 \end{pmatrix} \xi,$$

and the eigenvalues of the matrix of the right-hand side are given by the solutions of

$$\begin{vmatrix} -\lambda & 1 \\ 2 & -1 - \lambda \end{vmatrix} = \lambda^2 + \lambda - 2 = (\lambda - 1)(\lambda + 2) = 0,$$

i.e. $\lambda = 1$ or $\lambda = -2$. The eigenvalues are real and of opposite sign, and this stationary point is still a saddle. The eigenvector corresponding to $\lambda_1 = 1$ is given by

$$\begin{pmatrix} -1 & 1 \\ 2 & -2 \end{pmatrix} \begin{pmatrix} v_1 \\ v_2 \end{pmatrix} = \mathbf{0},$$

and therefore is $\mathbf{v}_1 = (1, 1)$; while the eigenvector corresponding to $\lambda_2 = -2$ can be found from

$$\begin{pmatrix} 2 & 1 \\ 2 & 1 \end{pmatrix} \begin{pmatrix} v_1 \\ v_2 \end{pmatrix} = \mathbf{0},$$

and therefore is $\mathbf{v}_2 = (1, -2)$.

Near the stationary point $(-1, 0)$ the linearisation is

$$\frac{d\xi}{dt} = \begin{pmatrix} 0 & 1 \\ -2 & -1 \end{pmatrix} \xi,$$

and the eigenvalues of the matrix are the solutions of

$$\begin{vmatrix} -\lambda & 1 \\ -2 & -1 - \lambda \end{vmatrix} = \lambda^2 + \lambda + 2,$$

which are

$$\lambda = \frac{-1 \pm \sqrt{1 - 8}}{2} = -\tfrac{1}{2}(1 \pm \sqrt{7}i);$$

this stationary point is a stable spiral.

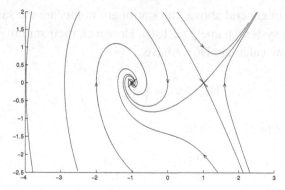

Fig. 34.6. The phase portrait for the dissipative system from Example 34.2.

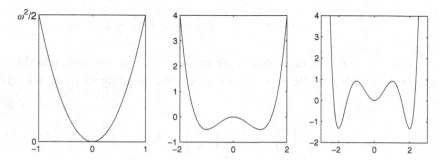

Fig. 34.7. From left to right, the potentials from parts (i), (ii) and (iii) of Exercise 34.1 plotted against x.

The phase portrait is shown in Figure 34.6. It is clear that now, with damping, there are no periodic orbits, and the particle will either come to rest at $x = -1$ or escape to $x = +\infty$.   □

Exercises

34.1 For the following choices of potential functions $V(x)$ write down the total energy for a particle of unit mass, and assuming that this is conserved write down a coupled system for x and $y = \dot{x}$. Draw the phase portrait and interpret the dynamics.
   (i) $V(x) = \frac{1}{2}\omega^2 x^2$;
   (ii) $V(x) = \frac{1}{2}x^4 - x^2$; and
   (iii) $V(x) = \frac{1}{6}x^6 - \frac{5}{4}x^4 + 2x^2$.
   (Pictures of these potentials are shown in Figure 34.7.)

34.2 For the functions $V(x)$ in parts (i) and (ii) of Exercise 34.1 write down the kinetic energy of a particle of unit mass moving on a wire whose height as a function of x is $V(x)$. Taking $g = 1$ write down the total energy, and hence derive the second order equation satisfied by x. Write down a coupled system for x and $y = \dot{x}$, and draw the phase portrait.

34.3 Write down the equation of motion for a particle of unit mass moving in each of the potentials in Exercise 34.1, when there is an additional damping force $-\dot{x}$ (in part (i) take $\omega = 1$). Draw the phase portrait for each case.

34.4 (T) A particle of unit mass moves on a wire whose height as a function of x is $V(x)$, and is subject to an additional damping force $-k\dot{x}$. Write down the equation of motion, and show that the behaviour of this system is qualitatively the same as that of

$$\ddot{x} = -V'(x) - k\dot{x}.$$

34.5 (C) Investigate the dynamics of the equations in exercise 34.1 both with and without damping, using the M-file `newtonplane.m`. The program asks for the level of damping k, and then a succession of initial conditions. The equation is specified in the file `newtonde.m`, currently set up for the example $V(x) = x - \frac{1}{3}x^3$ in the main text. By changing this file you should be able to consider all the examples in Exercise 34.1, and also the equivalent problems for a ball rolling on a wire.

35

The 'real' pendulum

We end our treatment of phase portraits by returning to the example of the simple pendulum.

35.1 The undamped pendulum

In Chapter 13 we derived the exact equation for the motion of an ideal pendulum,

$$\frac{d^2\theta}{dt^2} = -\omega^2 \sin\theta, \tag{35.1}$$

(where $\omega^2 = g/L$) but we then approximated this by $\ddot\theta = -\omega^2\theta$ in order to apply the methods we had just learned for linear equations.

Here we will use phase plane methods to understand the nonlinear equation (35.1). For simplicity we will choose $\omega = 1$ and consider the equation

$$\frac{d^2\theta}{dt^2} = -\sin\theta.$$

In order to look at this as a set of coupled first order equations we set $x = \theta$ and $y = \dot\theta$ and then

$$\begin{aligned} \dot x &= y \\ \dot y &= -\sin x. \end{aligned} \tag{35.2}$$

Note that the direction field (shown in Figure 35.1) repeats itself every 2π in the horizontal direction. This should not be a surprise, since the x coordinate represents the angle of the pendulum to the vertical (θ in our original equation), and the value $\theta = x + 2\pi$ corresponds to the same position of the pendulum as $\theta = x$. So we should consider (x, y) and $(x + 2\pi, y)$ as representing the same physical

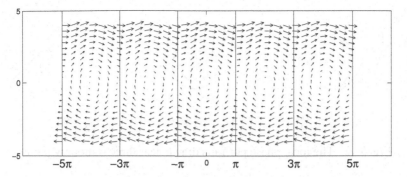

Fig. 35.1. The direction field for the pendulum equation.

state of the system. The natural way to present our phase diagrams, then, is to restrict to a range of x corresponding to one particular choice for the angle θ, $-\pi < x \le \pi$, say. We will show a slightly longer range of x, since this may be helpful in understanding the phase portraits, but you should remember that the figures 'wrap around' from one side to the other. We return to this at the end of the chapter.

The first step in the phase plane analysis is always to find all the stationary points. If (x^*, y^*) is a stationary point then we need

$$y^* = 0 \qquad \text{and} \qquad -\sin x^* = 0.$$

The first equation tells us that y^* must be zero at any stationary point, while the second implies that $x^* = k\pi$ for some (positive or negative) integer k.

Thus any point $\mathbf{x}^* = (k\pi, 0)$ with k an integer is a stationary point. In fact there are just two distinct stationary points here: $(0, 0)$, which corresponds to the pendulum hanging vertically downward; and $(\pi, 0)$ which corresponds to the pendulum being precariously balanced vertically upward.

To check the stability of these stationary points we have to consider the eigenvalues of the matrix of partial derivatives

$$\mathbf{Df}(\mathbf{x}) = \begin{pmatrix} 0 & 1 \\ -\cos x & 0 \end{pmatrix}$$

near the stationary points.

Near the origin (and all stationary points $(2k\pi, 0)$) the linearised equation is

$$\frac{d\xi}{dt} = \begin{pmatrix} 0 & 1 \\ -1 & 0 \end{pmatrix} \xi; \tag{35.3}$$

the eigenvalues of this matrix are $\pm i$, so the origin is a (linearised) centre, round which the orbits travel clockwise (when $\xi = 0$ and $\eta > 0$, $\dot{\xi} = \eta > 0$).

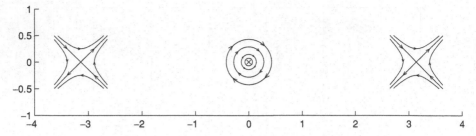

Fig. 35.2. The local phase portraits drawn near three stationary points.

Near $(\pi, 0)$ (and all the stationary points $((2k + 1)\pi, 0)$) the linearisation is

$$\frac{d\xi}{dt} = \begin{pmatrix} 0 & 1 \\ 1 & 0 \end{pmatrix} \xi; \tag{35.4}$$

the eigenvalues are ± 1 with corresponding eigenvectors $(1, \pm 1)$.

We can now put this information on a phase diagram, at present just drawing some trajectories close to the stationary points. Figure 35.2 shows the local phase portraits near the three stationary points closest to the origin, $(\pm\pi, 0)$ and $(0, 0)$ itself.

To complete the picture it is helpful to use the fact that the energy is constant,

$$\tfrac{1}{2}y^2 - \cos x = E = \text{constant}.$$

On each solution curve

$$y^2 = 2(E + \cos x) \qquad \Rightarrow \qquad y = \pm\sqrt{2(E + \cos x)}$$

and the type of curves traced out by trajectories depend on the value of E. It is clear that if $E > 1$ then we can solve for y for every value x; there will be one curve that has $y > 0$ and one with $y < 0$. If $E < 1$ then we can only solve for a range of x values, and this will lead to a closed curve passing through $y = 0$. The critical value is $E = 1$. Curves of constant E are shown in Figure 35.3, and the phase portrait is shown in Figure 35.4.

Note that

(i) for $E < 1$ there are a collection of closed curves circling the origin;
(ii) the curves corresponding to $E = 1$, shown as bold lines in Figure 35.4, form the stable and unstable manifolds of the saddle points at $(\pm\pi, 0)$, and connect these two points;
(iii) for $E > 1$ the value of x is either always increasing ($y > 0$) or always decreasing ($y < 0$).

These translate to the following behaviour of the original pendulum:

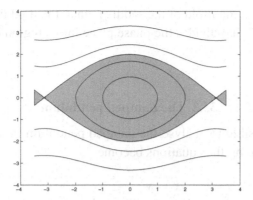

Fig. 35.3. Curves on which $E = \frac{1}{2}y^2 - \cos x$ is constant. In the shaded region $E < 1$.

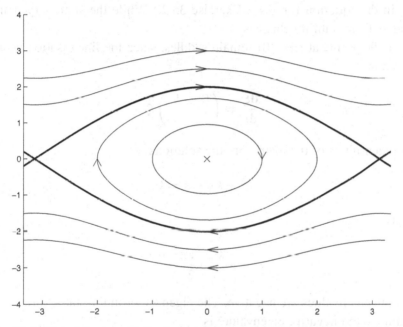

Fig. 35.4. The phase portrait for the simple pendulum. The bold lines indicate the 'separatrix', dividing a region in which the pendulum oscillates about the downward vertical, and a region in which the pendulum whirls around through $360°$.

(i) If the pendulum has less than a critical amount of energy ($E = 1$) then it performs oscillations back and forth about $\theta = 0$, i.e. its rest position pointing downwards.

(ii) If the pendulum has the critical amount of energy then it can make exactly one revolution from the upright position ($\theta = -\pi$) back to the upright position ($\theta = \pi$) (or vice versa) and no further.

(iii) If the pendulum has more energy ($E > 1$) then it can whirl around forever.

Note that the stable manifold of the saddle points (the bold lines in Figure 35.4) form the 'separatrix' that divides the phase plane into a region of oscillation and a region of 'whirling'.

35.2 The damped pendulum

It is, of course, possible to analyse the damped pendulum in a similar way. With the addition of damping the equations become

$$\dot{x} = y$$
$$\dot{y} = -\sin x - ky, \tag{35.5}$$

where $k > 0$. (For the behaviour of the system with a quadratic damping term $-ky|y|$ in the equation for $\dot{y}$ see Exercise 35.2.) While the stationary points are unchanged, their stability changes.

The saddle points at $(\pm\pi, 0)$ remain saddles, since the linearisation near these points is now

$$\frac{d\xi}{dt} = \begin{pmatrix} 0 & 1 \\ 1 & -k \end{pmatrix} \xi$$

and the eigenvalues of the matrix are the solutions of

$$\lambda^2 + k\lambda - 1 = 0,$$

which gives

$$\lambda = \frac{-k \pm \sqrt{k^2 + 4}}{2},$$

so that both eigenvalues are negative. The eigenvector in the stable direction (corresponding to the negative eigenvalue) is

$$\begin{pmatrix} 2 \\ \sqrt{k^2 + 4} - k \end{pmatrix},$$

and the eigenvector in the unstable direction (corresponding to the positive eigenvalue) is

$$\begin{pmatrix} 2 \\ -k - \sqrt{k^2 + 4} \end{pmatrix}.$$

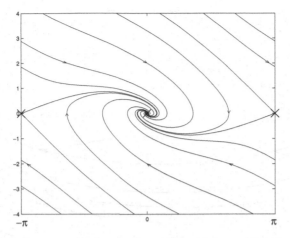

Fig. 35.5. The phase portrait for the damped pendulum. All trajectories tend to $(0, 0)$. Remember that $x = \pi$ and $x = -\pi$ correspond to the same angle, therefore as trajectories leave the diagram on the left they reappear on the right (and vice versa).

The origin becomes a stable node or spiral; the linearisation is now

$$\frac{d\xi}{dt} = \begin{pmatrix} 0 & -1 \\ 1 & -k \end{pmatrix} \xi,$$

and so the eigenvalues of the matrix are the solutions of

$$\lambda^2 + k\lambda + 1 = 0,$$

which are

$$\lambda = \frac{-k \pm \sqrt{k^2 - 4}}{2}.$$

The phase portrait is shown in Figure 35.5 for the choice $k = 1$ when

$$\lambda = \frac{-1 \pm i\sqrt{3}}{2}$$

and the origin is a stable spiral.

The behaviour predicted by this phase diagram accords with our physical intuition; from any initial condition (apart from being precariously balanced vertically upwards) the pendulum will eventually come to rest hanging vertically downward.

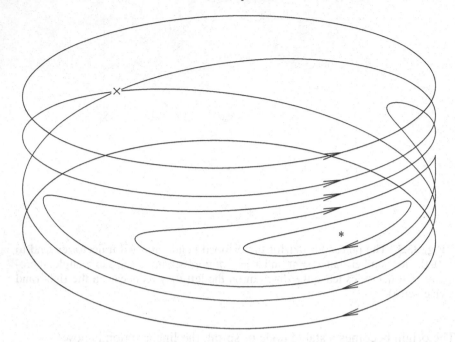

Fig. 35.6. The 'phase cylinder' for the pendulum. The '*' represents the pendulum pointing vertically downwards, and the '×' is the cylinder vertically upwards. The angular coordinate is θ (or x modulo 2π, the angle of the pendulum to the vertical), and the vertical coordinate is $\dot{\theta}$ (or y in our coupled equation).

35.3 Alternative phase space

Since the point $(x + 2\pi, y)$ corresponds to the same state of the system as the point (x, y), it would really be more sensible to draw a phase diagram that reflected this. Figure 35.6 shows the phase diagram drawn on the 'phase cylinder', where the coordinate on the circle represents the angle θ, and the vertical coordinate the angular velocity $\dot{\theta}$. The M-file `cylinder.m` draws animated trajectories of the pendulum equation moving on this surface.

It is possible to cast the whole theory of dynamical systems in a general setting that enables the use of such phase spaces (which are manifolds) in a very natural way.

Exercises

35.1 Draw the phase portrait for the damped pendulum equations in (35.5) when $k = 2$ and when $k = 3$.

35.2 Consider the equation for a pendulum with a quadratic damping term

$$\dot{x} = y$$
$$\dot{y} = -\sin x - ky|y|.$$

Show that if $E = \frac{1}{2}y^2 - \cos x$ then

$$\frac{\mathrm{d}E}{\mathrm{d}t} = -ky^2|y|. \tag{E35.1}$$

Show that the point $(0, 0)$ is a centre for the linearised equation, but using (E35.1) deduce that for the nonlinear equation it behaves like a stable spiral, and hence draw the phase diagram. (Remember that 'linearised centres' do not have to be centres for the nonlinear equation.)

36

*Periodic orbits

We have already seen that showing the existence of periodic solutions is much more difficult than showing that there are stationary points, and that the 'joining up' of trajectories that is required for a periodic orbit is a sensitive thing. In this brief chapter we look at two results, one that excludes the possibility of there being any periodic orbits, and one guaranteeing that there is at least one.

36.1 Dulac's criterion

Dulac's criterion is a way of showing that there cannot be any periodic orbits within some region of the phase space. Suppose that we are considering trajectories of the differential equation

$$\dot{x} = f(x, y)$$
$$\dot{y} = g(x, y).$$

Then given a region $\Omega \subset \mathbb{R}^2$, if we can find a smooth function $h(x, y)$ such that

$$\frac{\partial}{\partial x}(hf) + \frac{\partial}{\partial y}(hg) \neq 0$$

for all $x, y \in \Omega$ then there are no periodic orbits contained wholly within Ω. The proof is straightforward, but relies on the divergence theorem.[1]

---

[1] In $\mathbb{R}^2$ the divergence theorem says that if Ω is a region with smooth boundary Γ, $h : \mathbb{R}^2 \to \mathbb{R}$ and $\mathbf{f} : \mathbb{R}^2 \to \mathbb{R}^2$ are continuously differentiable functions, then

$$\int_\Omega \nabla \cdot (\mathbf{f} h) \, d^2\mathbf{x} = \int_\Gamma h(\mathbf{n} \cdot \mathbf{f}) \, ds, \qquad (36.1)$$

where $\mathbf{n}$ is the unit outward normal to Γ, see Figure 36.1.
   Now, if we can find a function h such that $\nabla \cdot (\mathbf{f} h) \neq 0$ in Ω, then this means that the sign of $\nabla \cdot (\mathbf{f} h)$ is constant throughout Ω. Then in particular we must have

$$\int_\Omega \nabla \cdot (\mathbf{f} h) \, d^2\mathbf{x} \neq 0.$$

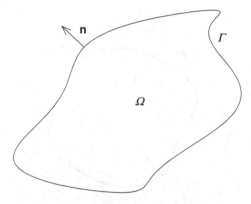

Fig. 36.1. A region Ω, its boundary Γ, and one of the outward normals **n**.

For example, we can easily show that for many choices of parameters there are no periodic orbits in the ecological models

$$\dot{x} = x(A - ax + by)$$
$$\dot{y} = y(B - cy + dx).$$

We will suppose that $a, c > 0$, but will say nothing about A, B, b and d. If we choose $h(x, y) = (xy)^{-1}$ then

$$\begin{pmatrix} hf(x, y) \\ hg(x, y) \end{pmatrix} = \begin{pmatrix} A/y - a(x/y) + b \\ B/x - c(y/x) + d \end{pmatrix}$$

and

$$\frac{\partial}{\partial x}\left[\frac{A}{y} - \frac{ax}{y} + b\right] + \frac{\partial}{\partial y}\left[\frac{B}{x} - \frac{cy}{x} + d\right] = -\frac{a}{y} - \frac{c}{x}.$$

Since $a, c > 0$ there can be no periodic orbits in the region $x, y > 0$, since this expression is always negative there.

36.2 The Poincaré–Bendixson Theorem

More positively, the Poincaré–Bendixson Theorem guarantees the existence of a periodic orbit under certain conditions, and limits the complexity of two-dimensional systems.

Theorem 36.1 *Let D be a bounded region that orbits enter and never leave and that contains no stationary points. Then any orbit entering D is attracted to a*

However, if Γ is a periodic orbit then it is everywhere tangent to the velocity field **f**, cf. Figure 32.3. This means that **n**, which is normal to the periodic orbit, must be normal to the velocity field, and so $\mathbf{f} \cdot \mathbf{n} = 0$. Thus the right-hand side of (36.1) must be zero, a contradiction.

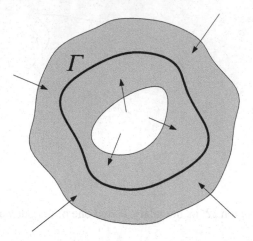

Fig. 36.2. If orbits enter and never leave the shaded region (D in the theorem), and this region contains no stationary points, then there is at least one periodic orbit Γ contained entirely within D.

periodic orbit; in particular there is at least one periodic orbit lying within D. (This is illustrated in Figure 36.2.)

The following corollary (which follows with a little additional work) says that the most complicated behaviour in two-dimensional systems is a periodic orbit.

Corollary 36.2 *If all orbits are bounded then each trajectory converges either to a stationary point or to a periodic orbit.*

Exercises

36.1 Use Dulac's criterion to show that periodic orbits in the equations

$$\dot{x} = y$$
$$\dot{y} = -ky - V'(x)$$

are only possible if $k = 0$.

36.2 Consider the coupled system[2]

$$\dot{x} = y + \tfrac{1}{4}x(1 - 2r^2)$$
$$\dot{y} = -x + \tfrac{1}{2}y(1 - r^2),$$

where $r^2 = x^2 + y^2$. First, show that the system has only one stationary point which

---

[2] This example is taken from P. A. Glendinning *Stability, instability, and chaos* (Cambridge University Press, 1994).

lies at the origin. Now, by finding the equation satisfied by r, show that trajectories enter (and do not leave) the region D, where

$$D = \{(x, y) : \tfrac{1}{2} \le r^2 \le 1\}.$$

Use the Poincaré–Bendixson Theorem to deduce that the system has a periodic orbit lying within D.

37

*The Lorenz equations

We saw at the very end of the last chapter that the most complicated dynamics that can occur in a system of two coupled equations (a two-dimensional system) is a periodic orbit.

In this chapter we discuss, briefly, a three-dimensional system of equations, the Lorenz equations, which demonstrates that much more complexity is possible once we have three variables. Without the help of numerical solutions and computer-aided visualisation our understanding of these equations would still be poor. This chapter contains many computer-generated images, and some samples of MATLAB output performing some tedious eigenvalue calculations. Most of this was generated by the M-file lorenz37.m, and it might be helpful to have this program running as you read.

Lorenz introduced his relatively simple system as a model in which to study various theoretical problems involved in meteorology, and in particular in weather prediction.[1] It is based on a model of convection (when a layer of fluid is heated from below), greatly simplified. The model is

$$\dot{x} = \sigma(-x + y)$$
$$\dot{y} = rx - y - xz \qquad (37.1)$$
$$\dot{z} = -bz + xy.$$

There are three parameters in the problem: b, r, and σ. Standard values have established themselves over the years: $b = 8/3$, $r = 28$, and $\sigma = 10$.

```
% parameters

>> sigma=10;  r=28;  b=8/3;
```

[1] E. Lorenz, Deterministic nonperiodic flow, *J. Atmos. Sci.* **20** (1963), 448–464.

The basic steps towards understanding the problem should be familiar from our analysis of two-dimensional systems; we find the stationary points and determine their stability.

For these parameter values there are three stationary points; one at the origin (it is easy to find this one) and two more at

$$(\pm\sqrt{b(r-1)}, \pm\sqrt{b(r-1)}, r-1).$$

```
% non-zero fixed points

>> x=sqrt(b*(r-1)); y=x; z=r-1; [x y z]

ans = 8.4853      8.4853      27.0000
```

In order to determine the stability type of the stationary points we have to look at the linearised equation near each one. The matrix of partial derivatives is now 3×3,

$$\mathbf{Df}(x, y, z) = \begin{pmatrix} -\sigma & \sigma & 0 \\ r-z & -1 & -x \\ y & x & -b \end{pmatrix}.$$

To determine the stability we look (as we did for systems of two equations) at the eigenvalues of this matrix. At the origin the matrix is

$$\begin{pmatrix} -\sigma & \sigma & 0 \\ r & -1 & 0 \\ 0 & 0 & -b \end{pmatrix},$$

and its eigenvalues are given by the solutions of

$$\begin{vmatrix} -\sigma - \lambda & \sigma & 0 \\ r & -1 - \lambda & 0 \\ 0 & 0 & -b - \lambda \end{vmatrix} = (-b - \lambda)[(-\sigma - \lambda)(-1 - \lambda) - \sigma r]$$

$$= -(\lambda + b)[\lambda^2 + (1 + \sigma)\lambda - \sigma(r-1)] = 0.$$

So $\lambda = -b$ or

$$\lambda = \frac{-(1+\sigma) \pm \sqrt{(1+\sigma)^2 + 4\sigma(r-1)}}{2}.$$

For the particular parameter values above this gives the three eigenvalues

$$-22.8277, \qquad 11.8277 \qquad \text{and} \qquad -8/3.$$

```
% eigenvalues and eigenvectors at the origin
% L gives values, V gives vectors
[V L] = eig([-sigma sigma 0; r -1 0; 0 0 -b])


V = -0.6148      -0.4165            0
     0.7887      -0.9091            0
          0            0       1.0000




L = -22.8277          0            0
           0    11.8277            0
           0          0      -2.6667
```

There are two stable directions at the origin, and one unstable one, so the origin is some kind of three-dimensional analogue of a saddle point, as can be seen if you look at the direction field near the origin (shown in Figure 37.1) in the right way.

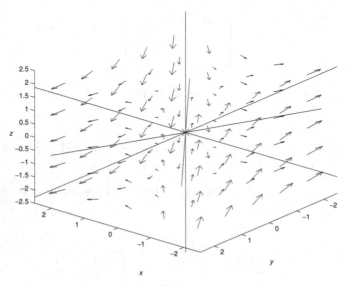

Fig. 37.1. The direction field near the origin.

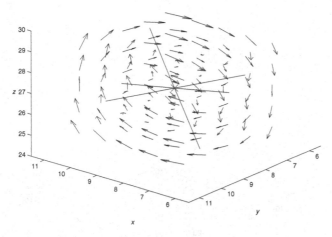

Fig. 37.2. The direction field near one of the non-zero stationary points.

The linearisations near the two non-zero stationary points is the same, and the eigenvalues are most easily found numerically:

```
% eigenvalues and eigenvectors at the non-zero points

>> [V L]=eig([-sigma sigma 0; r-z -1 -x; y x -b])

V =   0.8557      -0.2779 - 0.2839i   -0.2779 + 0.2839i
     -0.3298       0.0089 - 0.5699i    0.0089 + 0.5699i
     -0.3988      -0.7186 + 0.0293i   -0.7186 - 0.0293i


L = -13.8546          0                    0
          0      0.0940 +10.1945i          0
          0           0            0.0940 -10.1945i
```

Near these points there is one stable direction, and a two-dimensional unstable manifold of 'spiral type'. It is quite hard to see all this structure in the direction field (shown in Figure 37.2), but you should be able to make out the 'rotating' behaviour.

The trajectories cannot just settle down to a stationary point, since all of them have unstable directions. There is also no obvious way to 'join the dots' of the local phase portraits, as should be clear from Figure 37.3, which puts the direction fields near the three stationary points on one figure.

However, the trajectories do not escape to infinity; if we consider

$$V(x, y, z) = x^2 + y^2 + (z - \sigma - r)^2$$

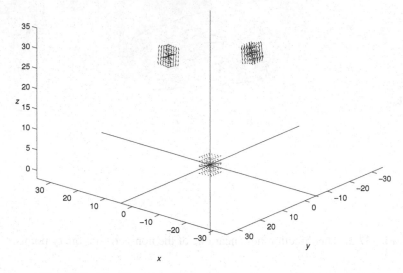

Fig. 37.3. The direction fields near the three stationary points.

(which is the square of the distance of the point (x, y, z) from $(0, 0, \sigma + r)$) then $V(t)$ changes according to

$$\frac{dV}{dt} = 2x\dot{x} + 2y\dot{y} + 2(z - \sigma - r)\dot{z}$$

$$= 2\sigma x(-x + y) + 2y(rx - y - xz) + 2(z - \sigma - r)[-bz + xy]$$

$$= -2\sigma x^2 - 2y^2 - 2z^2 + 2b(r + \sigma)z$$

$$= -2\sigma x^2 - 2y^2 - b(z - r - \sigma)^2 - bz^2 + b(r + \sigma)^2$$

$$\leq -\alpha V + b(r + \sigma)^2,$$

where $\alpha = \min(2\sigma, 2, b)$ (which is 2 for our choice of parameters). This shows that V decreases when $V > b(r + \sigma)^2/\alpha$. It follows that eventually all trajectories will have $V < 2b(r + \sigma)^2/\alpha$, and lie at a bounded distance from $(0, 0, r + \sigma)$.

So what happens to the trajectories? One thing we might try, looking at the solutions as functions of time, is unhelpful (see Figure 37.4).

Even though a picture of the trajectory traced out by a typical solution gives a very complicated looking picture, this approach is much more fruitful. Figures 37.5 and 37.6 give two different views of the same trajectory after a small period of time.

Essentially, these figures show the famous 'Lorenz attractor'. Unless you start on the stable manifold of one of the stationary points, if you trace out the trajectory in the three-dimensional phase space you will end up with pictures that look something like these. You can do this for various choices of initial conditions using the M-file `lorenzdraw.m`.

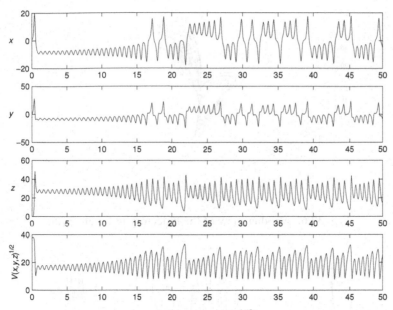

Fig. 37.4. Graphs of x, y, z, and $V(x, y, z)^{1/2}$ against t on one solution.

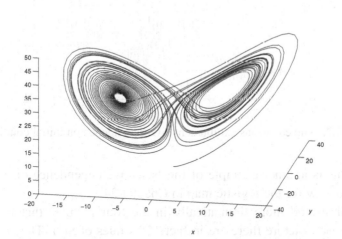

Fig. 37.5. A trajectory of the Lorenz equations.

However, suppose that you start two trajectories very close together. The result of doing this with two initial conditions that are extremely close (the initial x coordinates differ by 0.0001) is shown in Figure 37.7 (the x coordinate only). Although the solutions look the same for a time, after a while they are completely

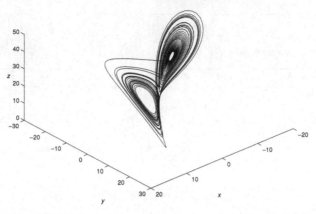

Fig. 37.6. Another view of the trajectory in Figure 37.5.

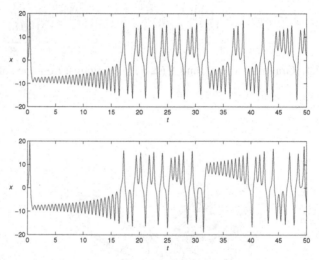

Fig. 37.7. The equations exhibit sensitive dependence on initial conditions.

different. This is another example of the 'sensitive dependence on initial conditions' that we saw for the logistic map in Chapter 24.

This is illustrated more dramatically in the four pictures that make up Figure 37.8. In each picture there are in fact 125 values of $\mathbf{x}(t)$. They start so close together that even at time $t = 18$, shown in the first picture, they still appear to be the same. However, in the following three pictures you can see them spreading apart, until at time $t = 34$ they are scattered all over the attractor. (The MATLAB program solvem.m performs all the integrations necessary to generate all the solutions, which you can then watch as they move using flies.m.)

A very small initial change makes a huge difference to the way the solutions eventually behave. This is the origin of the notion of the 'butterfly effect'; if a

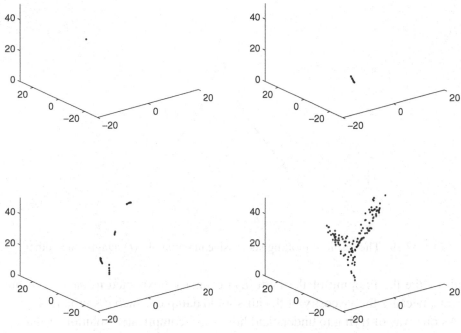

Fig. 37.8. 125 solutions of the Lorenz equations, at times $t = 18$ (top left); $t = 21$ (top right); $t = 24$ (bottom left); and $t = 34$ (bottom right). The axes are orientated similarly to those in Figure 37.6.

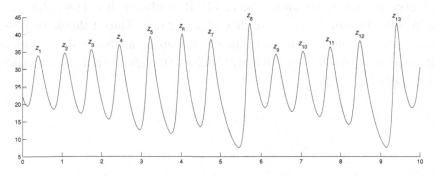

Fig. 37.9. Graph of a solution $z(t)$ and its successive maxima.

butterfly flaps its wings on one side of the world it could cause a storm on the other. In other words, very small changes to the initial condition can produce widely different effects after some time. In terms of weather prediction we know this very well; forecasts are reasonably accurate for the next day, but not for one week later. Of course, the 'butterfly effect' is a very colourful description of the idea. Atmospheric models will take into account the fact that variations on very small

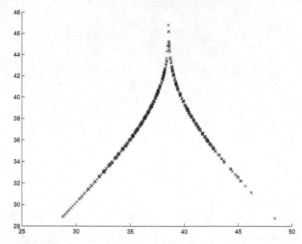

Fig. 37.10. The result of plotting successive maxima of $z(t)$ against each other.

scales (like the flapping of the butterfly) cannot be expected to have such huge effects, because the viscosity of the air should damp out such disturbances.

As one way of trying to understand how these complicated solutions behave, in his original paper Lorenz suggested concentrating on the successive local maxima of $z(t)$. If z_n is the nth local maximum, see Figure 37.9, Lorenz suggested plotting the next local maximum z_{n+1} against z_n. A priori there is no reason why this should not just produce points all over the plot.

However, Figure 37.10 shows the result. Remarkably, it appears that z_{n+1} is given by a function of z_n, say f, so that $z_{n+1} = f(z_n)$. Thus it should be possible to investigate the solutions of the Lorenz equations by analysing the dynamics of this iterated map, which is surprisingly similar to the tent map that we considered at the end of Chapter 24.

38

What next?

In this book we have covered all of the basic methods for finding the explicit solutions of simple first and second order differential equations, along with some
qualitative methods for coupled nonlinear equations. We have also discussed difference equations, and seen how complicated the dynamics of even very simple
iterated nonlinear maps can become.

There are two ways in which to proceed further with the material developed
here. One arises from turning first to the study of partial differential equations,
while the other essentially continues from where we have left off.

38.1 Partial differential equations and boundary value problems

Partial differential equations model systems that have spatial as well as temporal
structure, for example the temperature throughout an object, the vibrations of a
string or a drum, or the velocity of a fluid.

In general linear partial differential equations are easier to solve. By using the
technique known as 'separation of variables' it is possible to convert such a problem into an ordinary differential equation. This was touched on briefly in Exercise 20.10, and the exercises in this chapter apply this method in more detail for
the example of the vibrating string.

Viewed in general this approach leads to the theory of Fourier series (the expansion of an arbitrary function as a sum of sines and cosines, see Exercises 38.7
and 38.8), and its extension via the Sturm–Liouville theory of boundary eigenvalue
problems

$$-\frac{\mathrm{d}}{\mathrm{d}x}\left(p(x)\frac{\mathrm{d}y}{\mathrm{d}x}\right) + q(x)y = \lambda w(x)y \qquad \text{with} \qquad y(a) = y(b) = 0.$$

These topics are treated at length in many differential equation textbooks, for example,

373

W. E. Boyce and R. C. DiPrima, *Elementary differential equations and boundary value problems*, 7th edition (John Wiley & Sons, 2001)

C. H. Edwards and D. E. Penney, *Differential equations and boundary value problems*, 2nd edition (Prentice Hall, 2000)

R. K. Nagle, E. B. Saff and A. D. Snider, *Fundamentals of differential equations and boundary value problems* (Addison-Wesley, 2000),

all of which also cover much of the material presented in this book.

The theory of Fourier series is treated rigorously in

H. A. Priestley, *Introduction to Integration* (Oxford University Press, 1997)

Sturm–Liouville theory provides one of the first concrete applications in introductory courses on functional analysis, see for example,

E. Kreyszig, *Introductory Functional Analysis with Applications* (Wiley, 1978)

M. Renardy and R. C. Rogers, *An introduction to partial differential equations*, in the series Texts in Applied Mathematics Volume 13 (Springer Verlag, 1992)

N. Young, *Hilbert Spaces* (Cambridge University Press, 1988)

38.2 Dynamical systems and chaos

We have investigated, at least numerically, the complicated dynamics of the Lorenz equations. Similarly we spent a chapter examining the behaviour of the iterated logistic map. Both of these examples pose problems whose solution is beyond the techniques presented in this book, and fall into the realm of the theory of dynamical systems. The subject received a large boost in the 1980s, when it was popularised under the media-friendly 'chaos' banner. One particularly readable popular account is

J. Gleick, *Chaos: making a new science* (Minerva, 1997)

while one of the early pioneers in the subject presents it in an accessible way in

E. N. Lorenz, *The essence of chaos* (University of Washington Press, 1994).

The theory is concerned with the qualitative behaviour of the solutions of difference and differential equations in very general situations in which we cannot hope to be able to find an explicit solution. Since we cannot base our understanding on explicit solutions, the subject relies on a series of very powerful general results (like the Hartman–Grobman Theorem we used to draw our phase portraits for nonlinear coupled equations) and inspired simplifications (like the substitution we used to turn the logistic map $x_{n+1} = 4x_n(1 - x)n)$ into the simpler 'tent map').

Books that treat the subject from a rigorous but accessible point of view, and could be seen as continuing naturally from the material we have covered here are

V. I. Arnol'd, *Ordinary Differential Equations*, 3rd edition (Springer Verlag, 1992)

R. L. Devaney, *An introduction to chaotic dynamical systems*, 2nd edition (Westview Press, 2003)

P. A. Glendinning, *Stability, instability, and chaos* (Cambridge University Press, 1994)

J. Guckenheimer and P. Holmes, *Nonlinear oscillations, dynamical systems and bifurcations of vector fields*, in the series Applied Mathematical Sciences 42 (Springer Verlag, 1983)

M. W. Hirsch and S. Smale, *Differential equations, dynamical systems, and linear algebra* (Academic Press, 1974)

D. W. Jordan and P. Smith, *Nonlinear ordinary differential equations*, 2nd edition (Oxford University Press, 1999)

Two classic advanced texts that concentrate more on the rigorous proof of fundamental results are

J. K. Hale, *Ordinary Differential Equations*, 2nd edition (Krieger, 1980)

P. Hartman, *Ordinary Differential Equations*, 2nd edition (SIAM, 1973)

Finally, I cannot end without recommending unreservedly the thoughtful and entertaining book

D. Ruelle, *Chance and Chaos* (Penguin Books, 1993)

Written by one of the foremost mathematicians working in the field, it combines a discussion of various topics from dynamical systems and modern theoretical physics with many insights into the life of the research mathematician, and life in general.

Exercises

This sequence of exercises treats the problem of the vibrating string using the method of separation of variables. This produces a simple boundary value problem, and serves to introduce the idea of Fourier series.

The equation for the vibrations of a string stretched between $x = 0$ and $x = 1$ and attached at both endpoints is

$$\frac{\partial^2 u}{\partial t^2} = c^2 \frac{\partial^2 u}{\partial x^2},$$
(E38.1)

with $u(x, t)$ representing the height of the string at position x at time t. Since the string is fixed at the endpoints, we should have $u(0, t) = u(1, t) = 0$ for all t. See Figure 38.1.

38.1　Show that the principle of superposition is valid: if two functions $u_1(x, t)$ and $u_2(x, t)$ satisfy the equation and the boundary conditions, then $u(x, t) = \alpha u_1(x, t) + \beta u_2(x, t)$ also satisfies both the equation and the boundary conditions.

38.2　Show that if we guess that a solution has the form $u(x, t) = X(x)T(t)$ then $X(x)$ and $T(t)$ must satisfy

$$\frac{1}{c^2 T} \frac{d^2 T}{dt^2} = \frac{1}{X} \frac{d^2 X}{dx^2}.$$

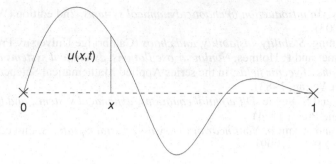

Fig. 38.1. A vibrating string, fixed at the endpoints $x = 0$ and $x = 1$.

Since the left-hand side is a function of t alone, and the right-hand side is a function of x alone, the only way that they can be equal is if they are both constant. If we choose this constant to be $-\lambda$ then we obtain two equations,

$$\frac{d^2 T}{dt^2} = -\lambda c^2 T \tag{E38.2}$$

and the boundary value problem

$$\frac{d^2 X}{dx^2} = -\lambda X \qquad \text{with} \qquad X(0) = X(1) = 0. \tag{E38.3}$$

38.3   Show that if $\lambda \leq 0$ then the only solution of (E38.3) is $X(x) = 0$ for all $x \in [0, 1]$. (You can find the general solution using the methods you have learned in this book, and then choose the constants in order to satisfy the boundary conditions.)

38.4   Show that if $\lambda > 0$ then we only have $X(x) \neq 0$ if we choose $\lambda = n^2 \pi^2$ for some integer n, and then

$$X(x) = A X_n(x), \qquad \text{where} \qquad X_n(x) = \sin n\pi x$$

and A is an arbitrary constant.

The values $\lambda_n = n^2 \pi^2$, and the corresponding solutions $X_n(x)$, are known as the eigen-values and eigenfunctions for the problem

$$\frac{d^2 X}{dx^2} = -\lambda X \qquad \text{with} \qquad X(0) = X(1) = 0. \tag{E38.4}$$

38.5   By requiring the solution of (E38.4) to be non-zero we have restricted the possible values of λ to the eigenvalues $\lambda_n = n^2 \pi^2$. Find the solution of (E38.2) when $\lambda = \lambda_n$, and hence show that one solution of (E38.1) is

$$u(x, t) = (A \sin n\pi ct + B \cos n\pi ct) \sin n\pi x. \tag{E38.5}$$

Use the principle of superposition to show that

$$u(x, t) = \sum_{n=1}^{\infty}(A_n \sin n\pi ct + B_n \cos n\pi ct) \sin n\pi x \qquad (E38.6)$$

solves (E38.1) for any choice of coefficients A_n and B_n.

38.6 Assuming that any solution of (E38.1) can be written in the form (E38.6), the problem becomes to determine the coefficients A_n and B_n. Show that if the initial position and velocity of the string, $u(x, 0)$ and $\partial u/\partial t(x, 0)$, are given then A_n and B_n must satisfy

$$u(x, 0) = \sum_{n=1}^{\infty} B_n \sin n\pi x \qquad (E38.7)$$

and

$$\frac{\partial u}{\partial t}(x, 0) = \sum_{n=1}^{\infty} n\pi c\, A_n \sin n\pi x.$$

An expansion of a function $f(x)$ as a sum of sine functions,

$$f(x) = \sum_{n=1}^{\infty} c_n \sin n\pi x \qquad (E38.8)$$

is known as a Fourier series expansion of f. It is one of the wonders of mathematics that any reasonably smooth function f that has $f(0) = f(1) = 0$ can be expanded in such a series. (If we also include cosine functions then we can remove the restrictions at the end-points.) Finding the coefficients c_n is also relatively straightforward, at least in principle.

38.7 Check that

$$\int_0^1 \sin n\pi x\, \sin m\pi x\, dx = \begin{cases} 0 & n \neq m \\ \frac{1}{2} & n = m. \end{cases}$$

(The functions $\sin n\pi x$ and $\sin m\pi x$ are orthogonal on $[0, 1]$.)

38.8 Multiply both sides of (E38.8) by $\sin m\pi x$ and, assuming that it is possible to integrate the series term-by-term, show that the coefficient c_m is given by

$$c_m = 2 \int_0^1 f(x) \sin m\pi x\, dx.$$

Sturm–Liouville theory treats the more general eigenvalue problem

$$-\frac{d}{dx}\left(p(x)\frac{dy}{dx}\right) + q(x)y = \lambda w(x)y \qquad \text{with} \qquad y(a) = y(b) = 0$$

(in equation (E38.3) we had $p(x) = w(x) = 1$, $q(x) = 0$, $a = 0$, and $b = 1$). There are, again, an infinite set of eigenvalues λ_n for which there is a corresponding non-zero eigenfunction $y_n(x)$. The eigenfunctions are 'orthonormal on $[a, b]$ with respect to the weight function $w(x)'$,

$$\int_a^b y_n(x)y_m(x)w(x)\,dx = \begin{cases} 0 & n \neq m \\ 1 & n = m. \end{cases} \tag{E38.9}$$

Furthermore, any function $f(x)$ satisfying the boundary conditions can be expanded as a generalised Fourier series using the eigenfunctions $y_n(x)$,

$$f(x) = \sum_{n=1}^{\infty} c_n y_n(x). \tag{E38.10}$$

38.9   Using the orthonormality relation in (E38.9) show that the coefficients c_m in (E38.10) are given by

$$c_m = \int_a^b f(x)y_m(x)w(x)\,dx.$$

Appendix A: Real and complex numbers

In this appendix we discuss some basic notation and properties of real and complex numbers. We use '$\in$' to denote 'is an element of'.

Real numbers

We use $\mathbb{R}$ to denote the collection of all the real numbers, so that $\alpha \in \mathbb{R}$ simply means that α is a real number. We use curved brackets to denote the open end of an interval, and square brackets to denote the closed end of an interval, so for example

$$x \in [a, b] \qquad \text{represents} \qquad a \leq x \leq b,$$

and

$$x \in (a, b] \qquad \text{represents} \qquad a < x \leq b.$$

One end of the interval is allowed to be $\pm\infty$, for example

$$t \in (-\infty, t^*) \qquad \text{represents} \qquad t < t^*.$$

Complex numbers

A complex number is a number z of the form $z = x + iy$, where $i = \sqrt{-1}$. Any complex number can be split into its real and imaginary parts, Re[z] and Im[z], where

$$\text{Re}[z] = x \qquad \text{and} \qquad \text{Im}[z] = y.$$

The rules for addition and multiplication of complex numbers follow from applying standard algebra and the fact that $i^2 = -1$; we have

$$(a + ib) + (c + id) = (a + c) + i(b + d)$$

379

and

$$(a + ib)(c + id) = (ac - bd) + i(ac + bd). \tag{A.1}$$

The complex conjugate of a complex number $z = x + iy$ is written z^* and is given by $z^* = x - iy$. Adding z and its complex conjugate yields twice the real part of z,

$$z + z^* = (x + iy) + (x - iy) = 2x = 2\,\mathrm{Re}[z],$$

while their difference gives i multiplied by twice the imaginary part of z,

$$z - z^* = (x + iy) - (x - iy) = 2iy = 2i\,\mathrm{Im}[z].$$

Also useful is the fact that the complex conjugate of a product is the product of the complex conjugates, $(wz)^* = w^*z^*$. In order to see this, note that

$$(a + ib)^*(c + id)^* = (a - ib)(c - id) = (ac - bd) - i(ac + bd)$$
$$= [(ac - bd) + i(ac + bd)]^* = [(a + ib)(c + id)]^*,$$

using (A.1).

One important identity involving complex numbers is Euler's formula,

$$e^{i\theta} = \cos\theta + i\sin\theta. \tag{A.2}$$

The easiest way to see this is to use the power series expansion of e^z,

$$e^z = \sum_{n=0}^{\infty} \frac{z^n}{n!}$$

(see Appendix C.) If $z = i\theta$ then we have

$$e^{i\theta} = \sum_{n=0}^{\infty} \frac{(i\theta)^n}{n!}.$$

Since $i^2 = -1$, the even powers in the expansion are real, while the odd powers are still imaginary, and we get

$$e^{i\theta} = \sum_{n=0}^{\infty} (-1)^n \frac{\theta^{2n}}{(2n)!} + i\sum_{n=0}^{\infty} (-1)^n \frac{\theta^{2n+1}}{(2n+1)!}.$$

The two sums on the right-hand side are just those for $\cos\theta$ and $\sin\theta$ respectively (see Appendix C again), and so $e^{i\theta} = \cos\theta + i\sin\theta$, as claimed.

This gives another, often convenient form in which we can write any complex number, known as *modulus and argument form*. This is when we express $z = x + iy$ as $z = re^{i\theta}$. This is illustrated in Figure A.1, which shows z plotted on

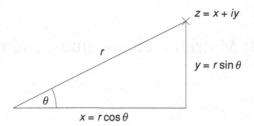

Fig. A.1. The complex number $z = x + iy$ written in modulus and argument form, $z = re^{i\theta}$.

the Argand diagram. (The Argand diagram shows the 'complex plane', where the real part of z gives the x coordinate and the imaginary part of z gives the y coordinate.)

It follows that

$$r = \sqrt{x^2 + y^2} \quad \text{and} \quad \theta = \tan^{-1}(y/x).$$

This is entirely analogous to using plane polar coordinates instead of standard cartesian coordinates, cf. (29.8).

Appendix B: Matrices, eigenvalues, and eigenvectors

This appendix covers the simple algebra of matrices, and some properties of eigenvalues and eigenvectors. The calculation of eigenvalues and eigenvectors is the main topic of Chapter 27.

Basic matrix algebra

For the most part, we will only need to consider the algebra of 2×2 matrices,

$$\begin{pmatrix} a & b \\ c & d \end{pmatrix}.$$

Addition of two matrices is component by component, so that

$$\begin{pmatrix} a_{11} & a_{12} \\ a_{21} & a_{22} \end{pmatrix} + \begin{pmatrix} b_{11} & b_{12} \\ b_{21} & b_{22} \end{pmatrix} = \begin{pmatrix} a_{11} + b_{11} & a_{12} + b_{12} \\ a_{21} + b_{21} & a_{22} + b_{22} \end{pmatrix},$$

while multiplication is given by

$$\begin{pmatrix} a_{11} & a_{12} \\ a_{21} & a_{22} \end{pmatrix} \begin{pmatrix} b_{11} & b_{12} \\ b_{21} & b_{22} \end{pmatrix} = \begin{pmatrix} a_{11}b_{11} + a_{12}b_{21} & a_{11}b_{12} + a_{12}b_{22} \\ a_{21}b_{11} + a_{22}b_{21} & a_{21}b_{12} + a_{22}b_{22} \end{pmatrix}.$$

We can write this more compactly by saying that

$$[\mathbb{AB}]_{ij} = a_{i1}b_{1j} + a_{i2}b_{2j} = \sum_{k=1}^{2} a_{ik}b_{kj},$$

where $[\mathbb{AB}]_{ij}$ is the entry in the ith row and jth column of the matrix $\mathbb{AB}$.

One special matrix is the identity matrix,

$$\mathbb{I} = \begin{pmatrix} 1 & 0 \\ 0 & 1 \end{pmatrix},$$

which has the property that $\mathbb{IA} = \mathbb{AI} = \mathbb{A}$ for any 2×2 matrix $\mathbb{A}$.

A matrix $\mathbb{A}$ is said to be *invertible*, or *non-singular*, if there is another matrix $\mathbb{A}^{-1}$ such that

$$\mathbb{A}^{-1}\mathbb{A} = \mathbb{A}\mathbb{A}^{-1} = \mathbb{I}.$$

The matrix

$$\mathbb{A} = \begin{pmatrix} a & b \\ c & d \end{pmatrix} \tag{B.1}$$

is invertible if and only if its determinant, $\det(\mathbb{A})$, given by

$$\det(\mathbb{A}) = \begin{vmatrix} a & b \\ c & d \end{vmatrix} = ad - bc$$

is not equal to zero, and then

$$\mathbb{A}^{-1} = \frac{1}{ad - bc} \begin{pmatrix} d & -b \\ -c & a \end{pmatrix}.$$

Matrices and vectors

Multiplication of vectors by matrices

In general we can calculate the product $\mathbb{A}\mathbb{B}$ when $\mathbb{A}$ is an $n \times m$ matrix and $\mathbb{B}$ is an $m \times k$ matrix (the columns of $\mathbb{A}$ have to match the rows of $\mathbb{B}$). In particular this allows us to calculate $\mathbb{A}\mathbf{x}$ if $\mathbb{A}$ is a 2×2 matrix and $\mathbf{x} = (x_1, x_2)$ is a two component vector,

$$\begin{pmatrix} a_{11} & a_{12} \\ a_{21} & a_{22} \end{pmatrix} \begin{pmatrix} x_1 \\ x_2 \end{pmatrix} = \begin{pmatrix} a_{11}x_1 + a_{12}x_2 \\ a_{21}x_1 + a_{22}x_2 \end{pmatrix}.$$

We can also write this more compactly as

$$[\mathbb{A}\mathbf{x}]_i = a_{i1}x_1 + a_{i2}x_2, \tag{B.2}$$

where $\mathbf{v}_i$ indicates the ith component of the vector $\mathbf{v}$. (Note that this means in particular that if $[\mathbf{v}_1 \ \mathbf{v}_2]$ is a matrix with columns made from the vectors $\mathbf{v}_1$ and $\mathbf{v}_2$ then for a 2×2 matrix $\mathbb{A}$ we have

$$\mathbb{A}[\mathbf{v}_1 \ \mathbf{v}_2] = [\mathbb{A}\mathbf{v}_1 \ \mathbb{A}\mathbf{v}_2],$$

which is used repeatedly in Chapters 28–30.)

Solution of simultaneous equations

The simultaneous linear equations

$$ax_1 + bx_2 = c_1$$
$$cx_1 + dx_2 = c_2$$

can be rewritten as the matrix equation

$$\mathbb{A}\mathbf{x} = \mathbf{c},$$

where $\mathbb{A}$ is defined as in (B.1), $\mathbf{x} = (x_1,\ x_2)$ and $\mathbf{c} = (c_1,\ c_2)$. This equation has a unique solution if and only if $\mathbb{A}$ is invertible, and then the solution is given by multiplying both sides by $\mathbb{A}^{-1}$,

$$\mathbf{x} = \mathbb{A}^{-1}\mathbf{c}.$$

It follows that $\mathbb{A}\mathbf{x} = \mathbf{0}$ can have a non-zero solution for $\mathbf{x}$ only if $\mathbb{A}$ is not invertible.

Eigenvalues and eigenvectors

If $\mathbf{v} \neq \mathbf{0}$ and

$$\mathbb{A}\mathbf{v} = \lambda\mathbf{v}$$

then λ is an eigenvalue of $\mathbb{A}$ and $\mathbf{v}$ is the corresponding eigenvector. The calculation of eigenvalues and eigenvectors for 2×2 matrices is treated in detail in Chapter 27.

Linear independence of eigenvectors

The eigenvectors corresponding to two distinct eigenvalues are linearly independent; if

$$\alpha\mathbf{v}_1 + \beta\mathbf{v}_2 = \mathbf{0} \tag{B.3}$$

then we can multiply both sides by $\mathbb{A}$ to obtain

$$\mathbb{A}(\alpha\mathbf{v}_1 + \beta\mathbf{v}_2) = \mathbf{0}.$$

Since $\mathbb{A}\mathbf{v}_j = \lambda_j\mathbf{v}_j$ we have

$$\alpha\lambda_1\mathbf{v}_1 + \beta\lambda_2\mathbf{v}_2 = \mathbf{0}. \tag{B.4}$$

While (B.3) requires that $\mathbf{v}_1 = -\beta\mathbf{v}_2/\alpha$, the second equation (B.4) says that $\mathbf{v}_1 = (\lambda_2/\lambda_1) \times (-\beta\mathbf{v}_2/\alpha)$. Since $\lambda_2 \neq \lambda_1$ these cannot both be true unless $\beta = 0$, in which case $\alpha = 0$ also, and $\mathbf{v}_1$ and $\mathbf{v}_2$ are therefore linearly independent.

The special case of symmetric matrices

The transpose of the matrix

$$\mathbb{A} = \begin{pmatrix} a_{11} & a_{12} \\ a_{21} & a_{22} \end{pmatrix},$$

written $\mathbb{A}^T$, is given by

$$\mathbb{A}^T = \begin{pmatrix} a_{11} & a_{21} \\ a_{12} & a_{22} \end{pmatrix};$$

i.e. $[\mathbb{A}^T]_{ij} = [\mathbb{A}]_{ji}$. A matrix is called *symmetric* if $\mathbb{A} = \mathbb{A}^T$, i.e. if $[\mathbb{A}]_{ij} = [\mathbb{A}]_{ji}$. A general 2×2 symmetric matrix is of the form

$$\begin{pmatrix} a & b \\ b & d \end{pmatrix}.$$

For such matrices all the eigenvalues are real, and eigenvectors corresponding to distinct eigenvalues are orthogonal.

To see that the eigenvalues are real, suppose that λ is an eigenvalue and $\mathbf{v} = (v_1, v_2)$ is the corresponding eigenvector. Then

$$\mathbb{A}\mathbf{v} = \lambda\mathbf{v}, \qquad \text{and} \qquad \mathbb{A}\mathbf{v}^* = \lambda^*\mathbf{v}^*,$$

where the second equation is the complex conjugate of the first. We take the inner (dot) product of the first equation with $\mathbf{v}^*$, and of the second with $\mathbf{v}$,

$$\mathbf{v}^* \cdot \mathbb{A}\mathbf{v} = \lambda|\mathbf{v}|^2 \qquad \text{and} \qquad \mathbf{v} \cdot \mathbb{A}\mathbf{v}^* = \lambda^*|\mathbf{v}|^2. \tag{B.5}$$

Now, the expression on the left-hand side of the first equation in (B.5) is

$$\mathbf{v}^* \cdot \mathbb{A}\mathbf{v} = \sum_{i=1}^{2} v_i^*[\mathbb{A}\mathbf{v}]_i = \sum_{i,j=1}^{2} v_i^* a_{ij} v_j$$

$$= \sum_{i,j=1}^{2} v_i^* a_{ji} v_j$$

$$= \sum_{i,j=1}^{2} v_j a_{ji} v_i^*$$

$$= \sum_{j=1}^{2} v_j[\mathbb{A}\mathbf{v}^*]_j = \mathbf{v} \cdot \mathbb{A}\mathbf{v}^*,$$

and so is the same as the expression on the left-hand side of the second equation in (B.5). It follows that

$$\lambda|\mathbf{v}|^2 = \lambda^*|\mathbf{v}|^2,$$

i.e. $\lambda = \lambda^*$ and so this eigenvalue is real.

To see that the eigenvectors corresponding to distinct eigenvalues are orthogonal, suppose that $\mathbb{A}\mathbf{v}^{(1)} = \lambda_1\mathbf{v}^{(1)}$ and $\mathbb{A}\mathbf{v}^{(2)} = \lambda_2\mathbf{v}^{(2)}$ with $\lambda_1 \neq \lambda_2$. Then

$$\mathbf{v}^{(1)} \cdot \mathbb{A}\mathbf{v}^{(2)} = \mathbf{v}^{(1)} \cdot \lambda_2\mathbf{v}^{(2)} = \lambda_2(\mathbf{v}^{(1)} \cdot \mathbf{v}^{(2)}).$$

Looking at the left-hand side of this we have

$$\mathbf{v}^{(1)} \cdot \mathbb{A}\mathbf{v}^{(2)} = \sum_i v_i^{(1)}[\mathbb{A}\mathbf{v}^{(2)}]_i = \sum_{i,j=1}^{2} v_i^{(1)} a_{ij} v_j^{(2)}$$

$$= \sum_{i,j=1}^{2} v_i^{(1)} a_{ji} v_j^{(2)}$$

$$= \sum_{i,j=1}^{2} v_j^{(2)} a_{ji} v_i^{(1)}$$

$$= \sum_{j=1}^{2} v_j^{(2)}[\mathbb{A}\mathbf{v}^{(1)}]_j = \mathbf{v}^{(2)} \cdot \mathbb{A}\mathbf{v}^{(1)}.$$

Now,

$$\mathbf{v}^{(2)} \cdot \mathbb{A}\mathbf{v}^{(1)} = \mathbf{v}^{(2)} \cdot \lambda_1\mathbf{v}^{(1)} = \lambda_1(\mathbf{v}^{(1)} \cdot \mathbf{v}^{(2)}),$$

and since $\mathbf{v}^{(1)} \cdot \mathbb{A}\mathbf{v}^{(2)} = \mathbf{v}^{(2)} \cdot \mathbb{A}\mathbf{v}^{(1)}$ we therefore have

$$\lambda_2(\mathbf{v}^{(1)} \cdot \mathbf{v}^{(2)}) = \lambda_1(\mathbf{v}^{(1)} \cdot \mathbf{v}^{(2)}),$$

i.e.

$$(\lambda_2 - \lambda_1)(\mathbf{v}^{(1)} \cdot \mathbf{v}^{(2)}) = 0.$$

Since $\lambda_2 \neq \lambda_1$ we must have $\mathbf{v}^{(1)} \cdot \mathbf{v}^{(2)} = 0$, i.e. the eigenvectors are orthogonal.

Appendix C: Derivatives and partial derivatives

This appendix covers the definitions and properties of ordinary and partial derivatives, Taylor expansions in one and two variables, and some properties of the critical points (turning points) of functions.

Functions of one variable: ordinary derivatives

We start by considering functions $f(x)$ of one variable, and their derivatives.

Definition and properties of the derivative

Let I be an interval. A function $f : I \to \mathbb{R}$ is *differentiable* at a point $x \in I$ if the limit

$$\lim_{h \to 0} \frac{f(x+h) - f(x)}{h} \tag{C.1}$$

exists, in which case the limit in (C.1) is the derivative of f at x, which we write as $(df/dx)(x)$ or $f'(x)$.

This basic definition implies the standard rules of differentiation. The product rule is

$$[fg]' = f'g + fg'$$

the quotient rule is

$$\left[\frac{f}{g}\right]' = \frac{gf' - fg'}{g^2};$$

and the chain rule, which allows us to differentiate functions of functions, is[1]

$$\frac{d}{dx}\left[f\big(g(x)\big)\right] = f'\big(g(x)\big)g'(x).$$

Taylor expansions

Taylor's Theorem allows us to expand a function f as a power series about a point x_0 using its derivatives. Suppose that f has $n+1$ derivatives, all of which are continuous functions, and that we use the notation

$$f^{(n)}(x) = \frac{d^n f}{dx^n}(x).$$

Then we can write

$$f(x) = f(x_0) + (x - x_0)f'(x_0) + \frac{(x - x_0)^2}{2!}f''(x_0) +$$
$$\cdots + \frac{(x - x_0)^n}{n!}f^{(n)}(x_0) + \frac{(x - x_0)^{n+1}}{(n + 1)!}f^{(n+1)}(y_n), \qquad \text{(C.2)}$$

for some point $y_n \in (x_0, x)$.

Provided that the remainder term

$$R_n = \frac{(x - x_0)^{n+1}}{(n + 1)!}f^{(n+1)}(y_n) \qquad \text{(C.3)}$$

tends to zero as n tends to infinity, we can write $f(x)$ as the power series

$$f(x) = \sum_{n=0}^{\infty} \frac{d^n f}{dx^n}(x_0)\frac{(x - x_0)^n}{n!},$$

known as the 'Taylor expansion' or 'Taylor series' for f.

Power series

We can use the Taylor expansion to find representations of common functions in terms of power series, i.e. an expression of the form

$$\sum_{n=0}^{\infty} a_n x^n.$$

For example, if we take $f(x) = e^x$ and $x_0 = 0$ then

$$\frac{d^n f}{dx^n} = e^x \qquad \text{for all} \qquad n = 0, 1, 2, \ldots.$$

---

[1] As written here we have confused the various functions that appear with their values at the point x. More correctly we should write $[f \circ g]'(x) = f'(g(x))g'(x)$, where $f \circ g$ is the composition of f and g, i.e. $(f \circ g)(x) = f(g(x))$. But such pedantry is probably unhelpful.

It follows that $d^n f/dx^n(0) = 1$ for all n, and so the Taylor series for e^x is

$$e^x = \sum_{n=0}^{\infty} \frac{x^n}{n!}.$$

For $f(x) = \sin x$, the derivatives are

$$\frac{d^{2n} f}{dx^{2n}}(x) = (-1)^n \sin x \qquad \text{and} \qquad \frac{d^{2n+1} f}{dx^{2n+1}} = (-1)^n \cos x.$$

It follows that $d^{2n} f/dx^{2n}(0) = 0$ and $d^{2n+1} f/dx^{2n+1} = (-1)^n$, so that

$$\sin x = \sum_{n=0}^{\infty} (-1)^{n+1} \frac{x^{2n+1}}{(2n+1)!}.$$

A similar calculation shows that

$$\cos x = \sum_{n=0}^{\infty} (-1)^n \frac{x^{2n}}{(2n)!}.$$

For a brief discussion of the convergence of such power series, see Chapter 20.

Turning points

A point x_0 is a *turning point* (or critical point) for f if $f'(x_0) = 0$. We can find out whether such a point is a local maximum, a local minimum or a point of inflection, by using the Taylor series expansion near x_0. If we assume that $f''(x_0) \neq 0$ and we keep only the first three terms from (C.2) then we have

$$f(x) \approx f(x_0) + \frac{(x - x_0)^2}{2!} f''(x_0),$$

where there is no second term since $f'(x_0) = 0$. (Because we know that the re-mainder term is $K(x - x_0)^3$ for some constant K (cf. (C.3)), we can be sure that sufficiently close to x_0 the last term that we have kept is larger than all the terms that we have neglected.) It is easy to see from here that if $f''(x_0) < 0$ then $f(x) < f(x_0)$ close to x_0, i.e. that f has a maximum at x_0; while if $f''(x_0) > 0$ then it follows that $f(x) > f(x_0)$ close to x_0 and so x_0 is a minimum of f.

If $f''(x_0) = 0$ and $f'''(x_0) \neq 0$ then we can take one further term of the Taylor expansion to find

$$f(x) \simeq f(x_0) + \frac{(x - x_0)^3}{3!} f'''(x_0).$$

Since $(x - x_0)^3$ changes sign near x_0, it follows that x_0 is now a point of inflection.

Functions of two variables: partial derivatives

We now treat similar topics for functions of two variables.

Partial derivatives and their properties

If $f(x, y)$ is a function of two variables then the partial derivative of f with respect to x is found by treating y as a constant and differentiating with respect to x,

$$\frac{\partial f}{\partial x}(x, y) = \lim_{h \to 0} \frac{f(x + h, y) - f(x, y)}{h};$$

similarly, $\partial f / \partial y$ is found by keeping x constant and differentiating with respect to y.

Since $\partial f / \partial x$ is in general another function of x and y it is possible to take partial derivatives again, e.g.

$$\frac{\partial^2 f}{\partial x^2} = \frac{\partial}{\partial x}\left(\frac{\partial f}{\partial x}\right), \qquad \frac{\partial^2 f}{\partial y \, \partial x} = \frac{\partial}{\partial y}\left(\frac{\partial f}{\partial x}\right).$$

One useful property of partial derivatives is that the order in which they are taken does not matter, so in particular we have

$$\frac{\partial^2 f}{\partial y \, \partial x} = \frac{\partial^2 f}{\partial x \, \partial y}, \tag{C.4}$$

i.e. it makes no difference if we take the partial derivative with respect to x and then with respect to y, or vice versa.

Obvious generalisations of the product and quotient rule apply to partial derivatives. When f is a function of x and y, and x and y depend on the same variable t, we have the following version of the chain rule. In this case, when $f(x(t), y(t))$ is in fact a function of the single variable t,

$$\frac{d}{dt} f\big(x(t), y(t)\big) = \frac{\partial f}{\partial x}\big(x(t), y(t)\big)\frac{dx}{dt}(t) + \frac{\partial f}{\partial y}\big(x(t), y(t)\big)\frac{dy}{dt}(t).$$

Taylor expansions

There is a two-dimensional version of Taylor's Theorem, which allows us to expand $f(x, y)$ near a point (x_0, y_0) as a series involving the partial derivatives of f at (x_0, y_0). Writing $\xi = x - x_0$ and $\eta = y - y_0$, and only giving the terms up to third order we have

$$f(x, y) = f(x_0, y_0) + \xi \frac{\partial f}{\partial x} + \eta \frac{\partial f}{\partial y} + \frac{1}{2!} \left(\xi^2 \frac{\partial^2 f}{\partial x^2} + 2\xi \eta \frac{\partial^2 f}{\partial x \partial y} + \eta^2 \frac{\partial^2 f}{\partial y^2} \right)$$

$$+ \frac{1}{3!} \left(\xi^3 \frac{\partial^3 f}{\partial x^3} + 3\xi^2 \eta \frac{\partial^3 f}{\partial x^2 \partial y} + 3\xi \eta^2 \frac{\partial^3 f}{\partial x \partial y^2} + \eta^3 \frac{\partial^3 f}{\partial y^3} \right) + \cdots,$$

where all the partial derivatives are evaluated at the point (x_0, y_0). (It is possible to check this by first doing a Taylor expansion with respect to x keeping y fixed, and then Taylor expanding each of these terms with respect to y.)

In Chapter 32 we will use the simplest non-trivial consequence of this expansion, where we keep only the terms that are linear in ξ and η,

$$f(x_0 + \xi, y_0 + \eta) = f(x_0, y_0) + \xi \frac{\partial f}{\partial x}(x_0, y_0) + \eta \frac{\partial f}{\partial y}(x_0, y_0) + \cdots$$

Critical points

In this section we will use the shorthand notation

$$f_x = \frac{\partial f}{\partial x}, \quad f_y = \frac{\partial f}{\partial y}, \quad f_{xy} = \frac{\partial^2 f}{\partial x \, \partial y},$$

etc.

A point (x_0, y_0) is called a critical point of f if $f_x = f_y = 0$, i.e. if

$$\frac{\partial f}{\partial x} = 0 \quad \text{and} \quad \frac{\partial f}{\partial y} = 0.$$

We can use the Taylor expansion of f near this point to see what kind of critical point we have. Keeping only terms up to second order (as we did for functions of one variable) we have

$$f(x, y) = f(x_0, y_0) + \xi f_x + \eta f_y + \tfrac{1}{2} \left(\xi^2 f_{xx} + 2\xi \eta f_{xy} + \eta^2 f_{yy} \right) + \cdots.$$

We can remove the terms linear in ξ and η (since $f_x = f_y = 0$ at the critical point), and then rewrite the quadratic terms using matrix notation to give

$$f(x_0 + \xi, y_0 + \eta) \approx f(x_0, y_0) + \tfrac{1}{2} (\xi \ \eta) \begin{pmatrix} f_{xx} & f_{xy} \\ f_{yx} & f_{yy} \end{pmatrix} \begin{pmatrix} \xi \\ \eta \end{pmatrix}. \qquad \text{(C.5)}$$

The matrix of second partial derivatives,

$$H = \begin{pmatrix} f_{xx} & f_{xy} \\ f_{yx} & f_{yy} \end{pmatrix},$$

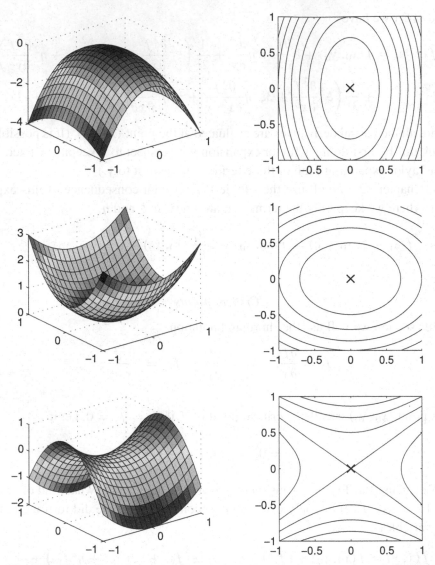

Fig. C.1. Plots of a surface whose height is $f(x, y)$, and contour plots of f, near a point (x_0, y_0) which is, from top to bottom, a maximum, a minimum and a saddle point.

is known as the *Hessian matrix*. Just as the type of the critical point x_0 for a function of one variable is determined by the sign of $f''(x_0)$, so the type of the critical point of $f(x, y)$ is determined by the properties of this matrix of second derivatives, and in particular by the sign of its eigenvalues.

Because $f_{xy} = f_{yx}$ (see (C.4)) the Hessian matrix is symmetric. It is a general result that a real symmetric matrix has real eigenvalues λ_1 and λ_2 and that if these eigenvalues are distinct then the corresponding eigenvectors $\mathbf{v}_1$ and $\mathbf{v}_2$ are orthogonal,

$$\mathbf{v}_1 \cdot \mathbf{v}_2 = 0,$$

see Appendix B. Suppose that we use this notation for the eigenvalues and eigenvectors of the Hessian matrix H, and choose the eigenvectors so that they have length 1, i.e. $|\mathbf{v}_j| = 1$. Now if we write the vector (ξ, η) in terms of the eigenvectors of the Hessian matrix,

$$\begin{pmatrix} \xi \\ \eta \end{pmatrix} = \alpha \mathbf{v}_1 + \beta \mathbf{v}_2,$$

the second term in (C.5) is

$$\tfrac{1}{2}[\alpha \mathbf{v}_1 + \beta \mathbf{v}_2] \cdot H[\alpha \mathbf{v}_1 + \beta \mathbf{v}_2] = \tfrac{1}{2}[\alpha \mathbf{v}_1 + \beta \mathbf{v}_2] \cdot [\alpha \lambda_1 \mathbf{v}_1 + \beta \lambda_2 \mathbf{v}_2]$$
$$= \tfrac{1}{2}[\alpha^2 \lambda_1 + \beta^2 \lambda_2].$$

So we have

$$f(x_0 + \xi, y_0 + \eta) \approx f(x_0, y_0) + \tfrac{1}{2}(\alpha^2 \lambda_1 + \beta^2 \lambda_2);$$

the behaviour of f near (x_0, y_0) does indeed depend on the eigenvalues of H. If both eigenvalues are positive then $f(x, y) > f(x_0, y_0)$ close to (x_0, y_0), so we have a minimum; if both eigenvalues are negative then $f(x, y) < f(x_0, y_0)$ close to (x_0, y_0) and this point is a maximum. If the two eigenvalues have opposite signs then we have a saddle point: f increases in one direction (the direction of the eigenvector corresponding to the positive eigenvalue), and decreases in another (the eigenvector for the negative eigenvalue). Plots of f against x and y, and the contour plots of curves of constant f in each of these three cases, are shown in Figure C.1. Note that near a maximum or minimum of f the curves of constant f are closed.

Index

Bold numbers indicate that the entry is the subject of a chapter beginning on that page, italic numbers indicate that the entry is the subject of a section starting on that page.

395

Printed in the United States
By Bookmasters